BIOTECHNOLOGY INTELLIGENCE UNIT

Integrated Biochips for DNA Analysis

Robin Hui Liu, Ph.D.
Osmetech Molecular Diagnostics
Pasadena, California, U.S.A.

Abraham P. Lee, Ph.D.
University of California at Irvine
Irvine, California, U.S.A.

LANDES BIOSCIENCE
AUSTIN, TEXAS
U.S.A.

SPRINGER SCIENCE+BUSINESS MEDIA
NEW YORK, NEW YORK
U.S.A.

INTEGRATED BIOCHIPS FOR DNA ANALYSIS

Biotechnology Intelligence Unit

Landes Bioscience
Springer Science+Business Media, LLC

ISBN: 978-0-387-76758-1   Printed on acid-free paper.

Please address all inquiries to the Publishers:
Landes Bioscience, 1002 West Avenue, 2nd Floor, Austin, Texas 78701, U.S.A.
Phone: 512/ 637 6050; Fax: 512/ 637 6079
www.landesbioscience.com

Springer Science+Business Media, LLC, 223 Spring Street, New York, New York 10013, U.S.A.
http://www.springer.com

Printed in the U.S.A.

9 8 7 6 5 4 3 2 1

Library of Congress Cataloging-in-Publication Data

Integrated biochips for DNA analysis / [edited by] Robin Hui Liu, Abraham P. Lee.
    p. ; cm. -- (Biotechnology intelligence unit)
  Includes bibliographical references and index.
  ISBN 978-0-387-76758-1
  1. DNA microarrays. 2. DNA--Analysis. 3. Biochips. I. Liu, Robin Hui. II. Lee, Abraham P. (Abraham Phillip) III. Series: Biotechnology intelligence unit (Unnumbered)
  [DNLM: 1. Oligonucleotide Array Sequence Analysis. 2. Sequence Analysis, DNA--methods. QU 450 I605 2007]
  QP624.5.D726I58 2007
  572.8'636--dc22
                            2007037845

About the Editors...

ROBIN HUI LIU, Ph.D., is the Director of Device Technology at Osmetech Molecular Diagnostics (OMD). He received his B.S. degree from South China University of Technology, his M.S. degree from Louisiana Tech University, and his Ph.D. degree in Mechanical Engineering from University of Illinois at Urbana-Champaign. Prior to joining OMD, Liu was the Senior Manager of Advanced Technology and principal investigator at CombiMatrix Corp. Prior to joining CombiMatrix, he was a Project Manager at Applied NanoBioscience Center, Arizona State University (2003), and a Principal Scientist and Group Leader at Motorola Inc. (2000-2003). He was a summer intern at Aclara Biosciences, Inc. (1999). Dr. Liu's research interests include development of commercial DNA biochip devices and systems, integrated microarray biochips, microfluidic systems for sample preparation and DNA analysis, and integrated immuno-sensors. Dr. Liu is an inventor of six US patents (issued and pending), and the author of over 30 peer-reviewed papers (including *Nature* and *PNAS*) and four book chapters. He served on the International Steering Committee of several international conferences, including the 26th Annual International Conference of IEEE Engineering in Medicine and Biology Society (Track Chair), the 2nd Annual International IEEE Conference on Microtechnology, Medicine, and the 2003 International Conference on Materials for Advanced Technologies in Singapore. He received two BRAVO awards and 22 Silver Quill Awards from Motorola. He is currently an Associate Editor for the *SPIE Journal of Microlithography, Microfabrication, and Microsystems.*

About the Editors...

ABRAHAM P. LEE, Ph.D., is a Professor in the Departments of Biomedical Engineering (BME) and Mechanical and Aerospace Engineering at the University of California at Irvine. He is also the associate chair of the BME Department and is the director the Micro/nano Fluidics Fundamentals Focus (MF3) Center, a Defense Advanced Research Projects Agency (DARPA) industry supported research center that includes 17 faculty and 17 companies. Prior to joining the UC Irvine faculty in 2002, he was a Senior Technology Advisor in the Office of Technology and Industrial Relations (OTIR) at the National Cancer Institute (NCI) in Bethesda, Maryland, and a program manager in the Microsystems Technology Office (MTO) of the DARPA. While at DARPA, Dr. Lee established the Bio-Fluidic Chips (BioFlips) program and was a founding co-manager of the Fundamental Research at the Bio: Info:Micro Interface program. At LLNL he was a Group Leader with projects on the treatment of stroke and CBW defense. Dr. Lee received his Ph.D. (1992) degree in Mechanical Engineering from the University of California at Berkeley. Professor Lee's current research is focused on the development of integrated and "digital" micro/nano fluidics (droplets, pumps, valves, sensors) for the following applications: molecular biosensors to detect environmental and terrorism threats, point-of-care diagnostics, "smart" nanomedicine for early detection and treatment, automated cell sorting based on electrical signatures, tissue engineering and stem cells, and the synthesis of ultra-pure materials. He currently serves as Editor for the *Journal of Microelectromechanical Systems* and International Advisory Editorial Board member of *Lab on a Chip*. Professor Lee has 32 issued US patents and has published over 60 peer-reviewed papers in journals and conferences.

Dedication

This book is dedicated to our families.
Dr. Liu's family: Fiona, Vick, Benjamin, Stanley, Shaowen and Zhongji
Dr. Lee's family: Chi, Carmel and Calvin

CONTENTS

EDITORS

Robin Hui Liu
Osmetech Molecular Diagnostics
Pasadena, California, U.S.A.
Email: robin.liu@osmetech.com
Chapters 1, 3, 4

Abraham P. Lee
University of California at Irvine
Irvine, California, U.S.A.
Email: aplee@uci.edu
Chapter 1

CONTRIBUTORS

Robert H. Austin
Department of Physics
Princeton University
Princeton, New Jersey, U.S.A.
Chapter 12

Shawn C. Baker
Illumina, Inc.
San Diego, California, U.S.A.
Chapter 2

David Barker
Illumina, Inc.
San Diego, California, U.S.A.
Chapter 2

Steven Barnard
Illumina, Inc.
San Diego, California, U.S.A.
Chapter 2

Marina Bibikova
Illumina, Inc.
San Diego, California, U.S.A.
Chapter 2

James P. Brody
Department of Biomedical Engineering
University of California at Irvine
Irvine, California, U.S.A.
Email: jpbrody@uci.edu
Chapter 10

Steven Y. Chou
Department of Electrical Engineering
Princeton University
Princeton, New Jersey, U.S.A.
Chapter 12

Eugene Chudin
Illumina, Inc.
San Diego, California, U.S.A.
Chapter 2

Edward Cox
Department of Molecular Biology
Princeton University
Princeton, New Jersey, U.S.A.
Chapter 12

David Danley
CombiMatrix Corp.
Mukilteo, Washington, U.S.A.
Chapter 3

Jian-Bing Fan
Illumina, Inc.
San Diego, California, U.S.A.
Chapter 2

Z. Hugh Fan
Department of Mechanical
    and Aerospace Engineering
University of Florida
Gainesville, Florida, U.S.A.
Email: hfan@ufl.edu
Chapter 6

Paolo Fortina
Kimmel Cancer Center
Thomas Jefferson University
Philadelphia, Pennsylvania, U.S.A.
Email: paolo.fortina@jefferson.edu
Chapter 13

H. Sho Fuji
CombiMatrix Corp.
Mukilteo, Washington, U.S.A.
Chapter 3

Michael Graige
Illumina, Inc.
San Diego, California, U.S.A.
Chapter 2

Piotr Grodzinski
National Cancer Institute
Bethesda, Maryland, U.S.A.
Email: grodzinp@mail.nih.gov
Chapters 1,4

Kevin L. Gunderson
Illumina, Inc.
San Diego, California, U.S.A.
Chapter 2

Mark Hansen
Illumina, Inc.
San Diego, California, U.S.A.
Chapter 2

Jong Wook Hong
Department of Mechanical Engineering
Auburn University
Auburn, Alabama, U.S.A.
Email: jwhong@end.auburn.edu
Chapter 8

Bob Kain
Illumina, Inc.
San Diego, California, U.S.A.
Chapter 2

Bahram G. Kermani
Illumina, Inc.
San Diego, California, U.S.A.
Chapter 2

Larry J. Kricka
Department of Pathology
    and Laboratory Medicine
University of Pennsylvania
Philadelphia, Pennsylvania, U.S.A.
Chapter 13

Kenneth M. Kuhn
Illumina, Inc.
San Diego, California, U.S.A.
Chapter 2

Ralf Lenigk
Applied NanoBioscience Center
Arizona State University
Tempe, Arizona, U.S.A.
Chapter 4

Shuang-Fang Lim
Department of Physics
Princeton University
Princeton, New Jersey, U.S.A.
Chapter 12

Limin Lin
Department of Biomedical Engineering
University of California at Irvine
Irvine, California, U.S.A.
Chapter 10

Mike Lodes
CombiMatrix Corp.
Mukilteo, Washington, U.S.A.
Chapter 3

Celeste McBride
Illumina, Inc.
San Diego, California, U.S.A.
Chapter 2

Tim McDaniel
Illumina, Inc.
San Diego, California, U.S.A.
Chapter 2

Andrew McShea
CombiMatrix Corp.
Mukilteo, Washington, U.S.A.
Chapter 3

Keith Morton
Department of Electrical Engineering
Princeton University
Princeton, New Jersey, U.S.A.
Chapter 12

Jason Y. Park
Department of Pathology
    and Laboratory Medicine
University of Pennsylvania
Philadelphia, Pennsylvania, U.S.A.
Chapter 13

Daniel A. Peiffer
Illumina, Inc.
San Diego, California, U.S.A.
Chapter 2

Christopher M. Puleo
Department of Biomedical Engineering
Johns Hopkins University
Baltimore, Maryland, U.S.A.
Chapter 11

Walter Reisner
Department of Physics
Princeton University
Princeton, New Jersey, U.S.A.
Chapter 12

Antonio J. Ricco
National Center for Space
    Biological Technologies
Stanford University
Stanford, California, U.S.A.
Chapter 6

Robert Riehn
Department of Physics
Princeton University
Princeton, New Jersey, U.S.A.
Email: rriehn@ncsu.edu
Chapter 12

Richard Shen
Illumina, Inc.
San Diego, California, U.S.A.
Chapter 2

Frank Steemers
Illumina, Inc.
San Diego, California, U.S.A.
Chapter 2

James C. Sturm
Department of Electrical Engineering
Princeton University
Princeton, New Jersey, U.S.A.
Chapter 12

Saul Surrey
Department of Medicine
Thomas Jefferson University
Philadelphia, Pennsylvania, U.S.A.
Chapter 13

Kai Tang
School of Biological Sciences
Nanyang Technological University
Singapore
Email: tkai@ntu.edu.sg
Chapter 9

Jonas O. Tegenfeldt
Lund University
Lund, Sweden
Chapter 12

Chih-kuan Tung
Department of Physics
Princeton University
Princeton, New Jersey, U.S.A.
Chapter 12

Victor M. Ugaz
Department of Chemical Engineering
Texas A&M University
College Station, Texas, U.S.A.
Email: ugaz@tamu.edu
Chapter 7

Joseph Wang
Departments of Chemical and Materials
    Engineering and Chemistry
Arizona State University
Tempe, Arizona, U.S.A.
Chapter 13

Tza-Huei Wang
Department of Mechanical Engineering
Johns Hopkins University
Baltimore, Maryland, U.S.A.
Email: thwang@jhu.edu
Chapter 11

Yan Mei Wang
Department of Physics
Princeton University
Princeton, New Jersey, U.S.A.
Chapter 12

Adam T. Woolley
Department of Chemistry
    and Biochemistry
Brigham Young University
Provo, Utah, U.S.A.
Email: awoolley@chem.byu.edu
Chapter 5

Jianing Yang
Applied NanoBioscience Center
Arizona State University
Tempe, Arizona, U.S.A.
Email: jianing.yang@asu.edu
Chapter 4

Joanne M. Yeakley
Illumina, Inc.
San Diego, California, U.S.A.
Email: jyeakley@illumina.com
Chapter 2

Hsin-Chih Yeh
Department of Mechanical Engineering
Johns Hopkins University
Baltimore, Maryland, U.S.A.
Chapter 11

Lixin Zhou
Illumina, Inc.
San Diego, California, U.S.A.
Chapter 2

PREFACE

The biochip field, fueled by contributors from both academia and industry, has been growing rapidly. DNA biochips are becoming a widespread tool used in life science, drug screening, and diagnostic applications due to the many benefits of miniaturization and integration. The term "DNA biochip" is used broadly and includes various technologies: DNA microarrays, microfluidics/Lab-on-a-Chip, and other biochips (such as integrated real-time PCR, mass spectrometry, and nanotechnology-based biochips). With the abundance of gene targets and combinatorial chemistry/biology libraries now available, researchers have the ability to study the effects of diseases, environmental factors, drugs, and other treatments on thousands of genes at once. Biochips can provide this information in a number of ways, depending on the type of chips and chosen design of the experiments. They can be used for pharmacogenomics that includes gene expression profiling, the measurement and analysis of regulated genes under various conditions, and genotyping, the detection of polymorphisms or mutations in a gene sequence. Another major application for DNA biochips is molecular diagnostics, which includes genetic screening (e.g., detection of mutations or inherited disorders), identification of pathogens and resistance in infections, and molecular oncology (e.g., cancer diagnosis). Biochips can also be used for high-throughput drug screening, food testing, chemical synthesis, and many other applications. Drawing steadily on expertise from engineering, biology, chemistry, and physics, integrated biochip devices are rapidly becoming sophisticated and affordable.

The objective of this book is to provide up-to-date coverage of some of the emerging developments in the field of integrated DNA biochips. Chapter 1 gives an overview of the biochip field, including its history, present developments, and future trends. Chapters 2-4 cover the latest developments in integrated microarray biochips. In Chapter 2, Yeakley et al describe a bead-based microarray platform that has been used in applications ranging from whole genome genotyping to whole transcriptome expression profiling. In Chapters 3 and 4, focus has been placed on integration of microfluidic technology with microarrays. Two different integrated biochip platforms are described: one is designed to automate and integrate the microarray process on a chip (Chapter 3) and the other is capable of performing sample preparation, DNA amplification, microarray hybridization, and electrochemical detection on a single device (Chapter 4).

Chapters 5 and 6 cover the topic of integrated microfluidic capillary electrophoresis (CE) biochips that have emerged as an especially promising approach for assaying genetic material. In Chapter 5, Woolley describes some of the highly parallel glass microchip CE devices and integrated PCR-CE systems that were developed by Mathies' group. In Chapter 6, Fan and Ricco detail the development of integrated plastic microfluidic CE devices that contain components for DNA amplification, microfluidic valving, sample injection, on-column labeling, separation, and detection.

Chapter 7 reviews the progress that has been made to date in the development of microfluidic devices capable of performing increasingly sophisticated PCR-based bioassays. In Chapter 8, Hong describes nucleic acid analyses on microfluidic chips with particular emphasis on parallel matrix architecture. Several polydimethysiloxane (PDMS)-based integrated systems developed by Quake's group are presented. In Chapter 9, Tang describes chip-based genotyping by mass spectrometry. One of the main advantages of mass spectrometry over other detection methods (e.g., microarrays and CE) is that it provides direct readout of allele types without the need for any labeling. Another label-free detection technology is a surface plasmon resonance based biosensor that is described in Chapter 10 for identification of DNA-protein interactions.

The focus of Chapter 11 is on single-molecule detection. Wang et al review different technologies for SMD, which has been demonstrated in quantitative studies of both specific DNA sequences and mutations and can potentially lead to amplification-free detection of genomic DNA. Both Chapters 12 and 13 describe the recent development of nanotechnology used for DNA analysis. In Chapter 12, Riehn et al provide a fairly complete "state of the art" report of how nanochannels can be used in molecular biology. For example, nanochannels can be used to analyze genomic length DNA molecules with very high linear spatial resolution. Instead of focusing on a particular nanostructure, Chapter 13 gives an overview of various nanotechnologies (including nanoparticles, nanopores, nanochannels, nanowires, nanotubes, etc.) and their applications in molecular diagnostics.

Undoubtedly further progress in the integrated DNA biochip field is expected with the help of microtechnology, nanotechnology, and modern molecular biology. This book will prove a useful source of information for researchers in the field of integrated DNA biochips and for those who are just entering the field of biochip research.

We wish to thank all of the contributing authors for their constantly high level of motivation and enthusiasm and for providing such good manuscripts. We are also thankful to our families (Liu's family including Fiona, Vick, Benjamin, Stanley, Shaowen, and Zhongji; Lee's family including Chi, Carmel, and Calvin) for their patience and support, and for sacrificing precious time that allowed us to launch into this promising project.

Robin Hui Liu
Abraham P. Lee

CHAPTER 1

Integrated DNA Biochips: Past, Present and Future

Piotr Grodzinski,* Robin Hui Liu and Abraham P. Lee

Early Developments

DNA biochip is becoming a widespread tool used in life science, drug screening and diagnostic applications due to its many benefits of miniaturization and integration. The term "DNA biochip" is used broadly and includes various technologies: DNA microarrays, microfluidics/Lab-on-a-Chip and other biochips (such as integrated real-time PCR, mass spectrometry and nanotechnology-based biochips). With the abundance of gene targets and combinatorial chemistry/biology libraries now available, researchers need the ability to study the effects of diseases, environmental factors, drugs and other treatments on thousands of genes at once. Biochips can provide this information in a number of ways, depending on the type of chips and chosen design of the experiment. They can be used for pharmacogenomics that includes gene expression profiling, the measurement and analysis of regulated genes under various conditions and genotyping, the detection of polymorphisms or mutations in a gene sequence. Another major application for DNA biochips is molecular diagnostics, which includes genetic screening (e.g., detection of mutations or inherited disorders), identification of pathogens and resistance in infections and molecular oncology (e.g., cancer diagnosis). Biochips can also be used for high-throughput drug screening, chemical synthesis and many other applications.

Microarrays are glass slides or other substrates with biomolecular probes immobilized on the surface to detect DNA targets in the sample solution through hybridization. The area of DNA microarrays enjoyed unprecedented growth in last two decades.[1,2] These arrays allow for a highly parallel analysis of a multitude of single-stranded DNA fragments. While ultra-high-density arrays are powerful in expression analysis, highly parallel low or medium density arrays find their use in many other applications such as clinical diagnostics and pharmacogenomic applications based on genotyping and SNP scoring. Early demonstration of the microarray concept was provided in Fodor's work.[1] Their method relied on photolithography and combinatorial chemistry allowing for parallel synthesis of probes: nucleotide-by-nucleotide in two-dimensional format. The use of photolabile chemistry provided for deprotection of the linker under the UV light exposure and attachment of the nucleotide. The advantage of this method is its high parallelism enabling fabrication of densely populated arrays (spot size in tens of microns). However, it is a technologically complex process associated with high number of lithographic steps needed to build even short oligonucleotide probes (~4 masks for each base). Nevertheless, the commercialization of this approach resulted in formation of Affymetrix, which is currently one of the largest suppliers of commercial DNA microarrays.

It is somewhat intriguing that the first demonstration of DNA microarray involved such a complex technological process, leveraging semiconductor industry experience, even before the

*Corresponding Author: Piotr Grodzinski—National Cancer Institute, Bethesda, Maryland U.S.A. Email: grodzinp@mail.nih.gov

Integrated Biochips for DNA Analysis, edited by Robin Hui Liu and Abraham P. Lee.
©2007 Landes Bioscience and Springer Science+Business Media.

synergy of molecular biology and engineering was exploited. Simpler designs, involving spotting of presynthesized oligonucleotides on the surface in predetermined two-dimensional pattern followed. The use of thermal, piezoelectric and solenoid-based spotters,[2] similar to those developed for ink-jet industry enabled deposition of minute amounts (in picoliter range) of oligonucleotides in humidity-controlled environment. This method allowed for an inexpensive approach to building custom-design, 'home-made' arrays in the research labs.

Second and parallel venue towards developing DNA analysis on the chip leveraged a concept of DNA fragment separation relying on electrophoresis and using microfluidic channels resulting in significantly reduced time of analysis.[3-5] Microfluidics technology involves movement and processing of samples and reagents inside the microchannels of a chip. The technology is descended from a field referred to as micro total analysis systems (μTAS), which first came about in the late 80s. The benefits of miniaturizing various lab equipment down to the size of a chip have been pursued for many years—the first such innovations took place as far back as the 60s and 70s with the development of microfabricated sensors for process monitoring and other applications.[6] This was the beginning of the micro-electromechanical systems (MEMS) field, which has branched into many other industries such as bioMEMS, optics and telecommunications.[6]

Microfluidics allows a reduction in the system size with a corresponding increase in the throughput of handling, processing and analyzing a biological sample. Other advantages of microfluidics include increased reaction rates, enhanced detection sensitivity and increasing level of device integration. Early devices[3-5] used glass chips allowing for the definition of a nL volume sample plug and subsequent separation of DNA fragments in the presence of a strong electric field. The applications of these miniaturized planar capillary electrophoresis (CE) biochips include nucleic acid analyses and protein assays for separating, sizing, quantifying and identifying the content of a sample of DNA, RNA or protein extracted from cells. These systems have been later expanded to 96-channel parallel microplate structures.[7] Microfluidics and DNA microarrays have been morphing into sophisticated concepts of 'lab-on-a-chip,' as we show in the subsequent parts of this chapter.

Formative Years

Maturation of DNA biochip technologies have subsequently occurred along a few parallel directions: (1) new methods towards microarray fabrication have been devised, (2) new approaches to detection (DNA binding recognition) have been assessed, (3) integration of, additional to hybridization based detection, functions such as sample preparation, mixing and sample transport using microfluidics technology and (4) development of rapid, high-throughput and fully integrated microfluidic separation analyses.

New Microarray Fabrication Methods

The use of on-chip in situ probe synthesis provides an attractive alternative to hybridization microarrays fabricated using presynthesized oligonucleotides and their physical deposition on the substrate. Chips with in situ probe synthesis are much more flexible to build and valuable in the experiments where new sequences need to be introduced frequently. Fodor's approach utilizes multiple-level photolithographic mask process.[1] Nuwaysir at Nimblegen has modified this technique using Digital Light Processor (DLP) to develop the Maskless Array Synthesizer (MAS) that delivers focused light beams to selected portions of the substrate, thus, eliminating the need for the photomasks.[8] Zhang's group has experimented with this technique further.[9] They added microfluidic assemblies to the chip to enable ease of solution manipulation and delivery to the synthesis site. They also varied the chemistry using reactants protected with acid-labile groups, which in turn can be cleaved with photogenerated acid (PGA).

Dill et al at Combimatrix introduced electrochemical in situ synthesis using electronic CMOS devices.[10,11] These chips contain high-density arrays of individually addressable microelectrodes. The use of phosphoramidite chemistry at the electrode sites, combined with sequential addressing of these electrodes in order to produce protons, results in the on-chip synthesis of oligonucleotides.[10]

CombiMatrix demonstrated in situ synthesis of 50-mers oligonucleotides; subsequently using these arrays for genotyping and gene expression assays.

Most of DNA microarrays mentioned above are fabricated on flat substrate surfaces. Illumina has developed a bead-based microarrray made by random positioning of 3 µm spherical silica beads with covalently attached oligonucleotide probes into wells manufactured in a solid surface.[12] This technique results in a high-density oligo microarray with a density of 500,000 bead types or more. The oligo probes on each set of beads are synthesized individually using standard chemistry, only full length oligos bearing a 5' modification are attached to beads and only beads that can hybridize to a target can be decoded (determining which randomly distributed bead is in which well). The decoding process involves iterative hybridizations with fluorescent oligo pools. The reassignment of fluors in the oligo pool for each hybridization step allows the unique identification of each bead by its particular sequence of fluors across a series of hybridization steps.

Alternate Detection Techniques

Early microarray demonstrations have utilized optical detection and fluorescence labeling of the target nucleotides that subsequently bind to surface immobilized probes and are imaged in the post-hybridization step. Fluorescence intensity compared with internal standards allows for assessment of hybridization rate. Multi-color labeling approaches in order to allow for multiplex recognition have also been developed.[13]

However, alternate electronic detection of DNA hybridization has been developed rapidly, since it offered potential towards further miniaturization of DNA diagnostic systems.[11,14,15] Use of electrochemical detection schemes obviates the need for large and expensive fluorescent scanners. The detection chips can be produced inexpensively using Si- or printed circuit board-based fabrication technologies.

Farkas et al have developed electrochemical detection based low-density hybridization arrays for diagnostic applications, in particular, detection of single-nucleotide-polymorphisms (SNPs).[16,17] The device is composed of a printed circuit board (PCB) consisting of an array of gold electrodes modified with a multi-component, self-assembled monolayer (SAM) that includes presynthesized oligonucleotide capture probes. Two ferrocene, electronic labels (capture probe and signaling probe) are used in the hybridization recognition process: they bind the target in a sandwich configuration. The presence of hybridized (double-stranded) DNA is detected using alternative current voltammetry (ACV) technique. Since incoming target itself is not labeled, the washing step (to remove excessive, nonbound target) prior to the signal collection is not required. A continuous monitoring of the binding process with a quantitative measurement of the target accumulation is possible.

Barton and coworkers used electron flow through a double helical DNA molecule to generate a current, which is dependent upon complementary base pairing within the helix. Perfect matches allow electrons to flow through the helix generating a measurable current, while single-base mismatches limit current flow.[18] Each position on the array has an independent current sensor at each register, such that parallel analysis is possible.

Departure from optical or electrochemical detection methods has also been accomplished through the development of Bead Array Counter (BARC) devices.[19,20] This system employs giant magneto-resistance (GMR) multilayer magnetic sensors for recognition of DNA hybridization events when target is labeled with magnetic beads. The detector size reduction can occur due to feasibility of integrating the magnetic sensor with electronics for signal acquisition on a single chip. Furthermore, this approach allows for improvement of reaction kinetics and control of hybridization stringency. The former is achieved through utilization of magnetic field for target preconcentration at the probe location and the latter through use of an AC field to repel mismatched targets.

DNA Microarray Biochips Integrated into Microfluidic Systems

Early developments of DNA microarrays were focused on the development and optimization of detection techniques. With the need of developing practical diagnostic approaches, the importance of building complete sample-to-answer systems emerged. The advances have been multi-directional and have addressed duration of the analysis, functional complexity, level of integration and on-chip sample preparation.

Microfluidic techniques have been employed to improve the kinetics of hybridization process since conventional DNA hybridization assays rely solely on the diffusion of target to the surface-bound probes. It has been recognized that mixing is important to achieve enhanced rates of hybridization and various methods were devised to accelerate this process. They include electronic enhancement of DNA hybridization,[21,22] dynamic DNA hybridization using paramagnetic beads,[23] rotation of the whole device,[24] the use of a micro porous three-dimensional biochip with the hybridization solution being pumped continuously through it,[25] shuttling the sample in microfluidic channel[26] and acoustic mixing.[27]

Other important works have been dedicated to the integration of sample collection and pretreatment with the DNA extraction, amplification and detection. Kricka and Wilding have demonstrated physical filters relying on separation of biological cells by size.[28] Anderson et al integrated monolithic genetic assay devices to carry out serial and parallel multistep molecular operations, including nucleic acid hybridization.[29] Liu and Grodzinski have developed plastic, disposable chips for pathogen detection, performing PCR amplification, DNA hybridization and a hybridization wash in a single device.[30] On-chip valving using phase change pluronics material was also implemented to facilitate for a separation of different stages of the assay. The level of integration was expanded further to a complete self-contained biochip capable of magnetic bead-based cell capture, cell preconcentration, purification, lysis, PCR amplification, DNA hybridization and electrochemical detection of hybridization events.[31] The device is completely self-contained and does not require external pressure sources, fluid storage, mechanical pumps or valves. Wilding, Kricka and Fortina have also developed a prototype of an integrated semi-disposable microchip analyzer for cell separation and isolation, PCR amplification and amplicon detection which is described with preliminary results.[32]

High-Throughput and Fully Integrated Microchip Separation Analyses

In addition to nucleic acid hybridization microarrays, microchip DNA separation is another important area of miniaturized DNA analysis. Significant progress has been made in the area of integrated miniaturized planar CE biochips. Devices with highly parallel arrays of micromachined separation channels that are integrated together to facilitate detection have been developed. The first reported microchip parallel DNA separation was performed in a 12-channel capillary array electrophoresis (CAE) microdevice to determine HFE gene variants from multiple individuals.[33] To increase the number of lanes that could be probed simultaneously, Mathies and coworkers developed 96-channel CAE microchips with a radial layout.[34] Samples were introduced near the perimeter of a circular substrate and confocal fluorescence detection occurred where the lanes converged near the center of the device. Demonstration of sequencing DNA from the phage M13mp18 using this CAE microplate showed a separation speed of 1.7 kb/min that corresponded to a 5-fold increase in separation throughput as compared with commercially available capillary array electrophoresis instrumentation. Later efforts to increase DNA sample capacity have led to the development of a 384-lane CAE microplate.[35] This system was used to simultaneously test 384 samples for a mutation in the HFE gene in just 7 min with 98.7% accuracy.

Not only has the microfluidic format proven advantageous for rapid, high-throughput CE analyses, but on-chip sample preparation steps including PCR have also been integrated into the microchip CE platform. PCR that amplifies a small number of copies of a DNA template to obtain a sufficient quantity for analysis is one of the most essential steps for preparing CE sample prior to electrophoresis separation and detection. Mathies and coworkers have developed a micromachined, integrated PCR/CE device that consisted of a 200-nL PCR chamber and a CE microchannel.[36] Integrated valves and hydrophobic vents were used to isolate the PCR chamber from the CE analyzer. Resistive heaters and resistance temperature detectors were photolithographically patterned on the device to provide heating and temperature monitoring during the PCR thermal cycling.

Advantages of this PCR/CE platform include small PCR reaction volume and efficient sample transfer. Integration of more front-end sample preparation has also been demonstrated. For example, Landers and coworkers developed a microchip device that performed DNA extraction prior to PCR amplification.[37] The device consists of a microchannel containing sol-gel-immobilized silica beads, which adsorbed DNA and other cellular components when a whole-blood sample was passed through. Undesired materials were removed with an isopropanol/water wash and the PCR-ready DNA sample was eluted with a low ionic-strength buffer. The devices reported by Burns et al are capable of metering aqueous reagents, mixing, amplifying, enzymatic digesting, electrophoretic separation and detection with no external lenses, heaters, or mechanical pumps.[38]

New Developments

Multiplex Tagging

New approaches devised by an emerging field of nanotechnology have found relevant applications and have provided for improvements in DNA hybridization assays.

Most of mature DNA microarray solutions have been built in planar configurations, where recognition sites with single-stranded probes are placed in a two-dimensional pattern on the flat surface. This approach can be modified to a tag-by-tag recognition in a flow system, providing that multiplex tagging of unknown molecules in a sample can be achieved. Highly multiplex tags have been developed based on quantum dots (QDs)[39] and metallic barcodes.[40] Tagging molecules with QDs have several advantages over standard fluorophore tags[41,42] since their absorption spectra are very broad, extending from the ultraviolet to a cutoff wavelength in the visible spectrum. Emission, on the other hand, is confined to a narrow band (typically 20-40 nm full width at half maximum) with a center wavelength being determined by the particle size. The differentiation of emission color, intensity and spectral width can lead to multiplex library of unique signatures. Nie's group, in their development of complex material microbeads built with embedded QDs, claimed theoretical multiplex capacity of one million.[39] The QD signatures can be recognized through optical means; Wang showed recently that they also could be used for coding in electrochemical detection schemes.[43]

Similarly, multi-metal nanorods with bar-coded stripes that can be recognized using reflectivity measurements have been demonstrated and can be used as molecular tags.[40] The nanorods are produced using a lithographic process, whereby different metal stripes (barcodes) are deposited on narrow silicon features. A subsequent read-out of such barcodes is performed using an optical microscope. The nanorod tagging can be performed in parallel with optical tagging, since the two detection methods do not interfere with each other and thus further add to the scale of the multiplexing capability.

Label-Free Detection

Most molecular recognition techniques rely on a binding event and subsequent interrogation of the optical,[1] electrochemical,[44] or magnetic tag[19,20] carried by the molecule involved in binding. An attractive and highly desirable alternative to this strategy would eliminate the tagging step and instead, rely on the detection of the change of an inherent property of the analyte or the molecular aggregate formed upon binding.

Lee's group explored the difference in dielectric properties of ssDNA and hybridized dsDNA fragments.[45,46] They developed nano-gap capacitors (50 nm electrode spacing) with ssDNA probe immobilized within the device. The change of dielectric constant upon hybridization can be measured through capacitance measurements. Such devices can be built into large two-dimensional arrays that could be used to provide capacitive, label-free simultaneous measurement of nucleic acid targets in a sample.

Nanomechanical deflections on micromachined silicon cantilevers have been also used to recognize the occurrence of DNA hybridization.[47] ssDNA is immobilized on the surface of a cantilever and subsequently target ssDNA is introduced. The binding on the surface contributes to the increased loading of the cantilever and its proportional deflection. The deflection is accurately measured using optical means.[48] An expansion of this method to multiplex recognition and prevention of nonspecific binding is still a challenge; however, the method is a clever demonstration utilizing the power of sophisticated microfabrication in assay applications.

A novel approach to sequencing of ssDNA fragments has been proposed based on electrophoretic transport of DNA chains through a single nanopore fabricated in a silicon nitride membrane[49] or through an alpha-hemolysis pore in lipid bilayers.[50] The pore dimensions were, in both cases, less than 10 nm in diameter. Measurements of the cross-pore current indicated that individual polynucleotides transported through the pore could provide a unique "signature" and could lead eventually to low-cost, rapid and direct methods for DNA sequence analysis.

Enhanced Sensitivity

Colloidal gold has been used in DNA assays with colorimetric read-out (color changes occur due to nanoparticle aggregation).[51-53] A subsequent combination of surface enhanced Raman spectroscopy (SERS) with gold labeling has enabled transformation of this method to a highly sensitive detection technique. Metal nanoparticles have distinct advantages over fluorescent dyes as labels. They have narrow Raman emission peaks and the surface plasmon resonance is directly related to nanoparticle size. Raman spectroscopy relies on detecting a signal that is only 0.0001% of the light that is scattered from the sample, thus the combination of a low signal intensity and fluorescence background could be problematic. SERS solves the problem of auto-fluorescence and background fluorescence through metal adsorption of the sample with metals like silver and gold. The resonant excitation of surface plasmon on the metal surface results in a Raman signal enhancement[54] compared to the Raman signal generated from the sample alone. SERS signal can be subsequently increased up to a trillion-fold using Ag enhancement.[55] SERS has been successfully used in DNA assays and allowed for detection sensitivity in the low zettomolar range.[55]

SERS in combination with bar-code technique led to more sophisticated multiplex DNA detection strategies. Mirkin's group developed the bar-codes made up of short sequences of DNA strands.[52,55] Gold nanoparticles carry the oligonucleotide/barcode DNA and are subsequently bound to magnetic particles carrying the complementary strand of DNA for hybridization DNA assays.[52,55] The bound gold nanoparticle—magnetic particle complexes are magnetically captured, the barcode DNA released and detected on arrays after oligonucleotide annealing to bound complements and detection using complementary oligonucleotide-bound gold nanoparticles.

Fully Integrated and Miniaturized Genetic Analysis Systems

Portable, integrated genetic analyzers have been developed for on-site pathogen detection.[56] Mathies et al developed an integrated, portable genetic analysis system measuring 8 × 10 × 12 in. The system included a microfluidic PCR/CE biochip that consisted of a 200-nL PCR chamber and microfabricated heaters and temperature sensors connected to a 7-cm CE microchannel. The biochip was filled with LPA sieving matrix and was loaded with sample and PCR reagents before being clamped into the instrument that provided pressure-driven pumping (using a rotary pump and two solenoids), high-voltage power supplies and electrical contacts for heating and electrophoresis separation. The instrument also included a confocal fluorescence detection unit that consisted of a solid-state laser, lenses, a dichroic beam splitter, excitation and emission filters and a photomultiplier tube. This portable PCR/CE system has been used for pathogen detection. Three *E. coli* strains were amplified with three sets of primers, allowing simultaneous identification of the bacterial species, testing for a specific serotype and probing for the shigatoxin I gene, which is present in enterohemorrhagic *E. coli*. Using this integrated system, a detection limit of 2 ~ 3 cells was achieved. Moreover, 10 *E. coli* O157: H7 cells in a mixture containing a background of *E. coli* K12 cells, which were 4 orders of magnitude more abundant, were also detected. The on-chip analysis was completed within 30 min.

Taylor and coworkers worked on a system with on-line PCR detection and showed on-chip spore preconcentration, lysing and DNA purification.[57] A similar prototype cartridge system that rapidly disrupts Bacillus spores by sonication, adds PCR reagent to the disrupted spores and dispenses the mixture into a PCR tube was reported by Belgrader et al.[58] The total time to automatically process the spores in the cartridge and then detect the spore DNA by real-time PCR was 20 min. This technology was commercialized by Cepheid and resulted in the GeneXpert® System (www.cepheid.com). The system automates and integrates all the steps required for real-time PCR-based DNA testing: sample preparation, DNA amplification and detection. The system consists of a microfluidic cartridge that incorporates a syringe drive, rotary drive and a sonic horn. The sonic horn delivers ultrasonic energy necessary to lyse the raw specimen and release nucleic acids contained within, while the combination of the syringe drive and rotary drive moves liquid between cartridge chambers in order to wash, purify and concentrate these nucleic acids. After the automated extraction is complete, the nucleic acid concentrate is moved into the cartridge reaction chamber where real-time PCR amplification and detection take place. The GeneXpert® module currently forms the core of the Biohazard Detection System deployed nationwide by the United States Postal Service for anthrax testing in mail sorting facilities.

Future and Path Forward

The DNA biochips carved a significant niche in today's biotechnological reality. With further development of DNA-based diagnostics, the push for the development of fully integrated biochip assays will continue. Similarly, high throughput genetic screening studies will fuel further developments of high-density microarrays. A transfer of these assay solutions to application space will, most likely, rely on more traditional solutions and fixed, two-dimensional arrays. At the same time, the desire for innovation will continuously explore new, more exotic approaches relying, for example, on nanowires,[59] carbon nanotubes,[60] as well as their large arrays.[61] The development of the future microfluidic lab-on-a-chip technologies will be continuously focused on addressing the following issues: high throughput of the analysis, increasing functional complexity, increasing level of integration, integration of on-chip sample preparation and low-cost fabrication and manufacturing.

All biochips developed to-date operated in invitro environment and analyzed the sample extracted from the patient: from the bodily fluids or through the biopsy. The next challenge will be to develop systems robust enough to operate in invivo environment and thus provide for earlier biomarker recognition at the disease site.

References

1. Fodor SPA, Read JL, Pirrung MC et al. Light-Directed, Spatially Addressable Parallel Chemical Synthesis. Science 1991; 251(4995):767-773.
2. Schena M. Microarray BiochipTechnology. Natick, MA: Eaton Publishing, 2000.
3. Harrison DJ, Fluri K, Seiler K et al. Micromachining a Miniaturized Capillary Electrophoresis-based Chemical Analysis System on a Chip. Science 1993; 261:895-897.
4. Manz A, Harrison DJ, Verpoorte E et al. Planar Chips Technology for Miniaturization and Integration of Separation Techniques into Monitoring Systems: Capillary Electrophoresis on a Chip. J of Chromatogr 1992; 593:253-258.
5. Woolley AT, Mathies RA. Ultra-High-Speed DNA Fragment Separations Using Microfabricated Capillary Array Electrophoresis. Proc Natl Acad Sci USA 1994; 91:11348.
6. Kovacs GTA. Micromachined Transducers Sourcebook. Boston: WCB McGraw-Hill, 1998.
7. Shi YN, Simpson PC, Scherer JR et al. Radial capillary array electrophoresis microplate and scanner for high-performance nucleic acid analysis. Anal Chem 1999; 71(23):5354-5361.
8. Nuwaysir EF, Huang W, Albert TJ. Gene Expression Analysis Using Oligonucleotide Arrays Produced by Maskless Photolithography. Genome Res 2002; 1749-1755.
9. Gao X, Yu P, LeProust E et al. Oligonucleotide Synthesis Using Solution Photogenerated Acids. J Am Chem Soc 1998; 120:12698-12699.
10. Oleinikov AV, Gray MD, Zhao J et al. Self-Assembling Protein Arrays Using Electronic Semiconductor Microchips and in vitro Translation. J Proteome Res 2003; 2:313.
11. Dill K, Montgomery DD, Ghindilis AL et al. Immunoassays and sequence-specific DNA detection on a microchip using enzyme amplified electrochemical detection. J Biochem Biophys Methods 2004; 59:181-187.
12. Gunderson KL, Kruglyak S, Graige MS. Decoding randomly ordered DNA arrays. Genome Res 2004; 14:870-877.
13. Fortina P, Delgrosso K, Sakazume T et al. Simple two-color array-based approach for mutation detection. EJHG 2000; 8:884-894.

14. Wang J. Electroanalysis and Biosensors. Anal Chem 1999; 71:328R-332R.
15. Wang J. From DNA biosensors to gene chips. Nucleic Acids Res 2000; 28(16):3011-3016.
16. Farkas DH. Bioelectronic detection of DNA and the automation of molecular diagnostics. J Assoc Laboratory Automation 1999; 4(20-24).
17. Umek RM, Vielmetter J, Terbrueggen RH et al. Electronic detection of nucleic acids: a versatile platform for molecular diagnostics. J Mol Diagn 2001; 3:74-84.
18. Drummond TG, Hill MG, J.K. B. Electrochemical DNA Sensors. Nat Biotechnol 2003; 10:1192-1199.
19. Edelstein RL, Tamanaha CR, Sheehan PE et al. The BARC biosensor applied to the detection of biological warfare agents. Biosens Bioelectron 2000; 14:805-813.
20. Miller MM, Sheehan PE, Edelstein RL et al. A DNA array sensor utilizing magnetic microbeads and magnetoelectronic detection. MAGMA 2001; 225(138-144).
21. Edman C, Raymond D, Wu D et al. Electric Field Directed Nucleic Acid Hybridization on Microchips. Nucl Acids Res 1998; 25:4907-4914.
22. Radtkey R, Feng L, Muralhidar M et al. Rapid, high fidelity analysis of simple sequence repeats on an electronically active DNA microchip. Nucleic Acids Res 2000; 28:E17.
23. Fan ZH, Mangru S, Granzow R et al. Dynamic DNA hybridization on a chip using paramagnetic beads. Anal Chem 1999; 71:4851-4859.
24. Chee M, Yang R, Hubbell E et al. Accessing genetic information with high-density DNA arrays. Science 1996; 274:610-614.
25. Cheek BJ, Steel AB, Torres MP et al. Chemiluminescence Detection for Hybridization Assays on the Flow-Thru Chip, a Three-Dimensional Microchannel Biochip. Anal Chem 2001; 73:5777-5783.
26. Lenigk R, Liu RH, Athavale M et al. Plastic biochannel hybridization devices: a new concept for microfluidic DNA arrays. Anal Biochem 2002; 311(1):40-49.
27. Liu RH, Lenigk R, Yang J et al. Hybridization Enhancement Using Cavitation Microstreaming. Anal Chem 2003; 75:1911-1917.
28. Wilding P, Kricka LJ, Cheng J et al. Integrated cell isolation and polymerase chain reaction analysis using silicon microfilter chambers. Anal Biochem 1998; 257(2):95-100.
29. Anderson RC, Su X, Bogdan GJ et al. A miniature integrated device for automated multistep genetic assays. Nucleic Acids Res 2000; 28(12):E60.
30. Liu Y, Rauch C, Stevens R et al. DNA Amplification and Hybridization Assays in Integrated Plastic Monolithic Devices. Anal Chem 2002; 74(13):3063-3070.
31. Liu RH, Yang J, Lenigk R et al. Self-Contained, Fully Integrated Biochip for Sample Preparation, Polymerase Chain Reaction Amplification and DNA Microarray Detection. Anal Chem 2004; 76:1824.
32. Yuen P, Kricka L, Fortina P et al. Microchip Module for Blood Sample Preparation and Nucleic Acid Amplification Reactions. Genome Res 2001; 11:405-412.
33. Woolley AT, Sensabaugh GF, Mathies RA. High-speed DNA genotyping using microfabricated capillary array electrophoresis chips. Anal Chem 1997; 69:2181-2186.
34. Paegel BM, Emrich CA, Weyemayer GJ et al. High throughput DNA sequencing with a microfabricated 96-lane capillary array electrophoresis bioprocessor. Proc Natl Acad Sci USA 2002; 99(2):574-579.
35. Emrich CA, Tian HJ, Medintz IL et al. Microfabricated 384-lane capillary array electrophoresis bioanalyzer for ultrahigh-throughput genetic analysis. Anal Chem 2002; 74(19):5076-5083.
36. Lagally ET, Medintz I, Mathies RA. Single-Molecule DNA Amplification and Analysis in an Integrated Microfluidic Device. Anal Chem 2001; 73:565-570.
37. Breadmore MC, Wolfe KA, Arcibal IG, al. e. Microchip-based purification of DNA from biological samples. Anal Chem 2003; 75:1880-1886.
38. Burns MA, Johnson BN, Brahmasandra SN et al. An integrated nanoliter DNA analysis device. Science 1998; 282(5388):484-487.
39. Gao X, al. e. Quantum-dot nanocrystals for ultrasensitive biological labeling and multicolor optical encoding. J Biomed Optics 2002; 7:532-537.
40. Nicewarner-Pena SR, al. e. Submicrometer metallic barcodes. Science 2001; 294:137-141.
41. Bruchez MJ, al. e. Semiconductor nanocrystals as fluorescent biological labels. Science 1998; 281:2013-2016.
42. Chan WC, al. e. Luminescent quantum dots for multiplexed biological detection and imaging. Curr Opin Biotechnol 2002; 13:40-46.
43. Wang J, al. e. Electrochemical coding technology for simultaneous detection of multiple DNA targets. J Am Chem Soc 2003; 125:3214-3215.
44. Wang J. From DNA biosensors to gene chips. Nucleic Acids Res 2000; 28(16):3011-3016.
45. Yi M, Jeong K, Lee LP. Theoretical and Experimental Study towards a Nanogap Dielectric Biosensor. Biosens Bioelectron 2005; 20:1320-1326.

46. Ionescu-Zanetti C, Nevill JT, Carlo DD et al. Nanogap Capacitors: Sensitivity to Sample Permittivity Changes. J Appl Phys 2006; 99:024305.
47. Fritz J, al. e. Translating biomolecular recognition into nanomechanics. Science 2000; 288:316-318.
48. Wu G, al. e. Bioassay of prostate-specific antigen (PSA) using microcantilevers. Nat Biotechnol 2001b; 19:856-860.
49. Li J, al. e. Solid state nanopore as a single DNA molecule detector. Biophys J 2003; 84:134A-135A.
50. Meller A, al. e. Rapid nanopore discrimination between single polynucleotide molecules. Proc Nat Acad Sci USA 2000; 97:1079-1084.
51. Bensch K, Gordon G, Miller L. Electron microscopic and cytochemical studies on DNA-containing particles phagocytized by mammalian cells. Trans NY Acad Sci 1966; 28(6):715-725.
52. Cao Y, Jin R, Mirkin CA. DNA-modified core-shell Ag/Au nanoparticles. J Am Chem Soc 2001; 123(32):7961-7962.
53. Elghanian R, Storhoff JJ, Mucic RC et al. Selective Colorimetric detection of polynucleotides based on the distance-dependent optical properties of gold-nanoparticles. Science 1997; 277:1078-1081.
54. Baltog I, Mihut L, Timucin V. Laboratory optics and spectroscopy. Natl Inst Mater Phys Ann Rep 2000; 58-60.
55. Nam JM, Stoeva SI, Mirkin CA. Bio-bar-code-based DNA detection with PCR-like sensitivity. J Am Chem Soc 2004; 126(19):5932-5933.
56. Lagally ET, Scherer JR, Blazej RG et al. Integrated portable genetic analysis microsystem for pathogen/infectious disease detection. Anal Chem 2004; 76:3162-3170.
57. Taylor MT, Belgrader P, Joshi R et al. Fully Automated Sample Preparation for Pathogen Detection Performed in a Microfluidic Cassette. In: Van den Berg A, ed. Micro Total Analysis Systems 2001. Monterey, CA: Kluwer Academic Publishers 2001:670-672.
58. Belgrader P, Okuzumi M, Pourahmadi F et al. A microfluidic cartridge to prepare spores for PCR analysis. Biosens Bioelectron 2000; 14:849-852.
59. Cui Y, Wei Q, Park H et al. Nanowire nanosensors for highly sensitive and selective detection of biological and chemical species. Science 2001; 293:1289-1292.
60. Chen RJ, Bangsaruntip S, Drouvalakis KA et al. Noncovalent functionalization of carbon nanotubes for highly specific electronic biosensors. Proc Natl Acad Sci USA 2003; 100(9):4984-4989.
61. Jin S, Whang D, McAlpine MC et al. Scalable interconnection and integration of nanowire devices without registration. Nano Lett 2004; 4:915-919.

CHAPTER 2

Integrated Molecular Analyses of Biological Samples on a Bead-Based Microarray Platform

Joanne M. Yeakley,* Daniel A. Peiffer, Marina Bibikova, Tim McDaniel, Kevin L. Gunderson, Richard Shen, Bahram G. Kermani, Lixin Zhou, Eugene Chudin, Shawn C. Baker, Kenneth M. Kuhn, Frank Steemers, Mark Hansen, Michael Graige, Celeste McBride, Steven Barnard, Bob Kain, David Barker and Jian-Bing Fan

Abstract

Molecular analyses of biological samples have traditionally been pursued in parallel, with those researchers studying genetic diversity having few technical approaches in common with those studying gene expression. Increasingly, scientists recognize the importance of integrating analytical technologies to further research, particularly into emerging fields such as epigenetics and the genetics of gene expression. In this chapter, we describe a suite of applications that take advantage of the Illumina® bead-based microarrays, all of which are read out on a single analytical instrument. The integration of whole genome genotyping, high throughput focused genotyping, whole transcriptome expression profiling, focused expression profiling of fresh or preserved tissues, allele-specific expression profiling and DNA methylation assays on the BeadArray™ Reader allows researchers to expand their perspectives, from whole genomes to single bases, from genetics to expression and on to epigenetics.

Introduction

Over the last decade or so, the use of microarray technologies to analyze nucleic acids from biological samples has revolutionized basic and applied biological research. Initial biological studies employing microarrays included detection of genomic differences[1] and gene expression monitoring.[2] Among these studies were analyses of differential expression in normal versus cancer tissue samples, allowing the identification of potential biomarkers for disease which would otherwise not have been readily achieved.[3-5] Coupled with highly parallel genomic assays, the extraordinary capability of this technology was easily recognized and rapidly adopted and has led to the expansion of microarray approaches to many other biological problems. Several recent reviews summarize the use of microarrays in biology and discuss the impacts they have had on experimental design and data analysis.[6-11]

While the types of microarrays and the assays deployed on them has exploded, DNA microarrays all rely on the underlying physical principle of the hybridization of complementary sequences. Fundamentally, DNA microarrays are comprised of DNA sequences immobilized on a surface,

*Corresponding Author: Joanne M. Yeakley—Illumina, Inc., 9885 Towne Centre Drive, San Diego, CA 92121-1975, U.S.A. Email: jyeakley@illumina.com

Integrated Biochips for DNA Analysis, edited by Robin Hui Liu and Abraham P. Lee. ©2007 Landes Bioscience and Springer Science+Business Media.

which are used to monitor the abundance of labeled complementary sequences generated by an assay of a biological sample. This concept is essentially the miniaturization and parallelization of traditional hybridization-based nucleic acid detection techniques.

Early microarrays employed spotting of DNAs derived from cloned fragments of genomic DNA (gDNA) or complementary DNA (cDNA) reverse transcribed from cellular RNA. These arrays were made using robots to spot small aliquots of known sequences at known locations on glass slides at low density.[1,2] To reduce the labeled biological material needed and to increase the number of sequence probes, microarrays evolved to high density positioning of oligonucleotide (oligo) probes. In a photolithographic approach developed by Affymetrix®, oligos are chemically synthesized on the surface of a silica chip, where the location of each probe is determined by masks used during oligo synthesis.[12] In contrast, Illumina high density microarrays are made by random positioning of microspheres bearing oligo probes into wells manufactured in a solid surface.[13,14] Both techniques result in a high density oligo microarray that can be used to perform genomic assays, though the strengths and limitations of the platforms differ. This chapter will describe some of the applications Illumina has developed for the bead-based technology.

The concept behind Illumina's BeadArray technology is straightforward: three-micron spherical silica beads with covalently attached oligonucleotides are randomly assembled into wells made in a solid substrate (Fig. 1A). On average, each bead carries 10^5 to 10^6 copies of the same oligo. The oligo sequence defines the "bead type". There are two substrate types currently offered: Sentrix® Array Matrix (SAM) arrays and Sentrix BeadChip arrays.

SAM arrays are manufactured by arranging ~1.4 mm diameter fiber optic bundles, each containing approximately 50,000 individual fiberoptic strands, into a 8 × 12 matrix that matches a standard 96-well microtiter plate (Fig. 1B). Because the core glass material of each individual fiber is designed to be softer than the cladding glass, wells can be formed on the ends of the fiber bundles by etching the softer glass with a mild acid. A pool that contains 1624 individual bead types is loaded onto the fiber bundle surface and the beads fall randomly into the wells. With 50,000 wells available, the loaded arrays contain about 30 beads of each type. This representation of 30 beads per bead type in each array allows redundancy in the assay fluorescence measurements, improving the robustness of the intensity data.

The Sentrix BeadChip (Fig. 1B), use micro-electromechanical systems (MEMS) technology to produce wells in a silicon slide. Each array section contains over one million wells. Thus, bead pools that contain thousands of bead types can be loaded onto each section while maintaining a representation of about 15 to 30 beads of each bead type. As shown in Figure 1B, BeadChips can be manufactured with various numbers of sections. Thus, the BeadChip format has the flexibility to create arrays of different dimensions to allow complexity of 500,000 bead types or more.

Among the advantages of this general approach are that oligo probes are synthesized individually using standard chemistry, only full length oligos bearing a 5' modification are attached to beads and only beads that can hybridize to a target can be decoded (determining which bead is in which well). Thus, the necessity of decoding the randomly distributed beads confers automatic quality control for the array. The decoding process involves iterative hybridizations with fluorescent oligo pools and has been described in detail by Gunderson et al.[13] Briefly, the reassignment of fluors in the oligo pool for each hybridization step allows the unique identification of each bead by its particular sequence of fluors across a series of hybridization steps.

The small feature size (the 3 micron bead) on the arrays imposes stringent array scanning requirements. The BeadArray Reader (Fig. 1C) is a laser confocal scanner capable of simultaneous measurements in two fluorescence channels at a resolution of greater than one micron.[14] The reader accommodates both the Sentrix Array Matrix and all Sentrix BeadChip formats. During a scan, all array images are automatically registered for fluorescence imaging and data are automatically extracted and associated with decoding data through the reader's BeadScan software. The resulting fluorescence intensity data files are recognized by the BeadStudio data analysis package, which contains modules for analyzing the different types of assays that are read out on Sentrix arrays.

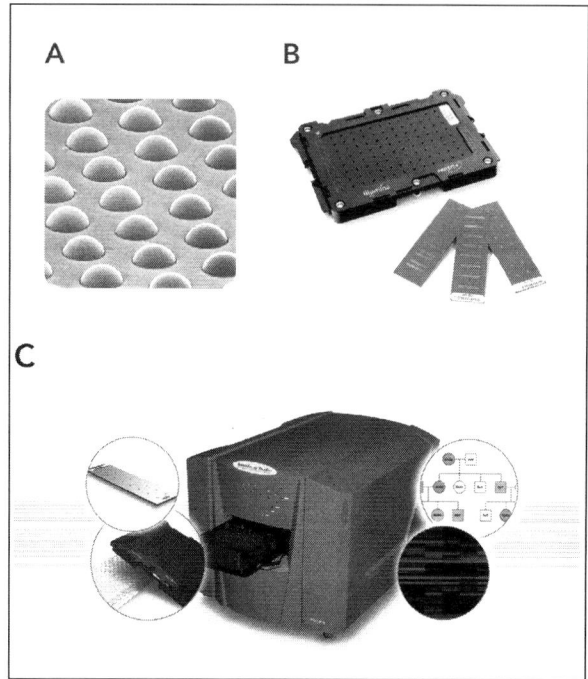

Figure 1. The Illumina BeadArray Platform. A) Beads in wells. A false color scanning electron microscopic image of the 3 micron beads positioned in the array wells. B) BeadArray formats. The randomly assembled arrays are manufactured on fiber-optic, bundle-based substrates for high throughput (above) or on silicon wafers for large feature sets (below). C) BeadArray Reader. The confocal scanning array reader accepts all BeadArray technology formats.

Thus, the BeadArray Reader becomes a point of integration for scientists exploring connections between the genome, the transcriptome and epigenomic regulation.

Universal Arrays and the GoldenGate® Genotyping Assay

To manage the diverse needs of different types of assays, beads assembled into Illumina arrays fall into two classes: those with short oligos, consisting of the address sequence used for decoding alone and those bearing long oligos, having an additional probe sequence concatenated to the address sequence. The address-only beads are used in arrays where the address sequence itself serves as a probe, which provides assay flexibility. That is, sample targets monitored with address-only beads can be revised by changing assay oligos, rather than changing array content. In contrast, beads with long oligos are used for assays requiring specific probes. This format is primarily used for arrays with a high density of features of fixed content, such as whole-genome arrays.

Because address-only arrays offer assay flexibility, they are considered "universal". That is, these arrays can be used to read out the results of several different kinds of genomic assays, so long as they are configured to produce labeled samples that hybridize to the address sequence. Illumina assays utilizing universal arrays include those monitoring genotypes and gene expression, as well as epigenetic phenomena such as allele-specific expression and DNA methylation, all through a common biochemical approach. For genotyping, this approach is called the GoldenGate Assay, illustrated in Figure 2A.[15,16]

While differences among individuals in genomic sequences have been established as biological dogma, the systematic identification of sequence differences has awaited the development of genomic assays for genotyping. In particular, single nucleotide differences among individuals are

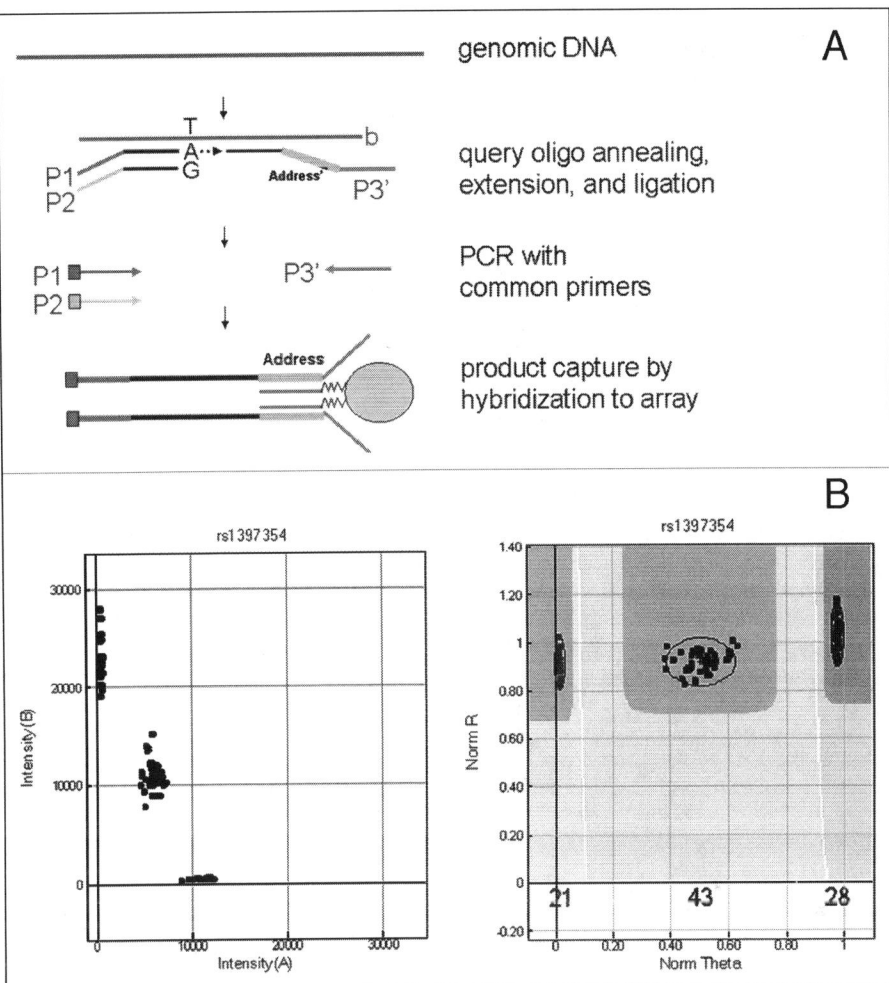

Figure 2. The GoldenGate Assay. A) Assay scheme. Genomic DNA is interrogated by chimeric query oligos. The upstream oligo matching the SNP is preferentially extended and ligated to the downstream oligo, forming an amplifiable template. Sample labeling occurs in the PCR through fluorescent primers P1 and P2. The address sequence contributed by the downstream oligo is used to hybridize the product to the array. B) An example genotyping plot. Intensities for a given locus are plotted in Cartesian coordinates on the left, where the x-axis indicates Cy3 intensity and the y-axis indicates Cy5 intensity. The 90 samples cluster into 3 categories, the 2 homozygote clusters, AA and BB and the heterozygote cluster, AB. On the right, the same data are normalized and transformed to polar coordinates to facilitate locus scoring. Samples falling outside the darkly shaded areas are not called with confidence.

thought to underlie complex diseases that have a genetic component. Such single nucleotide polymorphisms (SNPs) abound in the genome, making customizable, high-density, high-throughput genotyping assays essential for establishing links between genomic sequence and phenotypic consequence, from disease susceptibility, to drug tolerance, to molecular classification of disease state, among many potential applications.[6,9-11]

Table 1. GoldenGate assay metrics

All Service Projects: Jan 2005-Jun 2006	Average Performance
Custom assay sevelopment success	
– Human only	92.8%
– All species	91.2%
DNA success rate	97.6%
Call rate	99.75%
Reproducibility	>99.99%
Heritability	>99.99%

In GoldenGate genotyping, SNPs are genotyped in gDNA samples using a high-throughput multiplexed assay based on oligos for detection (Fig. 2A). In this approach, a SNP is targeted by a set of three chimeric oligos. These oligos consist of two oligos complementary to the genomic sequence just upstream of the SNP and another complementary to a genomic sequence downstream. The upstream allele-specific oligos (ASOs) are identical in the genomic sequence, but differ at their 3' terminal nucleotide, so that only one ASO matches a given SNP. The two ASOs also differ in the sequences fused at their 5' ends, where they bear different PCR primer landing sites. The downstream locus-specific oligo (LSO) is not polymorphic and has a sequence complementary to the genome, an array address sequence and a third PCR primer landing site fused to the 3' end of the address sequence.

Once annealed to gDNA, an allele-specific primer extension (ASPE) reaction occurs and the ASO matching the SNP is preferentially extended over the nonmatching ASO to meet the LSO. The extended ASO is now ligated to the LSO, forming an amplifiable template. Differential labeling of the two alleles is accomplished during PCR amplification by two fluorescent PCR primers, through the two primer landing sites that are associated with each allele. The amplified product is hybridized to the universal array and the difference in fluorescence indicates the original genotype of the sample, with homozygotes either red or green and heterozygotes yellow.

This oligo-based approach has several useful features. The requirement that separate oligos anneal at the same genomic site results in increased specificity. Because the primer landing sites are shared among all the oligo sets, the assay can be multiplexed to the number of address sequences available on the array (currently 1536), while avoiding the usual primer incompatibilities that limit multiplexing in PCR. In addition, activating the gDNA with biotin allows capture of the gDNA with the annealed oligos and removal of unbound oligos or enzymatic inhibitors, decreasing background and increasing efficiency. Further, because each SNP is associated with a particular address sequence, different SNPs can be targeted by reassigning address sequences and synthesizing a new oligo pool.

The GoldenGate Assay has proven to be quite robust in analyzing a wide variety of DNA samples,[17,18] including those subjected to whole-genome amplification. Indeed, this platform was used to generate about 70% of the genotypes for the first phase of the international Haplotype Mapping (HapMap) program.[19] Examples of genotyping plots are shown in Figure 2B, where the intensities of the two fluors and their normalized angles ("Norm Theta") is shown in both Cartesian and polar coordinates, respectively. In the polar coordinate plot, intensities are normalized as well ("Norm R"). For any given SNP, a sample is a member of one of the three possible genotype clusters. Samples falling outside the darkly shaded areas are not called with confidence. The performance of the GoldenGate Assay is summarized in Table 1. The metrics given for call rate, reproducibility and heritability are well above 99%, values all the more remarkable for their being derived from a wide variety of custom SNP genotyping projects, not selected for best data.

Gene Expression Analysis with the DASL® Assay

By changing the input to RNA instead of gDNA, the oligo-based genotyping assay can be adapted to monitor gene expression by targeting nonpolymorphic sequences. The result is the DASL Assay (cDNA-mediated annealing, selection, extension and ligation), illustrated in Figure 3A.[20] Biotinylated cDNA is made by reverse transcription of total RNA with biotinylated primers,

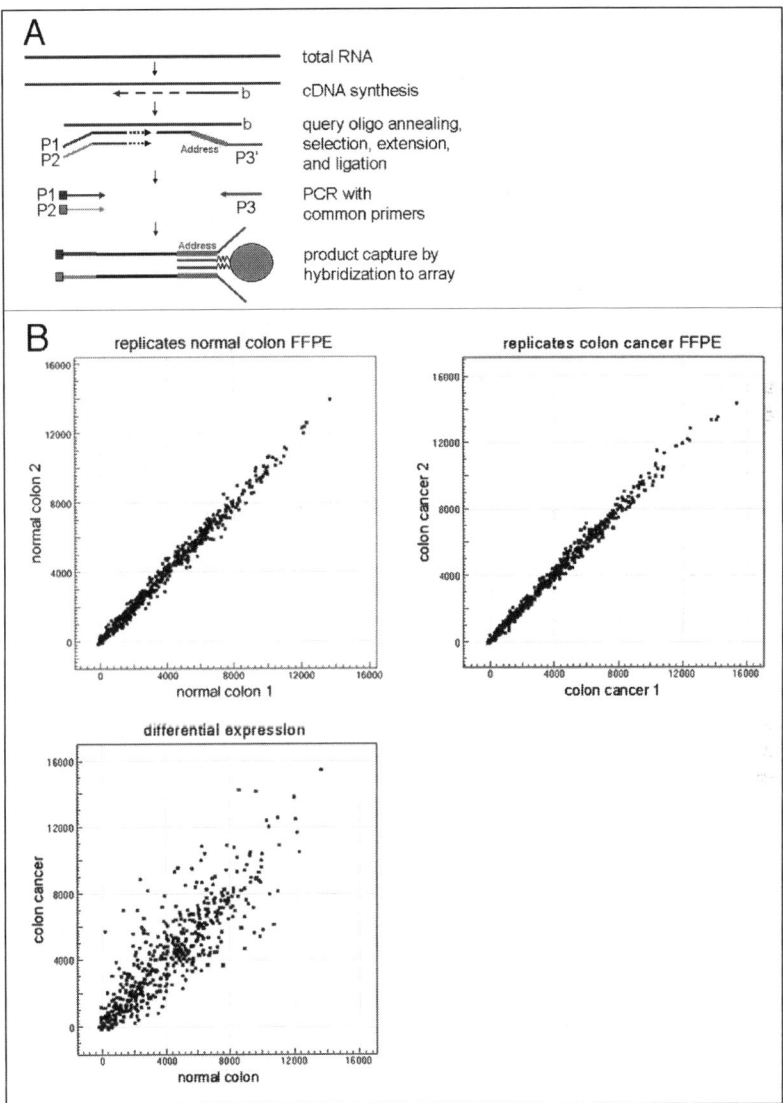

Figure 3. The DASL gene expression assay. A) Assay scheme. Total RNA is biotinylated by conversion to cDNA with biotinylated primers. Chimeric query oligos target nonpolymorphic sequences and are extended and ligated to form amplifiable templates, as in Figure 2A. B) Assay performance on FFPE-derived RNA samples with intensity data for technical replicates of RNA from normal colon (above, left) or colon adenocarcinoma (above, right). Comparison of normal and cancer samples shows differential expression of one-quarter of the 506 genes monitored (below).

both oligo-d(T) and randomers. The use of randomers frees the assay from the requirement for an intact poly-A tail and permits the analysis of gene expression in partially degraded RNAs. Another attractive feature of the DASL Assay is the ability to target sequence features such as alternative splicing.[21-23] All other GoldenGate Assay protocols and reagents are identical in the DASL Assay.

The capability of the DASL Assay to monitor expression in degraded RNAs means that RNAs from formalin-fixed, paraffin-embedded (FFPE) tissues can be analyzed.[24] These samples represent an extraordinary resource for retrospective analyses, but their use in expression studies has been limited by extensive RNA degradation.[25] The main technique commonly employed for these RNAs has been real-time RT-PCR, but this approach suffers from a high false negative rate, with many primer pairs unusable. Because the DASL Assay targets a smaller cDNA sequence than real-time RT-PCR (about 50 nucleotides), RNAs derived from FFPE tissues exhibit robust performance so long as the RNAs have a size distribution peaking at about 200 nucleotides (nt) long. Shorter RNAs of about 100 nt have shown poorer reproducibility in our experience, presumably due to the lower likelihood that both DASL Assay query oligos will land on the same molecule of cDNA.

Examples of DASL Assay data monitoring 502 genes in FFPE normal and cancerous colon tissues are shown in Figure 3B. Good assay reproducibility is observed for FFPE samples, typically giving a linear R^2 value of over 0.97 at the probe level. Including multiple probes per gene increases this number, with 3 probes per gene showing R^2 values from 0.98-0.99. Differential expression between normal and tumor samples is also evident and expression differences have been validated in several ways. For example, appropriate differential expression was observed in DASL Assay data for tissue and cancer-specific markers that were originally selected from whole-transcriptome profiling of fresh-frozen samples.[24] In addition, appropriate cell type-enriched expression was detected in microdissected FFPE samples as predicted by DASL Assay data from undissected samples.[26] Further, when normal and diseased samples were profiled in both fresh-frozen samples and FFPE samples of the same tissues, the correlation (R^2) in their differential expression p-value-based scores was ~ 0.7, suggesting that differential expression in FFPE-derived RNAs parallels that observed in intact RNAs.[24] Using this method to monitor gene expression in FFPE prostate cancer samples, a 16-gene profile has been shown to accurately predict patient relapse.[27]

Allele-Specific Expression Analysis

Allele-specific expression (ASE), where gene expression is dominated by one of the two chromosomal copies of a gene, has recently been recognized as surprisingly widespread.[28,29] The consequences of such a phenomenon are apparent if the two alleles differ in sequence in the same cell, presenting the possibility of expressing two transcripts with different protein coding potential or mRNA stability. Two alleles of a particular gene can be expressed at different levels due to gene imprinting, X chromosome inactivation, differential local promoter activity, or regulatory polymorphisms affecting transcription. Because ASE in a heterozygote results in sequence differences in RNA transcripts, it can be detected by genotyping cDNAs.

As shown in Figure 3A, the DASL Assay is configured with 2 upstream oligos, just as the GoldenGate Assay is. Although this arrangement is not necessary for monitoring nonpolymorphic sites, it does provide for assay flexibility in adapting the assay to target polymorphic sites in cDNAs. In the case where an individual is heterozygous for a SNP within a gene, expression from the two alleles can be monitored by comparing the fluorescence in 2 channels in a cDNA sample relative to the gDNA sample from the same individual, where both alleles are equally represented.[15]

Figure 4 illustrates an example of ASE. As in GoldenGate genotyping, the fluorescence intensities are expressed as "Norm R" and the relation between signals due to the two alleles is expressed as a "Norm Theta" value. The data shown were taken for matched gDNA and RNA samples from normal kidney or lung tissue samples. Among the 1390 SNPs monitored simultaneously in this case, the well-known imprinted gene SNRPN showed clear evidence of ASE, where the Theta values for two DNA samples were intermediate (heterozygous), but Theta for the matched RNA samples suggested only one allele was expressed. In addition to the SNP shown, ASE was also evident for a

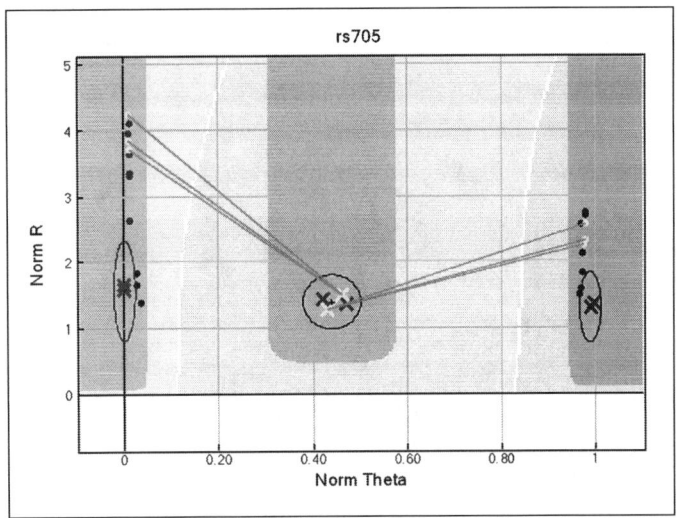

Figure 4. Allele-specific expression assay. Example of allele-specific expression in human kidney and lung tissues. Crosses represent allelic ratios in genomic DNA and dots, equivalent ratios in cDNA. Blue lines connect gDNA reference to corresponding value in cDNA, indicating full imprinting for the analyzed SNP.

series of imprinted genes in these samples. In other experiments, ASE was clearly shown for many SNPs, a finding that was also validated by sequencing of cDNA clones.[30]

DNA Methylation Analysis

Genomic DNA methylation is widespread and plays a critical role in the regulation of gene expression in development, differentiation and aging and in diseases such as multiple sclerosis, diabetes, schizophrenia and cancer.[31,32] It is often associated with gene silencing when it occurs in promoter areas. DNA methylation has recently received a great deal of attention for its potential use in classifying disease tissues and in cancer detection.[33,34] DNA methylation status has also been shown to be highly robust in distinguishing among closely related cells such as different human embryonic stem cell lines.[35]

The same GoldenGate Assay genotyping approach can be extended to analyze DNA methylation status by taking advantage of the resistance of methylated cytosines to deamination by bisulfite.[34] Treating a DNA sample with bisulfite results in a change in sequence from C to U for unmethylated cytosines, while methylated cytosines remain unchanged. Therefore, the methylation status of a given cytosine base in a CpG site can be interrogated using a genotyping assay for a C/T polymorphism after bisulfite treatment (see Fig. 5A). In adapting the oligo-based detection approach to bisulfite treated DNA, we have developed methods that minimize the impact of the greatly diminished sequence complexity caused by the loss of most cytosines in the genome when designing GoldenGate Assay query oligos.

For DNA methylation, the status of an interrogated locus is determined by calculating the "beta" value (or methylation fraction), which is defined as the un-normalized ratio of the fluorescent signal from the methylated allele to the sum of the fluorescent signals of both methylated and unmethylated alleles. The beta-value provides a continuous measure of levels of DNA methylation in samples. Good reproducibility of beta values is observed in the methylation assay ($R^2 >$ 0.98) and differences in methylation status are readily detectable. As shown graphically in Figure 5B, differential methylation is highly effective for classifying experimental samples in different biological states. In the data shown, a series of normal and cancer cell types were analyzed and

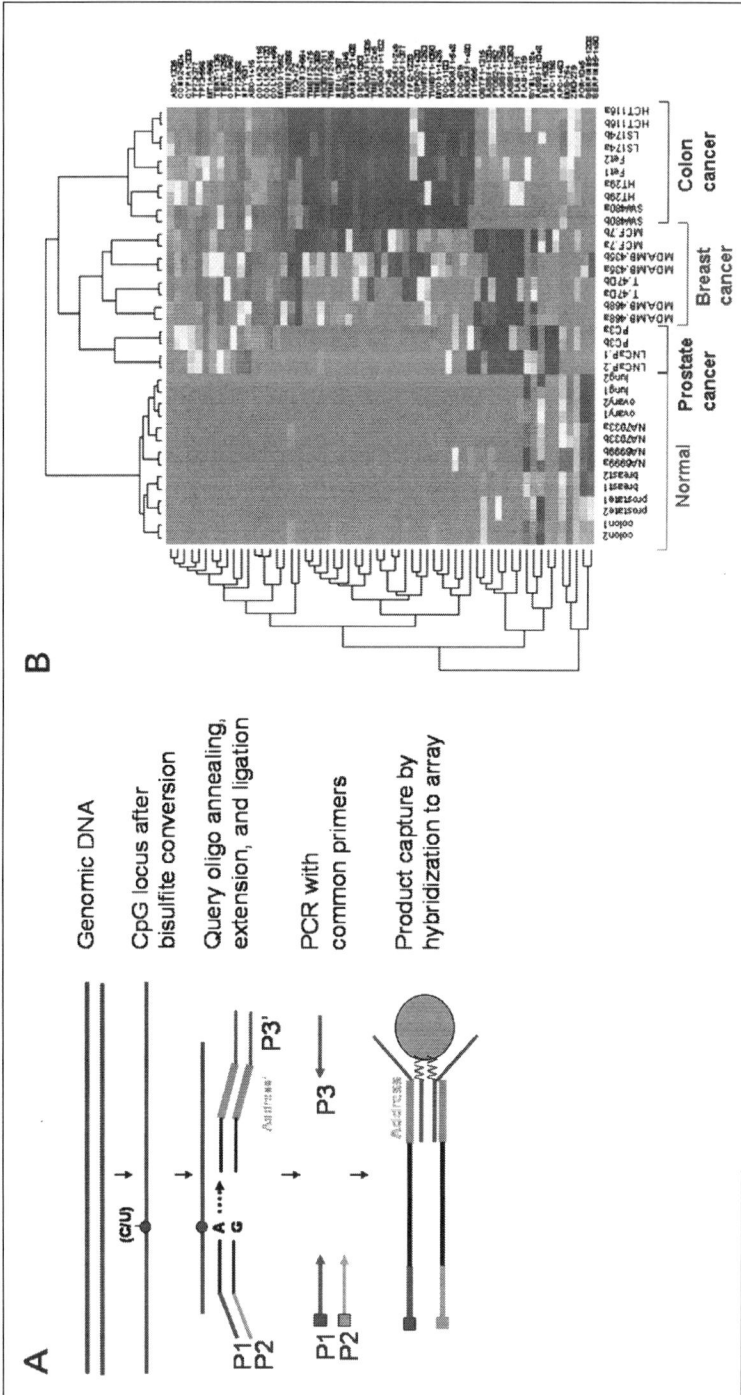

Figure 5. Methylation assay. A) Assay scheme. Bisulfite treatment of gDNA results in conversion of unmethylated cytosines to uracil, creating a polymorphism that can be monitored as in Figure 2A. B) Example of a heatmap representing differential methylation data. All cancer samples were correctly separated from normal samples using agglomerative clustering and highly specific methylation signatures were obtained for each cancer type. Green, yellow and red colors represent low, medium and high methylation levels, respectively. A color version of this figure may be found at www.eurekah.com

the samples were clustered based on probes that were selected for effectiveness in classifying these samples. Differences in methylation were readily apparent between analyzed cancer samples and they correctly segregated according to tissue of origin. Also noteworthy is how all the normal samples clustered together regardless of tissue, suggesting that differential methylation status may be an exceptionally sensitive measure of disease progression.

Whole-Genome Gene Expression

As mentioned above, one of the earliest applications of microarray technology was to profile gene expression. These sorts of studies have allowed the derivation of possible biomarkers for monitoring disease progression, diagnostics and prognostics. One drawback to early microarrays for gene expression was the necessity of running single samples at a time, limiting the number of samples in a typical experiment and thus the statistical power available in analyzing data. Because of the flexibility of the BeadChip platform, configurations are available to profile 6 or 8 samples in parallel at the whole transcriptome level. To expand to genome-wide analyses, we have turned to the longer capture probes mentioned above. For expression profiling, probes are designed as 50-mers that target transcript sequences accessible with standard in vitro transcription-based sample preparation (Fig. 6A).[36]

Six or eight RNA samples (human or mouse) can be profiled simultaneously on a single BeadChip monitoring ~48,000 or ~24,000 transcripts, respectively. The 24,000 genes in common between these sample number formats are primarily derived from the National Center for Biotechnology Information's RefSeq database and the supplementary genes in the 6-sample BeadChips include predicted genes. As shown in Figure 6B, these arrays are highly reproducible, with correlations (R^2) routinely exceeding 0.99 for labeling replicates. The assay performs very well, with a limit of detection of less than 0.25 pM and over 3 logs of dynamic range.[36,37]

In the data shown, genes that were differentially expressed between control mouse tissue samples and an experimental sample in this experiment (p <0.001) are shown as blue dots on all three plots. Differentially expressed genes were called by the Gene Expression module of the Illumina BeadStudio software package, which takes advantage of the ~30 fold redundancy of measurements per bead type for statistical power. Whole genome expression monitoring on these arrays has been used to characterize embryonic stem cell differentiation, among other studies.[38-41] In addition, these arrays have been used in combination with genotyping arrays to examine changes in expression that are associated with genetic variation.[42,43]

Whole-Genome Genotyping

The compilation of ever more SNPs in the human genome as a result of the HapMap and other studies has allowed researchers to initiate higher density genotype/phenotype association studies. To assess the genome at high enough SNP density to allow efficient linkage disequilibrum-based association studies requires as large a feature set as practical. Using the 50-mer capture probes and an entire chip for one sample, we have developed genotyping BeadChips that can monitor from ~109,000 up to ~650,000 SNPs. In principle, any SNP can be targeted for analysis and the number of SNPs per slide is only limited by feature density, so the assay used for this purpose was named the "Infinium™" Assay.[44,45]

The Infinium Assay procedure has four automated steps: (1) whole genome amplification, (2) hybridization to an oligonucleotide probe array, (3) an array-based SNP scoring assay and (4) signal amplification. For the Infinium I Assay, ASPE is used to score the SNP site, requiring two beads for each SNP where the perfectly matched bead type preferentially extends over the mismatched bead type. For the Infinium II Assay, one bead type is used and the allele is scored by single base extension (SBE) using labeled terminators (see Fig. 7A).

Illumina manufactures several high-density SNP genotyping arrays, including the Sentrix Human-1 BeadChip (109 k SNPs, exon-centric), the HumanHap300 (317 k tag SNPs), the HumanHap550 (550 k tag SNPs) and the HumanHap650Y (650 k tag SNPs) (for a list of current whole genome genotyping arrays, please see www.illumina.com). The Human-1 BeadChip utilizes

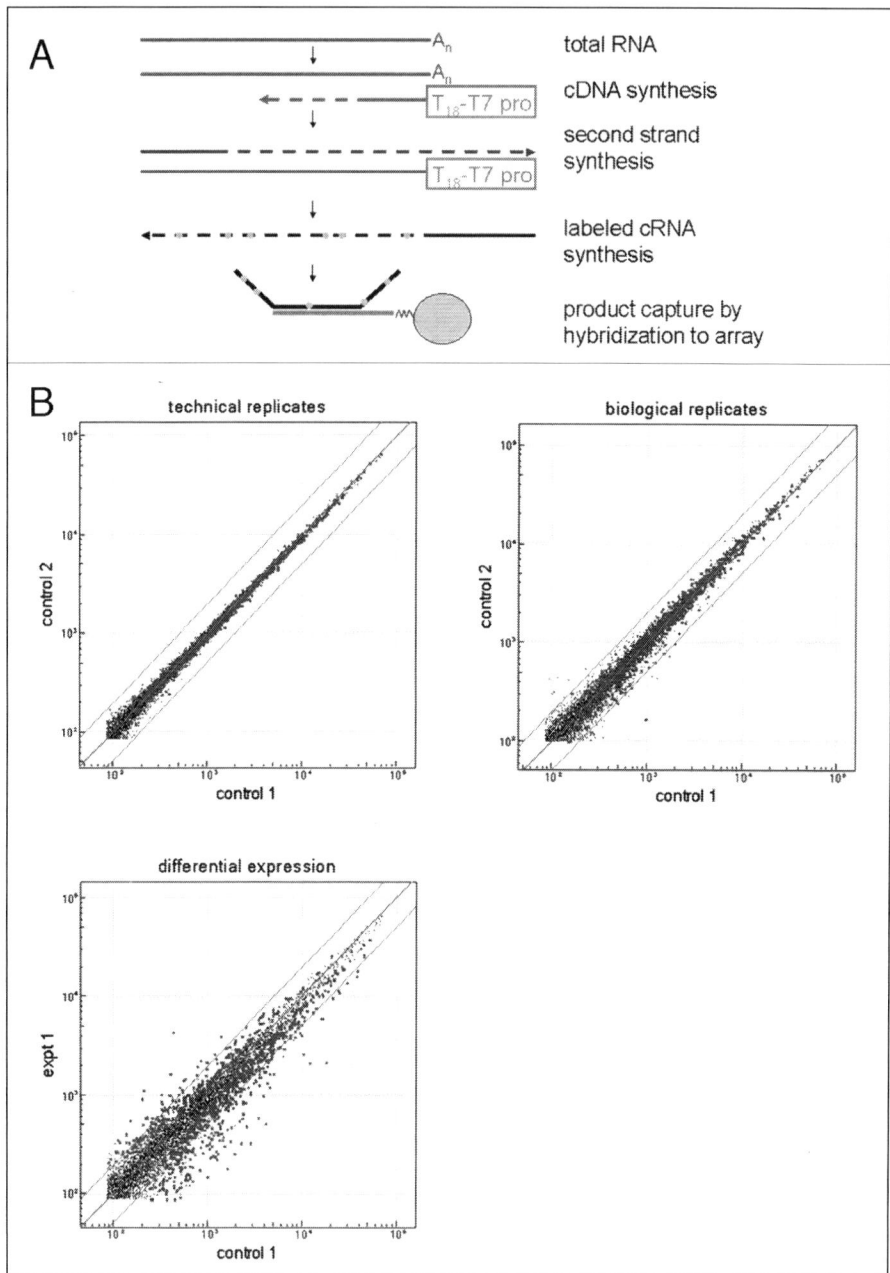

Figure 6. Whole Genome Gene Expression Assay. A) Labeling and probe scheme. Body labeled cRNA is prepared in the standard way, priming total RNA with a T7 RNA polymerase-oligo d(T) oligo. B) Technical replicates (top left), biological replicates (top right) and differential expression analysis (bottom) of mouse total RNAs are shown, analyzed for ~24,000 mouse RefSeq genes. Two fold difference lines are shown in red and genes that were differentially expressed (p <0.001) are shown in blue.

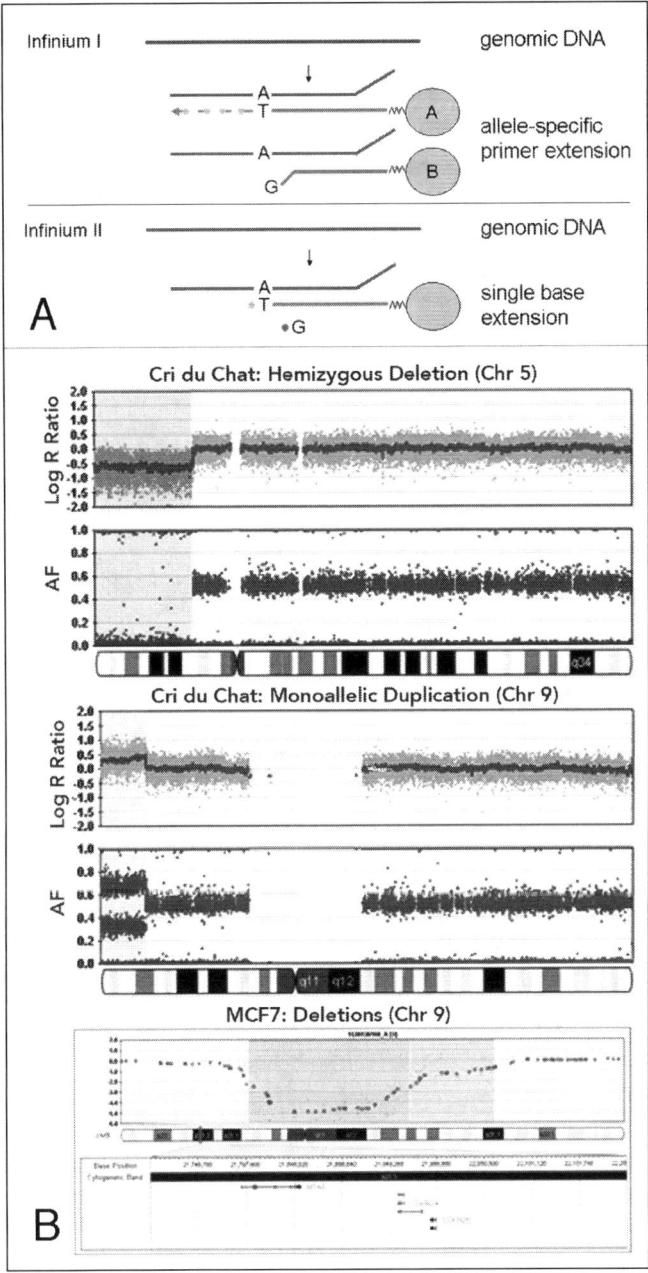

Figure 7. Whole genome genotyping assays and copy number detection. A) Infinium Assay Schemes. B) Single Sample Analysis. In a Cri du Chat sample, a hemizygous deletion on Chromosome 5 is noted by a deflection in the log R ratio and a loss of heterozygotes in the AF plot (top panel). A monoallelic duplication on Chromosome 9 is also present in Cri du Chat (middle panel). High resolution analysis of MCF7 cells showing a homozygous and hemizygous deletion in a region on Chromosome 9 containing MTAP, CDKN2A (p16) and CDKN2B (p15) genes.

the ASPE Infinium I Assay. The HumanHap300, HumanHap550 and HumanHap650Y BeadChips utilize the SBE Infinium II Assay to genotype over 317 k tag SNPs, ~550 k tag SNPs and ~650 k tag SNPs selected from the HapMap phase I and II data. The HumanHap300 has a 9 kb mean SNP spacing enabling an effective resolution of ~90 kb (10 SNP smoothing), the HumanHap550 increases resolution down to ~50 kb. By maximizing coverage of the human genome, these arrays can be used for whole genome association studies to dissect the genetics of complex diseases and pharmacogenomic drug responses.[46,47]

The allelic quantitation of genotyping also allows the detection of allelic imbalances. As such, a change in copy number associated with loss of heterozygosity (LOH) can be detected by a shift in theta values combined with a decrease in signal intensity. Although the arrays were not originally designed with this aim, the high density of whole genome SNP arrays has proven highly effective in the detection of DNA copy number changes, LOH and other chromosomal aberrations that are characteristically found in cancers and congenital disorders.[48,49]

Traditional array-comparative genome hybridization (CGH) has been used for detection of chromosomal aberrations and is based on the comparison of normalized intensities between a reference and subject sample. This approach was the foundation for the discovery of many well-known tumor suppressor genes.[50] In contrast to array-CGH, aberrations can be detected with high density genotyping arrays (SNP-CGH) using a combination of two genotyping parameters; a normalized intensity measurement and an allelic ratio. Together, these parameters provide a more sensitive and precise profile of aberrations as well as providing genetic information (haplotypes) of the locus undergoing aberration. Moreover, SNP-CGH has the ability to detect copy-neutral LOH events, which cannot be detected with array-CGH.[51]

For SNP-CGH, there are two modes of data analysis. The first is a single-sample mode in which reference values are derived from canonical clusters created from clustering on many normal reference samples. The other is a paired-sample mode in which direct intensity comparisons between a subject sample and its corresponding matched pair are performed. Genomic plots of $\log_2 (R_{subject}/R_{reference})$ (referred to as the log R ratio) and the allele frequency (AF) form the basis of detecting chromosomal aberrations.[52]

Figure 7B illustrates the ability of the HumanHap550 Genotyping BeadChip to detect chromosomal aberrations in the single sample mode. A ~33 Mb hemizygous deletion (2 copies to 1 copy) on the p-arm of Chromosome 5 was detected in a Cri du Chat patient sample as a downward deflection in the log R ratio profile and a loss of heterozygotes in the AF profile. On Chromosome 9, a monoallelic duplication (trisomy) of Chromosome 18 manifested itself as an increase in the log R ratio and split of the heterozygotes into two states; one at 0.67 (2:1 ratio) AF and another at 0.33 (1:2 ratio) AF. In another example, two adjacent homozygous and hemizygous deletions were observed in the breast cancer cell line MCF7. We were able to rapidly determine that these aberrations corresponded to homozygous deletions of the MTAP and CDKN2A genes and a hemizygous deletion of CDKN2B using the gene annotation provided in the Illumina BeadStudio data analysis package.[52]

High-density SNP-CGH arrays can profile chromosomal aberrations including LOH, deletions, duplications and amplifications. At unparalleled resolution, SNP-CGH arrays can also detect copy neutral genetic anomalies such as uniparental disomy (UPD) and mitotic recombination, events undetectable by conventional array-CGH. This application of a genotyping array is particularly relevant to cancer therapeutics because certain high-level amplifications are known to be correlated with clinical outcome.[53]

Conclusion

At a time of enormous growth in genome-level basic and applied research, microarray technologies are being developed and adapted to analyze ever more diverse aspects of biology. The applications described in this chapter are the first that have been developed for use on the Illumina BeadArray technology, but there is every likelihood that the bead-based microarray approach will continue to adapt to new assays that take advantage of the flexibility of this technology. Whether the throughput is a single sample or 96 samples at a time, all the assays are read on a single analytical instrument capable of rapid confocal scanning of these very small features. This point of integration means that fundamentally different types of analyses can be carried out on the same biological samples, allowing a researcher to correlate gene expression, for example, with genotype and/or DNA methylation status. This degree of flexibility and suite of complementary assays suggests that the aim of integrated analyses of biological specimens is now achievable.

References

1. Nelson SF, McCusker JH, Sander MA et al. Genomic mismatch scanning: a new approach to genetic linkage mapping. Nat Genet 1993; 4(1):11-18.
2. Schena M, Shalon D, Davis RW et al. Quantitative monitoring of gene expression patterns with a complementary DNA microarray. Science 1995; 270:467-470.
3. Golub TR, Slonim DK, Tamayo P et al. Molecular classification of cancer: class discovery and cleass prediction by gene expression monitoring. Science 1999; 286:531-537.
4. Perou CM, Sorlie T, Eisen MB et al. Molecular portraits of human breast tumors. Nature 2000; 406:747-752.
5. Welsh JB, Sapinoso LM, Su AI et al. Analysis of gene expression identifies candidate markers and pharmacological targets in prostate cancer. Cancer Res 2001; 61:5974-5978.
6. Fan JB, Chee MS, Gunderson KL. Highly parallel genomic assays. Nat Rev Genet 2006; 7(8):632-644.
7. Heller MJ. DNA microarray technology: devices, systems and applications. Annu Rev Biomed Eng 2002; 4:129-153.
8. Peeters JK, Van der Spek PJ. Growing applications and advancements in microarray technology and analysis tools. Cell Biochem Biophys 2005; 43(1):149-166.
9. Weeraratna AT, Nagel JE, de Mello-Coelho V et al. Gene expression profiling: from microarrays to medicine. J Clin Immunol 2004; 24(3):213-224.
10. Bucca G, Carruba G, Saetta A et al. Gene expression profiling of human cancers. Ann NY Acad Sci 2004; 1028:28-37.
11. Gibson, G. Microarray analysis: genome-scale hypothesis scanning. PloS Biol 2003; 1(1):E15 Epub 2003.
12. Lipshutz RJ, Fodor SP, Gingeras TR et al. High density synthetic oligonucleotide arrays. Nat Genet 1999; 21(1 Suppl):20-24.
13. Gunderson KL, Kruglyak S, Graige MS et al. Decoding randomly ordered DNA arrays. Genome Res 2004; 14:870-877.
14. Barker DL, Theriault G, Che D et al. Profiling alternatively spliced mRNA isoforms for prostate cancer. BMC Bioinform 2006; 7:202.
15. Fan JB, Oliphant A, Shen R et al. Highly parallel SNP genotyping. CSHL Symp Quant Biol 2003; 68:69-78.
16. Shen R, Fan JB, Campbell D et al. High-throughput SNP genotyping on universal bead arrays. Mutation Res 2005; 573(1-2):70-82.
17. Saarela J, Kallio SP, Chen D et al. PRKCA and multiple sclerosis: association in two independent populations. PLoS Genetics 2006; 2(3):e42 Epub 2006.
18. Zhang C, Cawley S, Liu G et al. Gene expression profiles in formalin-fixed, paraffin-embedded tissues obtained with a novel assay for microarray analysis. Clin Chem 2004; 50(12):2384-2386.
19. Altshuler D, Brooks LD, Chakravarti A et al. A survey of the human genome. Nature 2005; 437(7063):1299-1320.
20. Fan JB, Yeakley JM, Bibikova M et al. Epigenetics in human disease and prospects for epigenetic therapy. Nature 2004; 429:457-463.
21. Li HR, Wang-Rodriquez J, Nair TM et al. High-throughput DNA methylation profiling using universal bead arrays. Genome Res 2006; 16:383-393.
22. Zhang C, Li HR, Fan JB et al. Profiling alternatively spliced mRNA isoforms for prostate cancer. BMC Bioinform 2006; 7:202.
23. Yeakley JM, Fan JB, Doucet D et al. Profiling alternative splicing on fiber-optic arrays. Nat Biotech 2002; 20:353-358.
24. Bibikova M, Talantov D, Chudin E et al. NTera2: A model system to study dopaminergic differentiation of human embryonic stem cells. Stem Cells & Devel 2005; 14:517-534.
25. Bova GS, Eltoum IA, Kiernan JA et al. Optimal molecular profiling of tissue and tissue components: defining the best processing and microdissection methods for biomedical applications. Methods Mol Med 2005; 103:15-66.

26. Bibikova M, Yeakley JM, Chudin E et al. Experimental comparison and cross-validation of the Affymetrix and Illumina gene expression analysis platforms. Nucl Acids Res 2005; 33(18):5914-5923.
27. Bibikova M, Chudin E, Arsanjani A et al. Gene Expression Profiles to Predict Relapse of Prostate Cancer. J Mol Diagn 2006 submitted.
28. Knight JC. Allele-specific gene expression uncovered. Trends Genet 2004; 20(3):113-116.
29. Pastinen T, Sladek R, Gurd S et al. A survey of genetic and epigenetic variation affecting human gene expression. Physiol Genomics 2003; 16:184-193.
30. Bibikova M unpublished results.
31. Egger G, Liang G, Aparicio A et al. Genome-wide associations of gene expression variation in humans. PLoS Genet 2005; 1(6):e78.
32. Robertson KD. DNA methylation and human disease. Nat Rev Genet 2005; 6(8):597-610.
33. Hoque MO, Feng Q, Toure P et al. Detection of aberrant methylation of four genes in plasma DNA for the detection of breast cancer. J Clin Oncol 2006; [Epub ahead of print].
34. Bibikova M, Lin Z, Zhou L et al. High-throughput DNA methylation profiling using universal bead arrays. Genome Res 2006; 16:383-393.
35. Bibikova M, Chudin E, Wu B et al. Human embryonic stem cells have a unique epigenetic signature. Genome Res 2006; [Epub ahead of print].
36. Kuhn K, Baker SC, Chudin E et al. A novel, high-performance random array platform for quantitative gene expression profiling. Genome Res 2004; 14:2347-2356.
37. Chudin E, Kruglyak S, Baker SC et al. A model of technical variation of microarray signals. J Comput Biol 2005; 1(6):e65.
38. Schwartz CM, Spivak CE, Baker SC et al. NTera2: A model system to study dopaminergic differentiation of human embryonic stem cells. Stem Cells & Devel 2005; 14:517-534.
39. Cai J, Chen J, Lui Y et al. Assessing self-renewal and differentiation in hESC lines. Stem Cells 2006; 24(3):516-530, published online 2005.
40. Barnes M, Freudenberg J, Thompson S et al. Experimental comparison and cross-validation of the Affymetrix and Illumina gene expression analysis platforms. Nucl Acids Res 2005; 33(18):5914-5923.
41. Debey S, Zander T, Brors B et al. A highly standardized, robust and cost-effective method for genome-wide transcriptome analysis of peripheral blood applicable to large-scale clinical trials. Genomics 2006; 87(5):653-654.
42. Stranger BE, Forrest MS, Clark AG et al. Genome-wide associations of gene expression variation in humans. PLoS Genet 2005; 1(6):e78.
43. Seal JL, Gornick MC, Gogtay N et al. Segmental uniparental isodisomy on 5q32-qter in a patient with childhood-onset schizophrenia. J Med Genet 2006; epub.
44. Gunderson KL, Steemers FJ, Lee G et al. A genome-wide scalable SNP genotyping assay using microarray technology. Nat Genet 2005; 37(5):549-554.
45. Steemers FJ, Chang W, Lee G et al. Whole-genome genotyping with the single-base extension assay. Nat Methods 2006; 3(1):31-33.
46. Gunderson KL, Kuhn KM, Steemers FJ et al. Whole-genome genotyping of haplotype tag single nucleotide polymorphisms. Pharmacogenomics 2006; 7(4):641-648.
47. Amos CI, Chen WV, Lee A et al. High-density SNP analysis of 642 Caucasian families with rheumatoid arthritis identifies two new linkage regions on 11p12 and 2q33. Genes Immun 2006; 7(4):277-86 Epub 2006.
48. LaFramboise T, Weir BA, Zhao X et al. Allele-specific amplification in cancer revealed by SNP array analysis. PLoS Comput Biol 2005; 1(6):e65.
49. Albertson DG, Collins C, McCormick F et al. Chromosome aberrations in solid tumors. Nat Genet 2003; 34(4):369-376.
50. Greenwald BD, Harpaz N, Yin J et al. Loss of heterozygosity affecting the p53, Rb and mcc/apc tumor suppressor gene loci in dysplastic and cancerous ulcerative colitis Cancer Res 1992; 52(3):741-745
51. Struski S, Doco-Fenzy M and Cornillet-Lefebvre P. Compilation of published comparative genomic hybridization studies. Cancer Genet Cytogenet 2002; 135(1):63-90.
52. Peiffer DA, Le JM, Steemers FJ et al. High-resolution genomic profiling of chromosomal aberrations using Infinium whole-genome genotyping. Genome Res 2006; [Epub ahead of print].
53. Schwab M. Oncogene amplification in solid tumors. Semin Cancer Biol 1999; 9(4):319-325.

CHAPTER 3

Integrated Microfluidic CustomArray™ Biochips for Gene Expression and Genotyping Analysis

Robin Hui Liu,* Mike Lodes, H. Sho Fuji, David Danley and Andrew McShea

Abstract

DNA microarray technology has become one of the most promising analytical tools in molecular biology. It has been widely used for studying mRNA levels and examining gene expression in biological samples. It is becoming a powerful tool in the arena of diagnostics and personalized medicine. In this chapter, we present a fully integrated and self-contained microfluidic biochip device that has been developed to automate the fluidic handling steps required to carry out microarray-based gene expression or genotyping analysis. The device consists of a semiconductor-based CustomArray™ chip with 12,000 features and a microfluidic cartridge. The CustomArray™ was manufactured using a semiconductor-based in situ synthesis technology. The oligonucleotides were synthesized on an array of electrodes on a semiconductor chip using phosphoramidite chemistry under electrochemical control. The microfluidic cartridge consists of microfluidic pumps, mixers, valves, fluid channels and reagent storage chambers. Microarray hybridization and subsequent fluidic handling and reactions (including a number of washing and labeling steps) were performed in this fully automated and miniature device before fluorescent image scanning of the microarray chip. Electrochemical micropumps were integrated in the cartridge to provide pumping of liquid solutions. A micromixing technique based on gas bubbling generated by electrochemical micropumps was developed. Low-cost check valves were implemented in the cartridge to prevent cross talk of the stored reagents. Gene expression study of the human leukemia cell line (K562) and genotyping detection and sequencing of influenza A subtypes have been demonstrated using this integrated biochip platform. For gene expression assays, the microfluidic CustomArray™ device detected sample RNAs with a concentration as low as 0.375 pM. Detection was quantitative over more than three orders of magnitude. Experiment also showed that chip-to-chip variability was low indicating that the integrated microfluidic devices eliminate manual fluidic handling steps that can be a significant source of variability in genomic analysis. The genotyping results showed that the device identified influenza A hemagglutinin and neuraminidase subtypes and sequenced portions of both genes, demonstrating the potential of integrated microfluidic and microarray technology for multiple virus detection. The device provides a cost-effective solution to eliminate labor-intensive and time-consuming fluidic handling steps and allows microarray-based DNA analysis in a rapid and automated fashion.

*Corresponding Author: Robin Hui Liu—Osmetech Molecular Diagnostics, Pasadena, California U.S.A. Email: robin.liu@osmetech.com

Integrated Biochips for DNA Analysis, edited by Robin Hui Liu and Abraham P. Lee.
©2007 Landes Bioscience and Springer Science+Business Media.

Introduction

DNA microarray technology has become one of the most promising analytical tools in molecular biology. It has been widely used for studying mRNA levels and examining gene expression in biological samples.[1-5] Investigators rely on data produced by microarray experiments to assess changes in gene expression levels among various experimental tissues and treatments. The applications of microarrays for gene expression profiling[1] include pathway dissection,[6] drug evaluation,[2,3] discovery of gene function,[4] classification of clinical samples,[7-9] and investigation of splicing events,[10] among many others.[5] The highly parallel nature of microarrays has made them invaluable tools for monitoring gene expression patterns of numerous genes simultaneously. Biological experiments have a number of inherent variables making it imperative that the microarray platform be extremely reproducible, both to provide confidence in the data collected and to accurately identify small changes in gene expression patterns. Because the most interesting genes are often expressed at the lowest levels in the sample, it is equally important to use a highly sensitive microarray system.

The development of new detection methods, simplified methodologies and broad application to molecular diagnostics are rapidly migrating microarray technologies into the arena of genetic-based diagnostics and personalized medicine.[11] Array-based methods are now being applied in a diagnostic setting and offer the possibility of analyzing multiple analytes and even multiple samples in a highly parallel and high-throughput manner.[12] Because of their sensitivity, specificity and accuracy, DNA Microarrays have become an acceptable technology for screening samples for the presence or absence of a large variety of viruses simultaneously and identifying the genotype of an unknown specimen.[13-16] Microarrays are particularly useful for molecular detection and identification of influenza viruses because of their genetic and host diversity and the availability of an extensive sequence database.[15-17] For example, DNA microarrays have been recently used for identification of influenza A hemagglutinin and neuraminidase subtypes.[18]

Rapid detection and identification of influenza virus is becoming increasingly important in the face of concerns over influenza pandemic. Timely acquisition of information on the influenza A subtypes that are circulating in human and animal populations is crucial for the global surveillance program to effectively monitor disease outbreaks.[19] Knowledge of the exact strain, origin of the strain and probable characteristics of the virus is essential for surveillance of a disease outbreak and preventing the spread of the disease. Identification of a virus subtype can be realized by molecular identification of the subtype of viral hemagglutinin (HA) and neuraminidase (NA) genes. Viruses with any combination of the 16 HA and 9 NA subtypes can infect aquatic birds while few subtypes have been found to infect humans.[20] However, interspecies transmission can occur after recombination or mixing of subtypes in birds or pigs.[21-23] In addition, new human strains of virus can arise by reassortment or antigenic shifts when two or more subtypes are circulating in the human population.[24,25] Maintenance of a subtype in the human population can also occur by antigenic drift,[25] which occurs when genetic mutations of the HA and NA genes create virons that escape immune surveillance. These mutations arise as a result of viral polymerase infidelity. Therefore, in addition to identification of the circulating subtype, specific knowledge of the genetic makeup of the virus is required in many situations. For example, the avian H5N1 virus ("Bird Flu") has significant potential for further recombination with common human strains (such as H3N2) or other nonhuman strains common in avian populations (H7 and H9 strains). The H5N1 subtype is also difficult to identify because of the lack of sensitivity and specificity of many of the commercial tests, such as viral detection (cell culture) and serological techniques.[26-28] In addition, genotype Z, the dominant H5N1 virus genotype, contains a mutation that is associated with resistance to amantadine and rimantadine. Because of the high susceptibility in humans and resistance to antivirals of this isolate, neuraminidase inhibitors must be given within 48 hours of onset of illness to be effective. Thus rapid and specific identification of this subtype and accurate sequence information are crucial for proper treatment.

Rapid subtype identification of flu is not always straightforward. Simple serological tests on infected individuals are an ineffective tool for monitoring viruses undergoing a high rate of mutation or rapid recombination. Reverse transcription-polymerase chain reaction (RT-PCR) assays have better sensitivity but are problematic in scenarios where new strains of virus emerge or mixtures of viruses exist. DNA Microarrays have recently become an acceptable technology for screening samples for the presence or absence of a large variety of viruses simultaneously and identifying the genotype of an unknown specimen.[13-16] Microarrays that contain several thousand different DNA sequences (probes) can theoretically identify several thousand different organisms.

There are various microarray technologies and numerous commercially available sources of microarrays. Microarrays can be produced either by physical deposition of presynthesized DNA[1,29,30] or by in situ oligonucleotide synthesis.[31,32] The former requires labor-intensive preparation (and hence very significant upfront cost) and record-keeping of DNA probes, whereas the latter only requires DNA sequence design. There are several methods for in situ oligonucleotide synthesis. The most commercially successful method is using a photolithographic method.[33] However, this method, similar to deposition of presynthesized material, lacks flexibility and has to be designed with specific gene content due to high cost and lengthy time spent in generating the photolithographic masks. For those microarray users whose work does not fit in the catalog arrays, it can be a very costly and time-consuming process to adopt the usefulness of microarray technology to their specific research needs. Other in situ oligonucleotide synthesis technologies, such as ink-jet printing,[34] micromirror devices,[35,36] and electrochemical synthesis,[37-39] are more flexible and thus suitable for manufacturing customized oligonucleotide arrays. We have utilized a semiconductor-based in situ synthesis technology that is straightforward to manufacture microarrays and has the potential to sense hybridization electrochemically (in addition to established fluorescent methods).[37,40] The oligonucleotides are synthesized on an array of electrodes on a semiconductor chip using phosphoramidite chemistry under electrochemical control. The electrochemical reaction activated at specific electrodes on the chip generates protons, which in turn remove the blocking group on the oligonucleotide strand undergoing synthesis on the electrodes, allowing subsequent DNA synthesis to take place.

Microarray assays typically involve multi-stage sample processing and fluidic handling, which are generally labor-intensive and time-consuming. In both gene expression and genotyping assay described here, the array needs to be washed thoroughly to remove nonspecific binding of biotinylated cRNA target following hybridization of target in the sample solution to its complementary oligonucleotides synthesized on the microarray chip surface. Different salt concentrations of washing buffers are used to ensure satisfactory stringency. For indirect labeling, a labeling step is subsequently performed. Another washing is performed to remove excessive labeling reagents before the slide is ready for scanning. All the above processes involve many manual steps (handling arrays, moving and agitating racks etc.) with frequent run-to-run and operator-to-operator variation. The combination of these factors can lead to variability in array results. Automation of these processes would improve robustness and reduce costs. Robotic workstations have been developed to automate the whole hybridization and posthybridization process but such benchtop instruments are generally expensive for most research and clinical diagnostic applications. It is therefore desirable to develop a cost-effective method to integrate and automate the microarray processing in a single and miniature device using microfluidic technology.

Microfluidics lab-on-a-chip technology has proven to be useful for integrated, high-throughput DNA analysis.[41] Microfluidic devices can offer a number of advantages over conventional systems, e.g., their compact size, disposable nature, increased utility and a prerequisite for reduced concentrations of sample reagents. Miniaturized assemblies can be designed to perform a wide range of tasks that range from on-chip liquid pumping and handling to detection of DNA. Integration of several assay functions on a single device leads to assay automation and elimination of operator involvement as a variable. Microfluidic systems provide a real potential for improving the efficiency of techniques applied in drug discovery and diagnostics. Various materials have been used in the fabrication of microfluidic lab-on-a-chip devices. Lithographic techniques, adapted from semiconductor technology, have been used to build chips in glass[42] and silicon.[43] Unconventional lithography techniques such as soft lithography[44] have been used to fabricate reproducible microstructures of biological materials offering a multitude of possibilities to explore as molecular diagnostic tools.[45]

Recently, with increasing emphasis on disposable devices, the use of plastics and plastic fabrication methods has become popular.[46-49] Most of the integrated microfluidic works demonstrated to date have been in the area of on-chip capillary electrophoresis (CE),[50-53] polymerase chain reaction (PCR)[54,55] and sample preparation,[56,57] among many others[41] and there are only a few reports on combining microfluidics with DNA microarrays.[56,58,59] The microfluidic lab-on-a-chip devices with capabilities of on-chip sample processing and detection provide a cost-effective solution to direct sample-to-answer biological analysis. Such devices will be increasingly important for rapid diagnostic applications in hospitals and in-field bio-threat detection.

In this chapter, we report on the development of a self-contained and fully integrated microfluidic microarray device that automates hybridization and posthybridization processes for microarray gene expression and genotyping assays that involve multi-stage sample processing and fluidic handling. The description of the semiconductor-based CustomArray™ platform is included. The integrated device design, fluidics and developments of the key microfluidic components, such as pumps, valves and mixers, are described. The demonstrations of gene expression study of the human leukemia cell line (K562) and genotyping detection and sequencing of influenza A subtypes using this integrated biochip platform are presented.

Semiconductor-Based DNA Microarray

A CombiMatrix microarray chip (CustomArray™) was coupled with a microfluidic plastic cartridge to form the microfluidic microarray device as shown in Figure 1.[60] The CustomArray™ is a 1 inch × 3 inch alumina slide with an 11 × 25 mm silicon chip affixed in a cavity in the ceramic package (Fig. 2). A key aspect of the CustomArray™ platform is a semiconductor based microar-

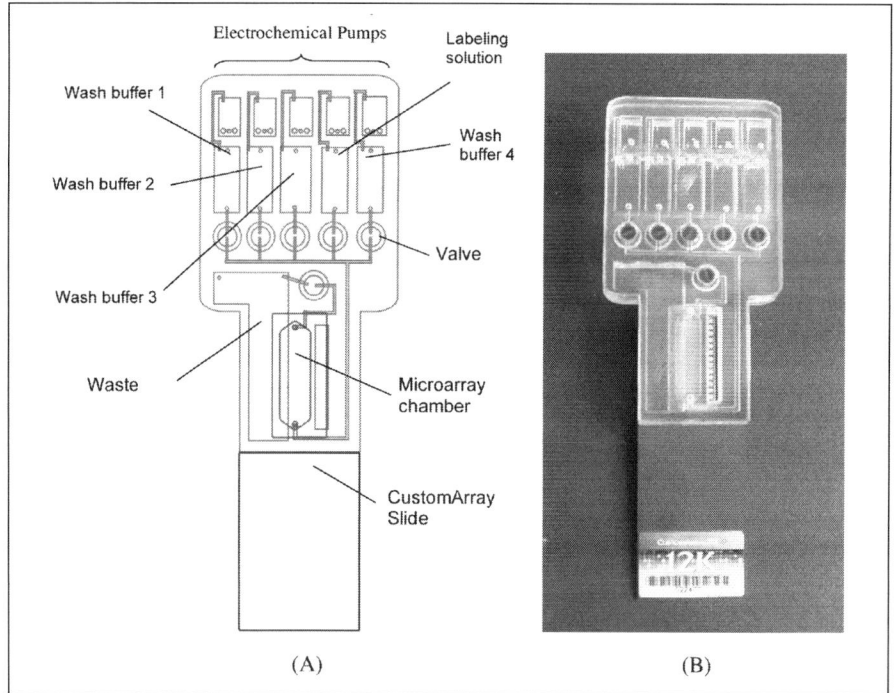

Figure 1. A) Schematic of the microfluidic array device. B) Photograph of the integrated device that consists of a plastic fluidic cartridge and a CombiMatrix CustomArray™ chip (reproduced with permission from ref. 60, copyright 2006 American Chemical Society).

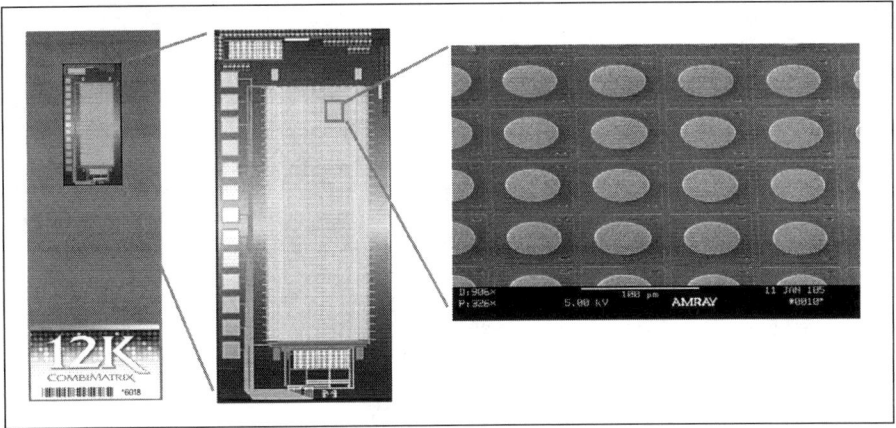

Figure 2. Photographs and SEM image showing the commercial 12K CustomArray™ with 12,544 microelectrodes mounted in a 1 × 3 inch slide. Each microelectrode has a size of 44 μm in diameter. The 13 silver pads on the left of the silicon array chip provide electrical connections required for communicating with the microelectrode array (large rectangle in the middle of the chip).

ray that allows the manufacture of high density microelectrode arrays that vary from a density of 1,000 to >100,000 electrodes/cm². Utilization of active circuit elements in the design permits the selection and parallel activation of individual electrodes in the array to perform in situ oligonucleotide synthesis of customized content on the chip.[37,40] The oligonucleotides were synthesized on an array of electrodes using phosphoramidite chemistry under electrochemical control. The electrochemical reaction generated at specific electrodes on the chip produced protons, which in turn removed the blocking group on the oligonucleotide strand undergoing synthesis on the electrodes, allowing subsequent DNA synthesis to take place.

The silicon chip was manufactured using a commercial mixed signal complementary metal oxide semiconductor (CMOS) process. The microarray chip used in this work has a 56 × 224 array of electrodes located in the center of the chip providing a total of 12,544 electrodes, each with a size of 44 μm in diameter (Fig. 2). All the electrodes on the chip are individually addressable, so that unique reactions can be carried out at each individual site. CMOS integrated circuit technology was utilized to create active circuit elements and digital logic on the chip that allowed complex functions to be implemented. These functions include a high-speed digital interface for efficient communication to the chip, data writing and reading from the electrode array and the setting of appropriate electrical states at the electrode to perform in situ oligonucleotide synthesis.

The oligonucleotide was in situ synthesized on the chip using phosphoramidite chemistry under electrochemical control.[37,40] The chip surface was coated with a proprietary membrane layer that facilitated the attachment and synthesis of biomolecules in a matrix above the platinum electrode surface. During DNA synthesis, the blocking DMT (dimethoxytrityl) group of the phosphoramidite on the selective electrode surface was removed by the acid (H^+) that was produced by the electrochemical reaction when these electrodes were turned on.[40] An activated nucleotide reagent was then introduced and reacted with the free hydroxyl groups on these electrodes. The chip was subsequently washed, followed by capping and then an oxidation step to stabilize the central phosphorous atom. The process continued with deprotection of certain electrodes and a coupling step. Using this in situ synthesis method, unique oligomers of 35-40 bases were synthesized at each electrode.

For the gene expression study, the array was designed with a variety of genes expressed by the K-562 leukemia cell line as well as a system of spiked-in control transcripts generated from segments of the *Escherichia coli* (*E. coli*) bacteriophage lambda genome (#NC_001416). The spiked-in control transcripts were used to determine sensitivity, reproducibility and dynamic range characteristics of microfluidic microarray devices. Probes were created against various genes involved with immune system pathways, as well as a number of housekeeping genes. The microarray was designed with four replicates of each probe distributed across the array to allow measurement of the variability within the array.

The design of the Influenza A subtyping array probes was based on the viral sequence data obtained from the GenBank database and from the Influenza Sequence Database (ISD).[61] For HA serotypes, 1,614 animal and 1,937 human isolates were selected; and for NA serotypes, 552 human and 831 animal isolates were selected. Both data sets were treated in the same way following a modification of the method of Wang, et al where probe uniqueness was based on subtype differences.[62] For each sequence, non-overlapping appended primers were made, tiling the entire sequence. These oligonucleotides were designed to have similar annealing stability as judged by a nearest neighbor thermodynamic model[63] and were designed to have a T_m of 50°C. Probes that had significant secondary structure ($T_m > 40°C$) were taken out of the set. Finally, probes from only the first 500 bp of sequence were used (bp 50 to 500). After tiling and culling, there were 23,568 HA probes and 15,191 NA probes left. Each sequence and probe was grouped and labeled by its serotype and databases were generated from the compiled sequences of HA and NA isolates. Probes were selected to be exclusive to a given sub-type as judged by pair-wise BLASTN.[64] The details of probe sequences and the number of probes selected for each subtype are described by Lodes et al.[18] A poly T_{10} spacer was added to the 3' end of all probes to avoid surface inhibition. Probe design files for array synthesis were generated with Layout Designer (CombiMatrix Corp., Mukilteo, WA). Oligonucleotide microarrays were synthesized on semiconductor microchips containing over 12,000 independently addressable electrodes. After the in situ synthesis of microarray, the oligonucleotide probes on the chip were phosphorylated with T4 polynucleotide kinase for 30 min at 37°C.

The sequencing microarray chips were designed using sequences that were representative of each subtype of interest.[18] For our experiments, we chose sequences that represent subtype H9N2 (GenBank accession numbers AF 156378 and AF 222654 respectively). Probes were tiled by one nucleotide to cover the sequences of interest and probes were designed with a T_m of approximately 55°C. Four probes were designed for each nucleotide to be examined that were identical except for the 5' nucleotides, which contained either an A, C, T, or G (see Fig. 3).[60] The chips with this design were hybridized with both HA and NA gene targets prepared from Influenza A isolates from infected Quail and Chicken (A/Quail/Hong Kong/G1/97 (H9N2) and A/Chicken/Hong Kong/NT17/99 (H9N2)).

Microfluidic Cartridge

Coupling with the microarray slide is a microfluidic plastic cartridge that consists of five micropumps, six microvalves, five chambers for the storage of different buffers and reagents, a microarray hybridization chamber and a waste chamber, as shown in Figure 1. The plastic cartridge measures $40 \times 76 \times 10$ mm and has channels and chambers that range from 500 μm to 3.2 mm in depth and 0.5 to 8.5 mm in width. The prototype of the plastic cartridge consists of multiple layers of acrylic materials that are laminated and assembled using double-sided adhesive tapes. All the layers, including 5 layers of acrylic sheets with various thicknesses ranging from 0.5 to 5.7 mm (MacMaster-Carr, Atlanta, GA) and 4 layers of double-sided adhesive tapes (Adhesive Research Inc., Glen Rock, PA, USA), were machined using a CO_2 laser machine (Universal Laser Systems, Scottsdale, AZ, USA). The 5.7 mm thick acrylic layer had six valve seats in which six duckbill check valves (Vernay Laboratories Inc., Yellow Springs, OH, USA) were glued using an epoxy. These check valves were normally close and could be open when the upstream pumping pressure exceeded the cracking pressure of the valves. The valves were used to retain the liquid solutions in their storage chambers and prevent cross talk of the solutions between two adjacent chambers. The implementation of these check valves did not require microfabrication process and their operation

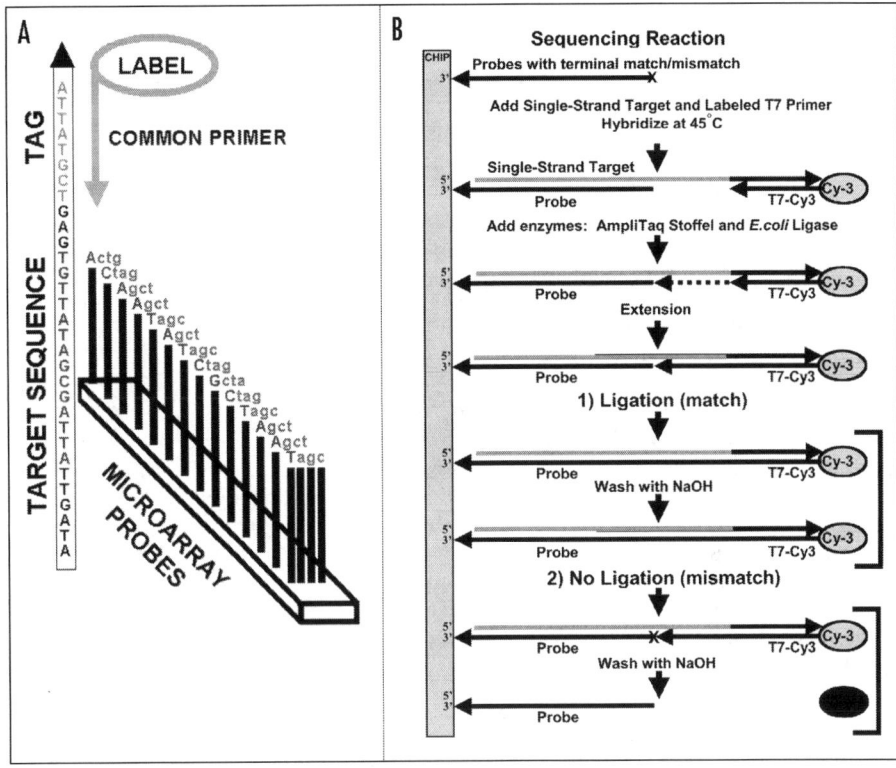

Figure 3. A) Diagram of the strategy for microarray DNA sequencing. Four target-specific (antisense) probes, for each base of sequence, are identical except for the 5' terminal residue (terminate in bases A, C, G and T). After hybridization of probes and target DNA, annealed Cy3-labeled primer, a mixture of enzymes, buffer and dNTPs were added to the array and incubated for 30 min at 37°C. The labeled primer (orange) was extended to matching probes and probes that are a perfect match to the target (blue) were ligated while mismatches (red) are not. The array was finally washed with 0.1 N NaOH and scanned for fluorescence. B) Diagram of the strategy for enzyme-based microarray sequencing. Sequencing probes as described above were hybridized with single-strand target that contains a 3' tag sequence and a labeled oligonucleotide that is complementary to the tag. After washing, an enzyme mix that contains Amplitaq polymerase Stoffel fragment and *E. coli* DNA ligase was added to the annealed complex. The labeled primer was extended on the target template to the probe. Ligation to the probe occurred when target and probe sequences match (1) and no ligation occurred when target and probe do not match (mismatch)(2). Stringent washing removed any signal that is not ligated to the probe (reproduced with permission from ref. 60, copyright 2006 American Chemical Society). A color version of this figure can be found at www.eurekah.com.

required no actuation. As a result, they are less expensive and easier to integrate and operate than most conventional microvalves.[65-68] After assembly, stainless steel wires with 0.5 mm diameter were inserted into the electrochemical pumping chambers followed by sealing with an epoxy. Each electrochemical pumping chamber was then loaded with 50 μL of 1M Na_2SO_4 solution to form an electrochemical pump. The electrolyte loading holes were subsequently sealed using an adhesive tape (Adhesive Research Inc., Glen Rock, PA). The electrochemical micropumps also served as an actuation source for micromixing in the array chamber. The venting hole of the waste chamber was sealed with a hydrophobic membrane vent (Sealing Devices, Lancaster, NY) that allowed gas molecules to pass through while the liquid solution was retained in the waste chamber. The plastic cartridge was then bonded with the microarray chip using a double-sided adhesive tape (Adhesive Research Inc., Glen Rock, PA). The tape with a thickness of 0.5 mm was machined with a pattern of the hybridization chamber using the CO_2 laser machine.

Fluidic Architecture and Operation

The operation of the microfluidic device is as follows. A sample solution was loaded in the array chamber using a pipette. Other solutions required for assays were separately loaded in different storage chambers. For gene expression and subtyping assays, these solutions included four washing buffer solutions and a labeling solution. For sequencing assays, the solutions included a ligase buffer, an enzyme solution containing T4 DNA ligase and Taq polymerase, a NaOH solution and a wash buffer solution. After sealing the loading ports using a sealing tape (Adhesive Research Inc., Glen Rock, PA), the device was then inserted into an instrument, which provided hybridization heating, temperature sensing and electrical power for liquid pumping and mixing. The instrument is measured $140 \times 200 \times 200$ mm. It consists of a clamping manifold, a printed circuit board and a power supply. The microfluidic device was inserted into the manifold where a thin-film heating element (Minco Corp., Minneapolis, MN) was physically pressed on the microarray slide of the device to provide the heating of the array chamber during the hybridization process. The thin-film heating element consists of a temperature sensor that provided the temperature feedback to the control circuit board. A flexible cable connector was used to connect the circuit board with the electrical pins for the electrochemical pumps in the cartridge. The board provides electronic control of the hybridization heating, temperature sensing and electrical power for liquid pumping and mixing. The on-chip assay process started with a hybridization step in the microarray hybridization chamber, followed by subsequent washing and posthybridization process. The pumping of liquid solutions was performed using the integrated electrochemical micropumps that operated with a DC current of 8.6 mA. The pump generated a pressure that was used to open the normally close check valve once the pressure exceeded the cracking pressure of the valve (i.e., 1 psi). During each pumping step, a mixing procedure as described in the following section was implemented. The device was then removed from the instrument. The microarray chip was detached from the microfluidic plastic cartridge before it was scanned using a commercial fluorescent microarray scanner. The fluorescent hybridization signals were detected on the chip and analyzed.

A key aspect of the microfluidic platform is that the device was placed vertically during operation in order to take advantage of the fluid gravity to remove the air bubbles from the system without the use of porous hydrophobic vents.[58] For example, the hybridization chamber was designed with a depth of 600 μm and a width of 6.5 mm and fluid volume was on the order of tens of μL. The Reynolds number for the fluid flow was less than 10.[69] Fluid gravity played an important role in fluidics in the hybridization chamber when the chamber was placed vertically. In this chamber where the liquid solutions and gas bubbles entered from the lower portion, buoyant force allowed gas bubbles to travel quickly to the upper portion of the chamber, leaving the chamber bubble-free.

All of the microfluidic components used in this integrated device, including micropumps, microvalves and micromixers, were designed to be simple, low-cost and easy to fabricate and integrate into the plastic cartridge, resulting in a cost-effective, manufacturable and disposable device. These micropumps, microvalves and micromixers are described in the following sections.

Micropumps

Micropump that can transport liquid solutions with volume of hundreds of μL is one of the most essential components required in this integrated microfluidic device. One of the key requirements of such a micropump is single use and low cost. Most traditional pressure-driven membrane-actuated micropumps did not meet the requirements since they generally suffer from complicated designs and fabrication and high cost.[70-72] In our device, we utilized integrated electrochemical pumps that rely on electrolysis of water between two electrodes in an electrolyte solution ($1M\ Na_2SO_4$) to generate gases when a DC current was applied. Electrolysis-based pumping techniques that

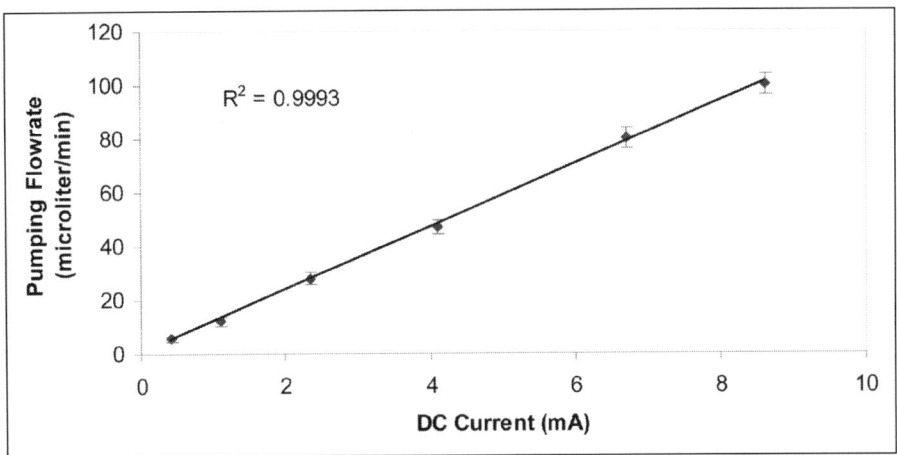

Figure 4. Measurement of the liquid pumping rate as a function of the applied DC current in an electrochemical pump that relies on the electrolysis of water between two stainless steel electrodes in an electrolyte solution (50 µL of 1M Na_2SO_4). Each data point is the mean value obtained from four pumping rate measurements with identical DC current. The pumping rate was determined by measuring the time required to pump a certain amount of liquid solution (e.g., 200 µL) from the storage chamber to the waste chamber at each DC current.

used the generated gases to displace fluids have been previously demonstrated.[59,73-75] In our device, stainless steel wires instead of platinum wires were used as electrodes, resulting in a reduced cost of the device. Note that the Na_2SO_4 solution was not allowed to come into contact with the array chip to prevent contamination of the hybridization with electrode breakdown products. The electrochemical pumps generated gas (H_2 and O_2) that was used to move liquid solutions from chamber to chamber in the device. Flow experiments demonstrated that the pumping flowrate, Q (µL/min), ranging from 5.5 to 100 µL/min, was in linear proportion with the DC current, i (mA), ranging from 0.43 to 8.6 mA. Six data points were used to determine the linear regression model. Each data point is the mean value obtained from four pumping rate measurements with identical DC current. The data fitted into the linear least squares regression equation $Q = 11.727$ i—0.0854, with correlation coefficient (R^2) of 0.9993, as shown in Figure 4. The pumping rate used in this work was 100 µL/min. It was observed that a yellow product was generated in the electrolyte solution during the electrolysis reaction, indicating that the stainless steel corroded. The corrosion did not pose any problem since the whole cartridge was disposed after use. This pumping mechanism did not require a membrane and/or check valves in the design. As a result, the fabrication and operation were simpler than most conventional micropumps.[70-72]

Micromixing

The electrochemical micropumps were also excellent sources to provide gas bubbles to enhance micromixing in the hybridization chamber during the washing and labeling steps. Since the Reynolds number for the fluid flow in the hybridization chamber was less than 10, fluid flows at such a low Reynolds number were predominantly laminar.[69] As a result, mixing of materials between streams in the array chamber was confined to molecular diffusion. The rapid mixing produced by turbulent flows is usually not available because the Reynolds number is below the critical value for transition to turbulence.[69] A pure diffusion-based mixing process can be very inefficient and often takes a long time, particularly when the solution streams contain macromolecules that have diffusion coefficients orders of magnitude lower than that of most liquids. Since the residence time of fluid elements in the flow was smaller than the diffusion time, pure diffusion failed to provide homogeneous mixing of confluent reagent solutions in the array chamber. Therefore, an efficient micromixer was required to enhance micromixing in this device during the washing and labeling steps. We have developed a simple and easy-to-operate mixing technique that is based on the continuous bubbling effect. During the mixing steps, gas bubbles generated from the electrochemical micropump entered into the hybridization chamber from the lower portion. Buoyant force allowed gas bubbles to travel quickly to the upper portion of the chamber. During this traveling process, the two-phase flow resulted in flow recirculation in the liquid solution around the bubbles.[58] As a result, the mixing was enhanced in the chamber. As shown in Figure 5, fluidic experiments with dye solutions demonstrated the efficiency of the mixing enhancement using this bubbling effect. The fluidic experiment showed that the flow recirculation around the continuously pumped bubbles produced homogeneous solutions in the chamber and thus facilitated a uniform reaction on the array surface during the reaction process (e.g., ligation, extension, or labeling) or washing steps. The assay experiments with DNA hybridization described in the following sections showed high sensitivity, low background signals and uniform hybridization signals on the microfluidic arrays, which suggested that on-chip microfluidic mixing was uniform and efficient and the bubbling mixing technique has a relatively low shear strain rate and is thus biofriendly to the hybridized DNA on the array. The use of gas bubbles to enhance mixing in the microfluidic chamber proved to be

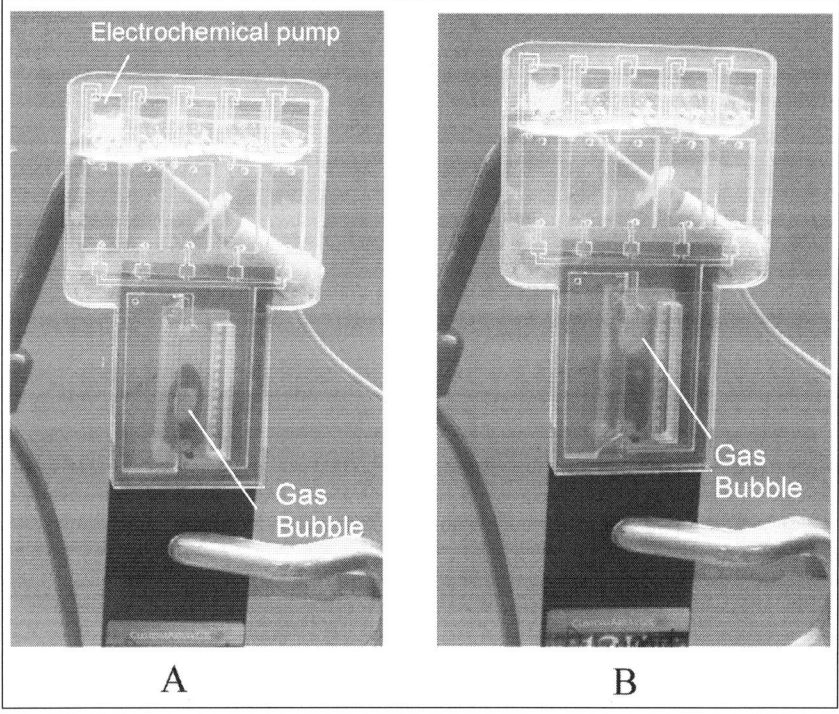

Figure 5. Photographs showing the bubbling mixing process in the hybridization chamber of a microfluidic array device that was placed vertically. A red color dye was used to visualize the mixing effect. A) A gas bubble generated using an electrochemical micropump was directed through a microchannel and introduced into the hybridization chamber from the lower portion of the chamber. The hybridization chamber was initially filled with DI water followed by injection of a small amount of red food dye at the lower portion of the chamber before the introduction of the bubble; B) When the gas bubble traveled to the upper portion of the chamber, flow recirculation was observed in the liquid solution around the bubble, resulting in a dye mixing enhancement in the chamber.

a simple but effective micromixing technique without the use of any external actuation methods such as acoustic agitation[59,76] or physical rotation.

Microvalves

It is necessary to incorporate microvalves in the cartridge to ensure robustness of the design and efficient isolation of liquid solutions in their storage chambers. Moreover, a normally closed valve between the array chamber and waste chamber could prevent evaporation of the hybridization solution during the one-hour hybridization process at 45°C, which would otherwise result in reduction of the hybridization solution and an increase in the salt concentration of the hybridization solution that would in turn lead to increased nonspecific adhesion of DNA to the array and loss of hybridization stringency. In the new design reported here, six commercially available duckbill check valves (VA 3426, Vernay Laboratory Inc., Yellow Springs, Ohio) were integrated. The Duckbill is a precision, one-piece elastomeric check valve that allows flow in one direction and checks flow in the opposite direction. These normally closed check valves are made of silicone and could be easily glued on the valve seats in the cartridge using epoxy. Once the upstream pumping pressure exceeded the cracking pressure of the valves, the valves were open. The implementation of these check valves did not require a microfabrication process and their operation required no actuation. As a result, they are less expensive and easier to integrate and operate than most conventional microvalves.[65-68]

Gene Expression Assay

The sample solution consisted of a complex background sample and spiked-in control transcripts. The complex background sample was prepared from Human Leukemia, Chronic Myelogenous (K-562 cell line) poly A + RNA (Ambion, Austin, TX) utilizing Ambion's MessageAmp aRNA Kit. Biotin was double incorporated using biotin-11-CTP (PerkinElmer, Boston, MA) and biotin-16-UTP (Roche Diagnostics, Mannheim, Germany). Varying concentrations of spiked-in biotin-cRNA control transcripts were combined with a constant amount (150 nM) of K-562 biotin-cRNA complex background such that final concentration of spiked-in control transcripts would range from 1 to 1000 pM in the hybridization. The biotin-cRNA mixtures were fragmented in a 1X fragmentation solution (40 mM Tris-Acetate, pH8.1, 100 mM KOAc, 30 mM MgOAc) at 95°C for 20 minutes. The fragmented cRNA sample was added to a hybridization solution (6X SSPE, 0.05% Tween-20, 20 mM EDTA, 25% DI Formamide, 0.05% SDS, 100 ng/µL sonicated Salmon Sperm DNA) and denatured for 3 minutes at 95°C. The sample was placed briefly on ice followed by centrifugation at 13,000 xg for 3 minutes. During the on-chip assay, 95 µL of the hybridization sample solution was loaded into the hybridization chamber. Other solutions including: (1) 200 µL of 3X SSPE, 0.05% Tween-20; (2) 200 µL of 0.5X SSPE, 0.05% Tween-20; (3) 200 µL of 2X PBST, 0.1% Tween-20; (4) 200 µL of a labeling solution and (5) 200 µL of 2X PBST, 0.1% Tween-20, were separately loaded in the storage chambers. The labeling solution contains streptavidin-Cy5 (Molecular Probes, Eugene, OR) that was diluted in a blocking solution (2X PBS, 0.1% Tween-20, 1% Acetylated BSA) to a final concentration of 1 µg/mL. Hybridization was carried out for 18 hours at 45°C.

Following hybridization, the array in the hybridization chamber was washed for 2 minutes with 3X SSPE, 0.05% Tween-20. On-chip washings continued with 0.5X SSPE, 0.05% Tween-20 for 2 minutes and 2X PBST, 0.1% Tween-20 for 2 minutes. The labeling solution was then pumped into the hybridization chamber and incubated for 30 minutes at room temperature. Note that the cartridge was protected from light using an external cover in the instrument to prevent photobleaching of the fluorescent dye. The final washing step was performed by flowing 2X PBST through the hybridization chamber. Subsequently, the cartridge was separated from the microarray chip with the use of a razor blade. The microarray chip was imaged on an Axon Instruments (Union City, CA) GenePix 4000B—5 µm resolution laser scanner. Imaging was performed while the array was wet with 2X PBST under a LifterSlip™ glass cover slip (Erie Scientific, Portsmouth, NH). Probe fluorescence on the microarray was analyzed and quantified using Microarray Imager software

(CombiMatrix Corp., Mukilteo, WA). To study chip-to-chip variability, gene expression assay with the same protocol was performed in three different microfluidic array devices.

Figure 6 shows the fluorescent scanning image of a section of the microfluidic array and the hybridization analysis of the phage lambda spiked-in control transcripts on the microfluidic array.

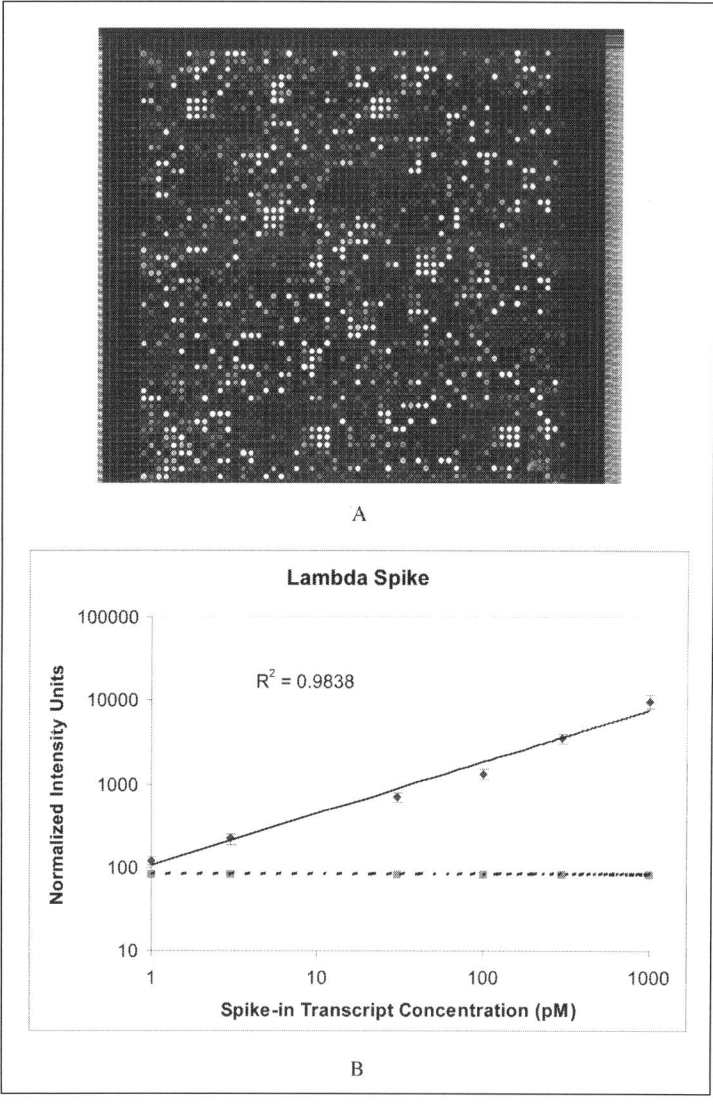

Figure 6. A) Fluorescent scanning image showing a section of the microfluidic array after hybridization and posthybridization processes; B) Hybridization analysis demonstrates the sensitivity and linear dynamic range of the microfluidic array in log scale. Each data point represents the mean of the normalized probe intensities for the spiked-in control transcripts across the array plotted against the corresponding concentrations. Error bars indicate the standard deviation across the array at each data point. Sensitivity was determined to be signal detectable above the average of the negative control signals (the bottom dot line) plus 3 standard deviations.

Each data point shown in Figure 6B represents the mean of the normalized probe intensities for the spiked-in control transcripts across the array plotted against the corresponding concentrations. Error bars indicate the standard deviation across the array at each data point. In determining the cutoff for sensitivity, signal was considered significant if greater than three standard deviations above the average of the negative control signals. The result showed that the dynamic range of the microfluidic platform covered 3 orders of magnitude. Further studies with lower spiked-in control transcript concentrations showed the sensitivity of the microfluidic array device achieved 0.375 pM (results not shown here). Since each measurement was made with replicate probes that were spaced across the array to allow measurement of the variability within the array, it was possible to get an accurate representation of reproducibility at each spiked-in concentration. Error bars in Figure 6B represent the standard deviation across the replicate probes and indicate that hybridization signals are uniform across the whole array. The low background signals and uniform hybridization signals suggest that on-chip microfluidic washing and labeling are uniform and efficient.

The result of cartridge-to-cartridge reproducibility study is shown in Figure 7. Given that three microfluidic array devices in this study received the same concentration of background target (K-562 cRNA), inter-array comparison could be demonstrated by comparing the probes specific to genes expressed by this sample. Scatter plots comparing these probe intensities (raw data) on three different arrays against their average intensities are demonstrated in Figure 7. The median covariance (CV) of the un-normalized probe intensities across these three arrays was 14%, indicating that the cartridge-to-cartridge variability is low. This inter-cartridge CV was calculated by averaging all the CVs of the un-normalized fluorescent intensities of identical probes on the three arrays. Similarly, the inter-chip CV of the un-normalized probe intensities across three conventional microarray chips with regular hybridization chambers that have no integrated microfluidic components was

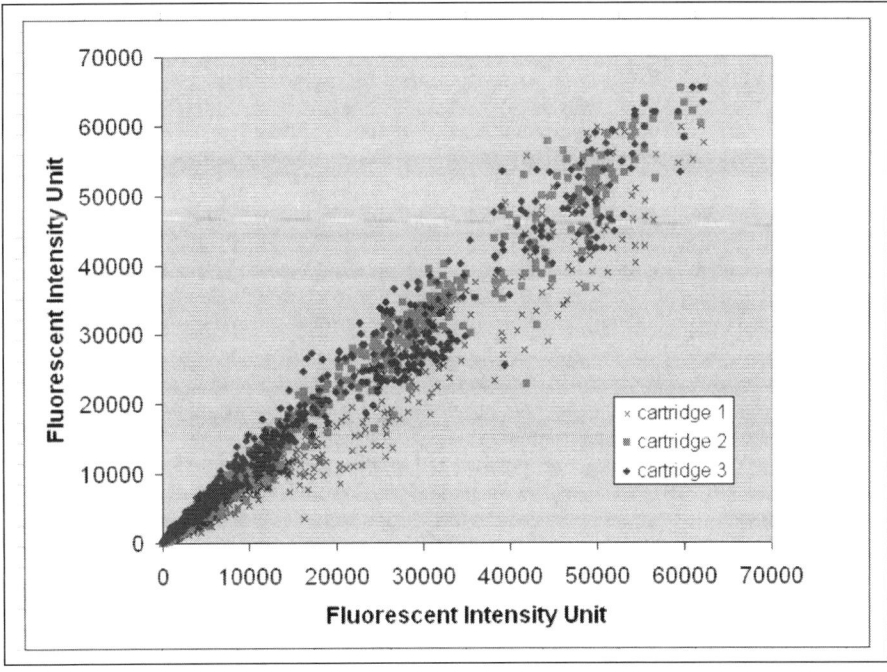

Figure 7. Scatter plot of the raw fluorescent intensities comparing data from three microfluidic array devices against their average intensities (x-axis) and showing cartridge-to-cartridge reproducibility. These three microfluidic array devices received the same concentration of the background target (K-562 cRNA). Both x- and y- axes are in fluorescent intensity unit.

approximately 14.5%. For these conventional microarray chips, all the fluidic handling and processes were carried out manually using pipettes by the same user and the sequence and composition of the buffers for manual processing were identical to those used for the automated microfluidic processing. The comparison results indicate that the cartridge-to-cartridge variability is slightly better than or equivalent to the manual chip-to-chip variability.

Subtyping Assay

The Influenza virus subtype reference samples were prepared as follows. The Influenza viruses were isolated using Madin-Darby canine kidney (MDCK) cells supplemented with 1 μg/mL L-(tosylamido-2-phenyl) ethyl chloromethyl ketone (TPCK)-treated trypsin. The samples were first added to monolayers of MDCK cells and incubated for 1 hr at 37°C to allow viral adsorption to the cells. The inoculum was decanted, Eagle's minimum essential medium supplemented with 0.2% bovine serum albumin was added and monolayers were incubated for 3-5 days at 37°C. After cytopathic effects appeared, influenza virus was confirmed by using hemagglutination of chicken erythrocytes and RT-PCR against the HA gene. A panel of reference influenza A virus RNA samples for 15 HA and 9 NA subtypes was developed by conducting hemagglutination inhibition (HI) assays with a panel of reference antisera against HA subtypes 1 through 15 and NA subtypes 1 through 9. Reference virus sequences were confirmed for each HA and NA subtype. Viral RNA was extracted from supernatants of cultures of infected cells by using the RNeasy Mini Kit (Qiagen, Chatsworth, CA) according to the manufacturer's instructions. Reverse transcription and PCR amplification were carried out under standard conditions by using influenza-specific primers. PCR products were purified with QIAquick PCR purification kits (Qiagen, Chatsworth, CA) and sequencing reactions were performed at the Hartwell Center for Bioinformatics and Biotechnology at St. Jude Children's Research Hospital (Memphis, TN).

The target for both subtyping and sequencing studies is a 500~600 bp amplicon that contains the 5' end of either the HA or NA gene and was amplified from first-strand cDNA. First-strand cDNA was produced from influenza virus RNA (5-20 ng/μL) with SuperScript II reverse transcriptase (Invitrogen, Carlsbad, CA) and a tagged universal primer, *TAATACGACTCACTATAGG*AGCAA AAGCAGG (tag sequence is italicized and universal influenza sequence is in bold). Amplifications were accomplished with a 10 μM forward tag primer [GCATCCTAATACGACTCACTATA GG] and specific reverse primer and 2 to 5 μL of first-strand cDNA per 100 μL reaction.[18] The reaction conditions consisted of a 5-min denaturation at 94°C, followed by 40 cycles of a 30-sec 94°C denaturation step; a 30-sec 55°C annealing step; and a 30-sec 72°C extension; and finally a 10-min extension at 72°C. The resulting PCR product was cleaned with a Qiagen QIAquick PCR purification kit and eluted in 100 μL of distilled water. A second, one-way amplification resulted in single stranded target. One-way amplifications were accomplished with the respective specific reverse primer only and 2 to 5 μL of cleaned amplification product from the first amplification. The reaction conditions were similar to those described above with 50 cycles of amplification. The resulting product was purified with a Qiagen QIAquick PCR purification kit and eluted in 100 μl of distilled water. This step resulted in tagged, single-stranded target for hybridization in sequencing assays. For subtyping assays, biotinylated single-strand target for standard hybridizations was produced as described above for one-way PCR, however, biotin-14-dCTP (Invitrogen, Carlsbad, CA) was incorporated into the product during amplification. Prior to hybridization, the single-stranded target was heated to 95°C for 10 min and then placed on ice. Ten × T4 ligase buffer was then added to bring the solution to 1 × concentration and, finally, a 5' labeled T7 oligonucleotide (Cy-3 or Cy-5, Integrated DNA Technologies, Inc., Coralville, IA) was added to a concentration of 1 μM.

During the on-chip subtyping assay, 95 μL of the hybridization sample solution was loaded into the array chamber in the cartridge. Other solutions including: (1) 200 μL of 6x SSPE, 0.05% Tween-20; (2) 200 μL of 3x SSPE, 0.05% Tween-20; (3) 200 μL of 2X PBST, 0.1% Tween-20; (4) 200 μL of a labeling solution and (5) 200 μL of 2X PBS, 0.1% Tween-20, were separately loaded in the storage chambers 1-5 (Fig. 1A). The labeling solution contains streptavidin-Cy5

(Molecular Probes, Eugene, OR) that was diluted in a blocking solution (2X PBS, 0.1% Tween-20, 1% Acetylated BSA) to a final concentration of 1 µg/mL. Hybridization was carried out for one hour at 45°C in the microarray chamber. Following hybridization, the array was washed with 6X SSPE, 0.05% Tween-20. On-chip washings continued with 3X SSPE, 0.05% Tween-20 and 2X PBS, 0.1% Tween-20, respectively. The labeling solution was then pumped into the hybridization chamber and incubated for 30 minutes at room temperature. The final washing step was performed by flowing 2X PBST through the hybridization chamber. During each pumping step, a bubbling mixing procedure was implemented. The microarray chip was then detached from the microfluidic plastic cartridge before it was scanned. The total on-chip processing time is approximately 1 hour and 50 minutes. Subtype identification was accomplished by averaging HA and NA subtype intensity values and graphing in Microsoft Excel.

The broad-scan microarray was designed to identify influenza A subtypes based on unique probe sets for HA subtypes 1 through 15 and NA subtypes 1 through 9. After hybridization of biotinylated H2 and N7-specific target and labeling with streptavidin-Cy5, high fluorescent signal intensity demonstrates that the subtype specific target DNA annealed to the correct probes, as shown in Figure 8.[60] By averaging subtype probe signal intensities and graphing the results, the correct influenza A subtypes were identified using the microfluidics device. In general, average positive probe signal was at least 4 times greater than average negative probe signal. Variable signal intensities from the negative probes are the result of cross hybridizations that are due to the close relationship among subtype sequences. Results using the microfluidics device were comparable to the manual control results (where the average signal intensities for H2 and N7 are 3,400 and 2,500, respectively and the negative probe and background signals are below 600), suggesting that on-chip microfluidic washing and labeling are efficient and compatible with manual washing and labeling processes. Since there were no replicate probes in the current subtyping array, it is difficult to study the hybridization uniformity. The Influenza A subtyping array used in this study has previously been shown to positively identify all 15 hemagglutinin and all 9 neuraminidase subtypes.[18] Previous study with gene expression analysis in the microfluidic array devices showed high detection sensitivity (375 fM) and uniform hybridization signals,[77] further indicating that

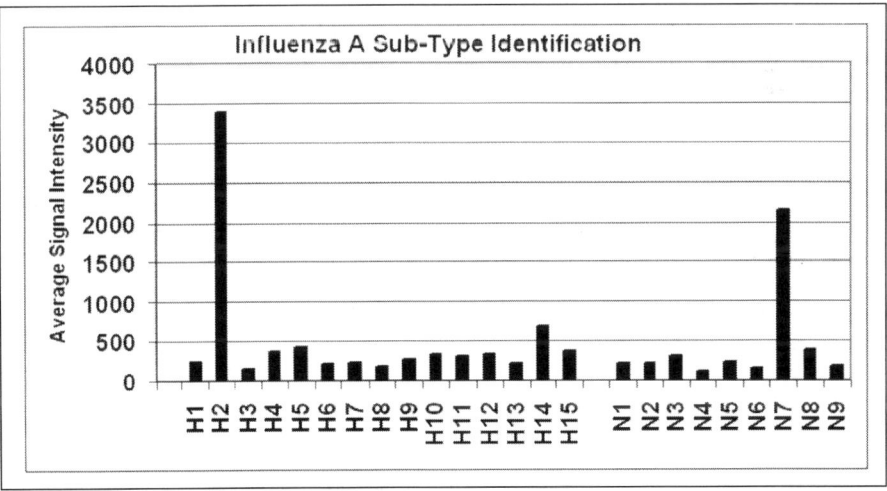

Figure 8. Influenza subtype analysis demonstrates the robustness of the microfluidic system. The correct Influenza A hemagglutinin (H2) and neuraminidase (N7) subtypes were identified with this automated system with results equal to or better than standard manual techniques (reproduced with permission from ref. 60, copyright 2006 American Chemical Society).

the on-chip fluidic handling (washing and reaction) is efficient and can be automated with no loss of performance.

Sequencing Assay

In many cases, the identification of an influenza A subtype is not sufficient for making decisions on vaccine development, patient treatment and general surveillance and specific sequence is needed. For example, mutations in sequence coding for the hemagglutinin receptor binding site of subtype H5N1 have been shown to be important for replication of avian viruses in humans.[78] In this study, viral HA and NA subtype sequencing was accomplished with a simple and rapid enzymatic assay in a microfluidics device. A mixture including DNA polymerase and ligase extended a labeled common primer on single-stranded target DNA to the 5' end of the hybridized HA and NA sequencing probes and then ligated the extended primer to probes that matched the target sequence (Fig. 3). A 0.1 N NaOH wash, a stringent washing procedure, removed any un-ligated signal and after scanning the array, intensity data was exported to an Excel worksheet. Sequence information was extracted with a routine designed to associate the correct base with the highest signal from sets of 4 probes that were tiled by one nucleotide to cover the sequence of interest (Fig. 3B). The preparation of the sequencing sample target has been described in the above section for the subtyping assay. During the on-chip sequencing assay, 95 μL of the hybridization sample solution was loaded into the array chamber in the cartridge. Other solutions include: 1) 200 μL of 1 × *E. coli* ligase buffer; 2) 200 μL of a mixture containing 155 μl of dH$_2$O, 18 μl of 10 × *E. coli* ligase buffer, 3 μL of 10 mM dNTP, 2 μL (20 units) of AmpliTaq DNA polymerase, Stoffel fragment (Applied Biosystems, Foster City, CA) and 2 μL (20 units) of *E. coli* ligase; 3) 200 μL of 0.1 N NaOH buffer; 4) 200 μL of 2X PBS, 0.1% Tween-20, were separately loaded in the storage chambers 1-4 (Fig. 1A). The 5th storage chamber was left empty in the device. Hybridization was carried out for one hour at 45°C. Following the hybridization process, the ligase buffer solution was first pumped through the array chamber, removing the sample mixture into the waste chamber and washing the array. The DNA ligase/polymerase mixture was subsequently pumped through the array chamber followed by a 30-minute incubation at 37°C. Once the extension and ligation were completed, the 0.1 N NaOH buffer was pumped through the array chamber at room temperature to denature and wash the array. The 2x PBST (phosphate-buffered saline-Tween 20) buffer was subsequently pumped through the array chamber to ensure a thorough washing. During each pumping step, a bubbling mixing procedure as described in the following section was implemented. The total on-chip processing time is approximately 1 hour and 45 minutes. The device was then removed from the instrument. The microarray chip was separated from the microfluidic plastic cartridge with the use of a razor blade before it was scanned using a commercial fluorescent scanner (GenePix 4,000B, Molecular Devices, Sunnyvale, CA). Imaging was performed while the array was wet with 2X PBST under a LifterSlip™ glass cover slip (Erie Scientific, Portsmouth, NH). Image intensities were analyzed and quantified using Microarray Imager software (CombiMatrix Corp., Mukilteo, WA). HA and NA subtype DNA sequence information was generated from sequencing array intensity data with a Microsoft Excel routine designed to interrogate units of 4 data points and then associate the most intense signal with the nucleotide represented by that probe. Sequence strings were then used to search the GenBank nonredundant database with BLASTN.

The results of hemagglutinin (H9) and neuraminidase (N2) subtype sequencing in the microfluidic cartridge is shown in Figure 9.[60] With such an array chip that has over 12,000 features, we were able to sequence over 500 nt, depending upon the length of the target DNA, of both the HA and NA genes with an accuracy of over 90%. In this study, the sequencing accuracies of the HA and NA genes using the microfluidic device are 91% and 94%, respectively, as compared to 95% for HA gene and 93% for NA gene using the conventional manually handling array (i.e., the control). The results indicated that the performance of the microfluidic array device was comparable to that of conventional manual handling. Inaccurate sequence calls were generally the result of difficulty to detect mismatches, such as A/G or T/G mismatches or secondary structure within the probe sequences. This artifact of the sequencing array can be resolved by including replicates

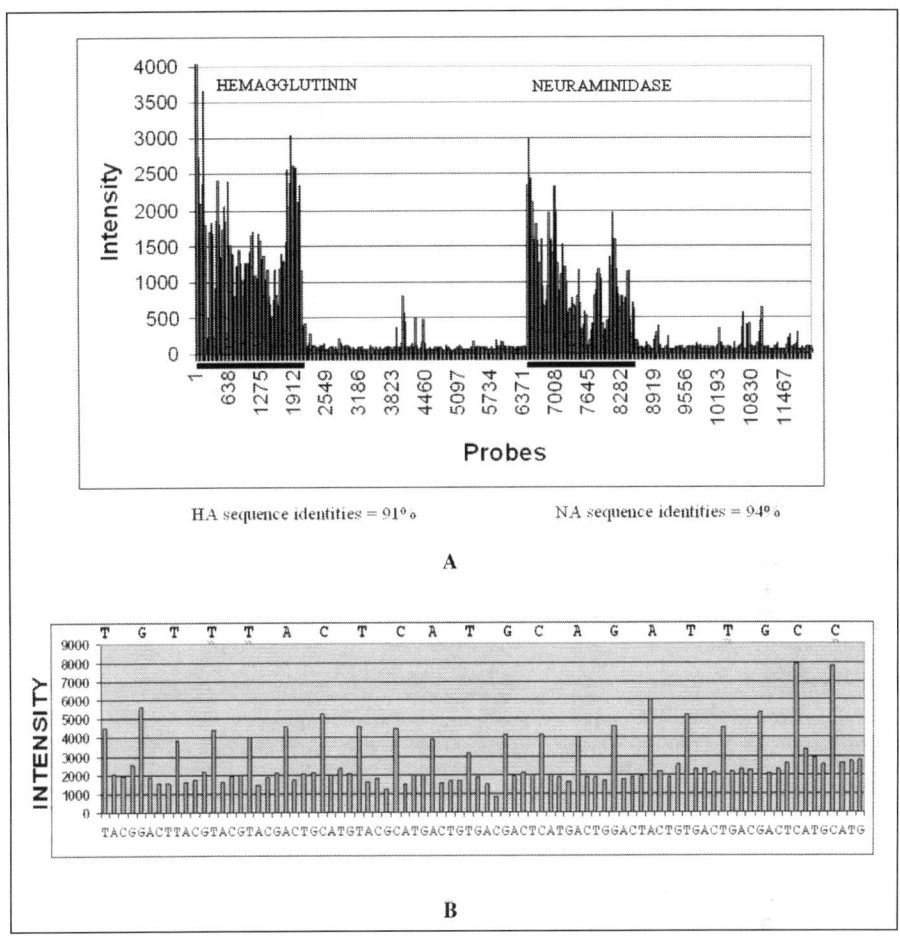

Figure 9. A) Probe hybridization intensity values for hemagglutinin H9 and neuraminidase N2 sequencing using the automated cartridge. Although the entire sequences of the HA (approximately 1,700 bp) and NA (approximately 1,400 bp) genes were encoded on the array, the targets were synthesized from and hybridized to the 5′ 500 nucleotides of the genes as demonstrated by the hybridization patterns (underlined). After extraction of data, resulting sequences for the HA and NA genes were 91% and 94% respectively identical to the correct GenBank sequence data. B) Sequence information is obtained by associating the DNA bases with the greatest signals from groups of 4 sequencing probes. An expanded section of the neuraminidase gene is shown here. DNA bases above the figure indicate the sequence at the 5′ end of probes and bases below the figure indicate the sequence extracted from the data (reproduced with permission from ref. 60, copyright 2006 American Chemical Society).

of the sequencing probes. The average values will reduce signal from ligated mismatches and other artifacts on the chip. This approach to sequencing is rapid, requiring less than 2 hours on chip to sequence a target.

Discussion

The integrated microfluidic array device automates microarray assays that involve multi-stage sample processing and fluidic handling that are in general labor-intensive and time-consuming and have significant potential to be error-prone.[79] Automation of the hybridization and posthybridization process allows more stringent processing control over the microarrays and eliminates variations in array data caused by subtle, day-to-day differences in protocol and manual handling.[77] There is more control over a variety of parameters including hybridization temperature and time, washing time and speed, mixing/agitation speed and labeling time. Although some commercial robotic workstations have also been developed to automate microarray processing, such instruments are generally too expensive and tend to limit the application of microarrays to high budget applications. These bulky workstations are complicated to operate and often require high cost for maintenance. In contrast, the self-contained and fully integrated microfluidic array devices reported here are disposable and require simple portable instruments for operation. The plastic microfluidic cartridges can be fabricated using injection molding (instead of laser machining reported here) that would lead to low-cost microfluidic devices. The cost of a microfluidic plastic cartridge component is small as compared to the cost of the microarray chip. In order to reduce the overall cost of the device (primarily of the microarray chip) for diagnostic applications, we are currently developing different methods or technologies. For example, the array is currently being redesigned to have 4 identical sub-array sectors so that 4 assays can be run on the same array chips. Moreover, the reuse of the array by stripping the hybridized DNA/RNA targets using chemicals before the next use is being explored. These strategies will reduce the per-assay costs so that the arrays are comparable in cost to other diagnostic technologies (e.g., real-time PCR). Furthermore, the fluorescent detection method used for the semiconductor-based microarrays reported in this work can be replaced with the electrochemical detection based method. Using an electrochemical detector (currently under development), we could further reduce the costs of equipment by eliminating the expensive optical scanning instrument.

The integration of microfluidics adds new significant functionalities to the conventional microarray platform. Although gene expression, sequencing and subtyping assays were demonstrated in this study, this integrated microfluidic platform can potentially be applied to many other assays. Although the sample and reagent consumption in the current microfluidic device design is tens to hundreds of µL, it is believed that the design could be further miniaturized so that the sample and reagent consumption could be significantly reduced. It is also possible to integrate front-end sample preparation and DNA amplification into the same platform, which would lead to a fully integrated system with direct sample-to-answer capability.[59]

The microfluidic components in the device, including electrochemical pumps, duckbill check valves and bubbling mixer, are simple in design, inexpensive and easy to fabricate and integrate into a complex microfluidic system, as compared with most of the existing micropumps, microvalves and micromixers. Because the power consumption of the electrochemical pumps is low (~ mW), hand-held field use of the integrated microfluidic components is feasible. The combination of integrated microfluidics with electrochemical detection based microarray that is currently under development would allow one to perform gene expression study or identify influenza subtypes and sequence samples of interest rapidly and cost-effectively. The read out will no longer be visual and rely on fluorescence that often suffers from photobleaching issues, but rather be electrochemical relying on the intrinsic electronic functionality of the silicon integrated circuit chip. The electrochemical detection will be a key differentiation of the integrated microfluidic array device since the chip will no longer need to be removed from the microfluidic cartridge for fluorescent scanning. The analysis workflow will also be improved as the stitching or templating of the fluorescent image will not be required. The integrated microfluidic platform provides a step towards fulfilling the promise of rapid, automated genetic analysis from complex sample fluids in cost-effective and portable instruments.

Conclusion

The use and applications of microarray technology have grown since its invention. More recent advances indicate that microfluidics can be integrated along with microarrays to fabricate multi-function devices. In this chapter, a self-contained and disposable microfluidic array device for integrated gene expression and genotyping assays is reported. The device automated and integrated hybridization, sample processing and fluidic handling steps that are considered labor-intensive and time-consuming in regular manual handling process. All microfluidic components such as micropumps, microvalves and micromixers are integrated on the microfluidic cartridge, but use simple and inexpensive approaches in order to reduce device complexity. Gene expression study of the human leukemia cell line (K562) and genotyping detection and sequencing of influenza A subtypes have been demonstrated using this integrated biochip platform. It is believed that this integrated microfluidic platform can potentially be applied to many other assays for DNA analysis.

Acknowledgements

The authors thank Tai Nguyen, Kevin Schwarzkopf, Tony Siuda, Alla Petrova, Kia Peyvan, Michael Bizak, Jeff Kemper, Al Pierce and Mike Slota for technical supports and useful discussions. This work has been sponsored by DoD contract #1999011104A.

References

1. Schena M, Shalon D, Davis RW et al. Quantitative monitoring of gene expression patterns with a complementary DNA microarray. Science 1995; 270:467-470.
2. Hughes TR, Marton MJ, Jones AR et al. Functional discovery via a compendium of expression profiles. Cell 2000; 102:109-126.
3. Gray NS, Wodicka L, Thunnissen AM et al. Exploiting chemical libraries, structure and genomics in the search for kinase inhibitors. Science 1998; 218:533-538.
4. Chu S, DeRisi J, Eisen M et al. The transcriptional program of sporulation in budding yeast. Science 1998; 282:699-705.
5. Schena M. Microarray BiochipTechnology. Natick, MA: Eaton Publishing 2000.
6. Roberts CJ, Nelson B, Marton MJ et al. Signaling and circuitry of multiple MAPK pathways revealed by a matrix of global gene expression profiles. Science 2000; 287:873-880.
7. Khan J, Simon R, Bittner M et al. Gene expression profiling of alveolar rhabdomyosarcoma with cDNA microarrays. Cancer Res 1998; 58:5009-5013.
8. Perou CM, Jeffrey SS, van de Rijn M et al. Distinctive gene expression patterns in human mammary epithelial cells and breast cancers. Proc Natl Acad Sci USA 1999; 96:9212-9217.
9. Golub TR, Slonim DK, Tamayo P et al. Molecular classification of cancer: class discovery and class prediction by gene expression monitoring. Science 1999; 286:531-537.
10. Hu GK, Madore SJ, Moldover B et al. Predicting splice variants from DNA chip expression data. Genome Res 2001; 11:1237-1245.
11. Dill K, McShea A. Recent advances in microarrays. Drug Discovery Today: Technologies 2005; 2(3):261-266.
12. Gershon D. DNA Microarrays. Nature 2005; 437:1195-1200.
13. Ellis JS, Zambon MC. Molecular diagnosis of influenza. Rev Med Virol 2002; 12(6):375-389.
14. Ivshina AV, Vodeiko GM, Kuznetsov VA et al. Mapping of genomic segments of influenza B virus strains by an oligonucleotide microarray method. J Clin Microbiol 2004; 42(12):5793-5801.
15. Kessler N, Ferraris O, Palmer K et al. Use of the DNA flow-thru chip, a three-dimensional biochip, for typing and subtyping of influenza viruses. J Clin Microbiol 2004; 42(5):2173-2185.
16. Li J, Chen S, Evans DH. Typing and subtyping influenza virus using DNA microarrays and multiplex reverse transcriptase PCR. J Clin Microbiol 2001; 39(2):696-704.
17. Sengupta S, Onodera K, Lai A et al. Molecular detection and identification of influenza viruses by oligonucleotide microarray hybridization. J Clin Microbiol 2003; 41(10):4542-4550.
18. Lodes MJ, Suciu D, Elliott M et al. Influenza A Subtype Identification and Sequencing with Semiconductor-based Oligonucleotide Microarrays. J Clin Microbiol 2006; 44:1209-1218.
19. Hay A, Gregory V, Douglas AR et al. The evolution of human influenza viruses. Phil Trans R Soc Lond 2001; B356:1861-1870.
20. Fouchier RA, Munster V, Wallensten A et al. Characterization of a novel influenza A virus hemagglutinin subtype (H16) obtained from black-headed gulls. J Virol 2005; 79(5):2814-2822.
21. Hoffmann E, Stech J, Guan Y et al. Universal primer set for the full-length amplification of all influenza A viruses. Arch Virol 2001; 146:1-15.
22. Lipatov AS, Govorkova EA, Webby RJ et al. Influenza: Emergence and control. J Virol 2004; 78(17):8951-8959.
23. Scholtissek C, Burger H, Kistner O et al. The nucleoprotein as a possible major factor in determining host specificity of influenza H3N2 viruses. Virology 1985; 147(2):287-294.

24. Mizuta K, Katsushima N, Ito S et al. A rare appearance of influenza A(H1N2) as a reassortant in a community such as Yamagata where A(H1N1) and A(H3N2) cocirculate. Microbiol Immunol 2003; 47(5):359-361.

25. Webby RJ, Webster RG. Emergence of influenza A viruses. Phil Trans R Soc Lond 2001; B356:1815-1826.

26. Allwinn R, Preiser W, Rabenau H et al. Laboratory diagnosis of influenza-virology or serology? Med Microbiol Immunol (Berl) 2002; 191(3-4):157-160.

27. Amano Y, Cheng Q. Detection of influenza virus: traditional approaches and development of biosensors. Anal Bioanal Chem 2005; 381(1):156-184.

28. Ueda M, Maeda A, Nakagawa N et al. Application of subtype-specific monoclonal antibodies for rapid detection and identification of influenza A and B viruses. J Clin Microbiol 1998; 36(2):340-344.

29. Ramakrishnan R, Dorris D, Lublinsky A et al. An assessment of Motorola CodeLink microarray performance for gene expression profiling applications. Nucleic Acids Res 2002; 30:e30.

30. Yue H, Eastman PS, Wang BB et al. An evaluation of the performance of cDNA microarrays for detecting changes in global mRNA expression. Nucleic Acids Res 2001; 29:e41.

31. Southern EM, Maskos U, Elder JK. Analyzing and comparing nucleic acid sequences by hybridization to arrays of oligonucleotides: evaluation using experimental models. Genomics 1992; 13:1008-1017.

32. Maskos U, Southern EM. Oligonucleotide hybridizations on glass supports: a novel linker for oligonucleotide synthesis and hybridization properties of oligonucleotides synthesized in situ. Nucleic Acids Res. 1992;20:1679-1684.

33. Fodor SPA, Read JL, Pirrung MC et al. Light-directed, spatially addressable parallel chemical synthesis. Science 1991; 251:767-773.

34. Hughes TR, Mao M, Jones AR et al. Expression profiling using microarrays fabricated by an ink-jet oligonucleotide synthesizer. Nat Biotechnol 2001; 19:342-347.

35. Singh-Gasson S, Green RD, Yue Y et al. Maskless fabrication of lightdirected oligonucleotide microarrays using a digital micromirror array. Nat Biotechnol 1999; 17:974-978.

36. Pellois JP, Zhou X, Srivannavit O et al. Individually addressable parallel peptide synthesis on microchips. Nat Biotechnol 2002; 20:922-926.

37. Dill K, Montgomery DD, Ghindilis AL et al. Immunoassays and sequence-specific DNA detection on a microchip using enzyme amplified electrochemical detection. Biochem Biophys Methods. 2004; 59:181-187.

38. Nittler MP, Hocking-Murray D, Foo CK et al. Identification of Histoplasma capsulatum Transcripts Induced in Response to Reactive Nitrogen Species. Mol Biol Cell 2005 (in press).

39. Maurer K, McShea A, Strathmann M et al. The Removal of the t-BOC Group by Electrochemically Generated Acid and Use of an Addressable Electrode Array for Peptide Synthesis. J Comb Chem 2005 (in press).

40. Oleinikov AV, Gray MD, Zhao J et al. Self-Assembling Protein Arrays Using Electronic Semiconductor Microchips and in Vitro Translation. J Proteome Res 2003; 2:313.

41. Kelly RT, Woolley AT. Microfluidic Systems for Integrated, High-Throughput DNA Analysis. Anal Chem 2005; 77:97A-102A.

42. Harrison DJ, Manz A, Fan Z et al. Capillary Electrophoresis and Sample Injection Systems Integrated on a Planar Glass Chip. Anal Chem 1992; 64:1926-1932.

43. Wilding P, Pfahler J, Bau HH et al. Manipulation and Flow of Biological-Fluids in Straight Channels Micromachined in Silicon. Clin Chem 1994; 40(1):43-47.

44. Xia YN, Whitesides GM. Soft lithography. Annual Review of Materials Science 1998; 28:153-184.

45. Piner RD, Zhu J, Xu F et al. "Dip-pen" nanolithography. Science 1999; 283(5402):661-663.

46. Alonso-Amigo MG, Becker H. Microdevices fabricated by polymer hot embossing. Abstr Pap Am Chem Soc 2000; 219:468-COLL.

47. Becker H, Dietz W, Dannberg P. Microfluidic Manifolds by Polymer Hot Embossing for Micro Total Analysis System Applications. Paper presented at: uTas 98, 1998; Banff, Canada.

48. Boone T, Fan ZH, Hooper H et al. Plastic advances microfluidic devices. Anal Chem 2002; 74(3):78A-86A.

49. Grodzinski P, Liu RH, Chen H et al. Development of Plastic Microfluidic Devices for Sample Preparation. Biomed Microdevices 2001; 3(4):275.

50. Harrison DJ, Fluri K, Seiler K et al. Micromachining a Miniaturized Capillary Electrophoresis-Based Chemical-Analysis System on a Chip Science 1993; 261(5123):895 897.

51. Burns MA, Johnson BN, Brahmasandra SN et al. An integrated nanoliter DNA analysis device. Science 1998; 282(5388):484-487.

52. Waters LC, Jacobson SC, Kroutchinina N et al. Microchip Device for Cell Lysis, Multiplex PCR Amplification and Electrophoretic Sizing. Anal Chem 1998; 70:158-162.

53. Woollery AT, Hadley D, Landre P et al. Functional integration of PCR amplification and capillary electrophresis in a microfabricated DNA analysis device. Anal Chem 1996; 68(23):4081-4086.
54. Kopp M, Mello AD, Manz A. Chemical amplification: Continuous-Flow PCR on a chip. Science 1998; 280:1046-1048.
55. Ibrahim MS, Lofts RS, Jahrling PB et al. Real-time microchip PCR for detecting single-base differences in viral and human DNA. Anal Chem 1998; 70(9):2013-2017.
56. Yuen PK, Kricka LJ, Fortina P et al. Microchip module for blood sample preparation and nucleic acid amplification reactions. Genome Res 2001; 11(3):405-412.
57. Taylor MT, Belgrader P, Furman BJ et al. Lysing bacterial spores by sonication through a flexible interface in a microfluidic system. Anal Chem 2001; 73(3):492-496.
58. Anderson RC, Su X, Bogdan GJ et al. A miniature integrated device for automated multistep genetic assays. Nucleic Acids Res 2000; 28(12):e60.
59. Liu RH, Yang J, Lenigk R et al. Self-contained, Fully Integrated Biochip for Sample preparation, PCR amplification and DNA microarray Detection. Anal Chem 2004; 76:1824-1832.
60. Liu RH, Lodes MJ, Nugyen T et al. Validation of A Fully Integrated Microfluidic Array Device for Influenza A Subtype Identification and Sequencing. Anal Chem 2006; 78:4184-4193.
61. Macken C, Lu H, Goodman J et al. The value of a database in surveillance and vaccine selection. In: A.D.M.E. Osterhaus NCAWH, ed. Options for the Control of Influenza IV. Amsterdam: Elsevier Science 2001; 103-106.
62. Wang D, Urisman A, Liu Y-T et al. Viral discovery and sequence recovery using DNA microarrays. PLoS Biology 2003; 1(2):257-260.
63. Allawi HT, Jr. JS. Nearest-neighbor thermodynamics and NMR of DNA sequences with internal A.A, C.C, G.G and T.T mismatches. Biochemistry 1999; 38:3468-3477.
64. Altschul SF, Madden TL, Schaffer AA et al. Gapped BLAST and PSI-BLAST: a new generation of protein database search programs. Nucleic Acids Res 1997; 25:3389-3402.
65. Beebe DJ, Moore JS, Bauer JM et al. Functional Structures For Autonomous Flow Control Inside Microfluidic Channels. Nature 2000; 404:588-590.
66. Liu RH, Yu Q, Beebe DJ. Fabrication and Characterization of Hydrogel-based Microvalves. J Microelectromechan Syst 2002; 11:45-53.
67. Jerman H. Electrically-activated, normally-closed diaphragm valves. J Micromech Microeng 1994; 4:210-216.
68. Ray CA, Sloan CL, Johnson AD et al. A Silicon-based Shape Memory Alloy Microvalve. Mater Res Soc Symp 1992; 276:161-166.
69. Liu RH, Stremler M, Sharp KV et al. A Passive Micromixer: 3-D C-shape Serpentine Microchannel. J Microelectromechan Syst 2000; 9(2):190-197.
70. Su YC, Lin LW, Pisano AP. A water-powered osmotic microactuator. J Microelectromechan Syst 2002; 11(6):736-742.
71. Zengerle R, Skluge S, Richter M et al. A Bidirectional Silicon Micropump. Sens Actuators A Phys 1995; 50:81-86.
72. Unger MA, Chou H, Thorsen T et al. Monolithic Microfabricated Valves and Pumps by Multilayer Soft Lithography. Science 2000; 288:113-116.
73. Bohm S, Olthuis W, Bergveld P. An Integrated Micromachined Electrochemical Pump and Dosing System. Biomed Microdevices 1999; 1(2):121-130.
74. Richter G. Device for Supplying Medicines. U.S. Patent, 3,894,538. 1975.
75. Munyan JW, Fuentes HV, Draper M et al. Electrically actuated, pressure-driven microfluidic pumps. Lab Chip 2003; 3:217-220.
76. Liu RH, Lenigk R, Yang J et al. Hybridization Enhancement Using Cavitation Microstreaming. Anal Chem 2003; 75:1911-1917.
77. Liu RH, Nguyen T, Schwarzkopf K et al. A Fully Integrated Miniature Device for Automated Gene Expression DNA Microarray Processing. Anal Chem 2006; 78:1980-1986.
78. Iwatsuki-Horimoto K, Kanazawa R, Sugii S et al. The index influenza A virus subtype H5N1 isolated from a human in 1997 differs in its receptor-binding properties from a virulent avian influenza virus. J Gen Virol 2004; 85:1001-1005.
79. Dobbin KK, Beer DG, Meyerson M et al. Interlaboratory comparability study of cancer gene expression analysis using oligonucleotide microarrays. Clin Cancer Res 2005; 11:565-572.

CHAPTER 4

Self-Contained, Fully Integrated Biochips for Sample Preparation, PCR Amplification and DNA Microarray Analysis

Robin Hui Liu,* Piotr Grodzinski, Jianing Yang and Ralf Lenigk

Abstract

Rapid developments in back-end detection platforms (such as DNA microarrays, capillary electrophoresis, real-time polymerase chain reaction and mass spectroscopy) for genetic analysis have shifted the bottleneck to front-end sample preparation where the 'real' samples are used. In this chapter, we present a fully integrated biochip device that can perform on-chip sample preparation (including magnetic bead-based cell capture, cell preconcentration and purification and cell lysis) of complex biological sample solutions (such as whole blood), polymerase chain reaction, DNA hybridization and electrochemical detection. This fully automated and miniature device consists of microfluidic mixers, valves, pumps, channels, chambers, heaters and DNA microarray sensors. Cavitation microstreaming was implemented to enhance target cell capture from whole blood samples using immunomagnetic beads and accelerate DNA hybridization reaction. Thermally actuated paraffin-based microvalves were developed to regulate flows. Electrochemical pumps and thermopneumatic pumps were integrated on the chip to provide pumping of liquid solutions. The device is completely self-contained: no external pressure sources, fluid storage, mechanical pumps, or valves are necessary for fluid manipulation, thus eliminating possible sample contamination and simplifying device operation. Pathogenic bacteria detection from ~ mL whole blood samples and single-nucleotide polymorphism analysis directly from diluted blood were demonstrated. The device provides a cost-effective solution to direct sample-to-answer genetic analysis and thus has a potential impact in the fields of point-of-care genetic analysis, environmental testing and biological warfare agent detection.

Introduction

Molecular approach to detect low abundance markers in biological tissue is becoming critical to assess genetic and environmental interactions. The miniaturization of biological assays to the chip level carries several advantages. On-chip assays use reduced volumes of reagents (2-3 orders of magnitude as compared to traditional bench approaches) and allow for reducing cost per reaction and improving reaction kinetics.[1-3] On-chip reactions are performed in miniature channels or chambers that can be distributed on the device wafer at high density. This high population of identical reaction paths allows for the development of highly parallel analytical systems with high

*Corresponding Author: Robin Hui Liu—Osmetech Molecular Diagnostics, Pasadena, California U.S.A. Email: robin.liu@osmetech.com

Integrated Biochips for DNA Analysis, edited by Robin Hui Liu and Abraham P. Lee.
©2007 Landes Bioscience and Springer Science+Business Media.

system throughput.[4,5] Furthermore, integration of several assay functions on a single chip leads to assay automation and elimination of operator involvement as a variable. The microfluidic lab-chip device with capabilities of on-chip sample processing and detection provides a cost-effective solution to direct sample-to-answer biological analysis. Such devices will be increasingly important for rapid diagnostic applications in hospitals and in-field bio-threat detection.

A fully integrated biochip needs to perform all functions including sample preparation, mixing steps, chemical reactions and detection in a miniature fluidic device. Most of the currently demonstrated microfluidic or microarray devices pursue single functionality and use purified DNA or homogeneous sample as an input sample. On the other hand, practical applications in clinical and environmental analysis require processing of samples as complex and heterogeneous as whole blood or contaminated environmental fluids. Due to the complexity of the sample preparation, most available biochip systems still perform this initial step off-chip using traditional benchtop methods. As a result, rapid developments in back-end detection platforms have shifted the bottleneck, impeding further progress in rapid analysis devices, to front-end sample preparation where the 'real' samples are used.

Since the early work of Harrison and coworkers[6] on chip-based capillary electrophoresis (CE), the advances of the microfluidic lab-on-a-chip technologies have been multi-directional and have addressed the following issues: high throughput of the analysis,[7] functional complexity,[8-11] level of integration,[12] on-chip sample preparation[8,13-15] and low-cost fabrication and manufacturing.[16,17] Several researchers have developed devices allowing for performance of multi-step assays using complicated channel networks, while pumps, valves and detectors were left off-chip and were built into the desktop test station.[13,14,18] Others argued for integrating all functional components into the chip and preferred portable solutions.[1] The latter efforts led to ingenious demonstrations of on-chip valving [12,19,20] and pumping schemes[21-24] in an attempt to depart from traditional Microelectro-mechanical Systems (MEMS) approaches that are complicated in fabrication and therefore expensive. Various materials have been used in the fabrication of lab-chip devices. Lithographic techniques, adapted from semiconductor technology, have been used to build chips in glass[25] and silicon.[26] Unconventional lithography techniques such as soft lithography[27] have been used to fabricate reproducible microstructures of biological materials offering a multitude of possibilities to explore as molecular diagnostic tools.[28] Recently, with increasing emphasis on disposable devices, the use of plastics and plastic fabrication methods has become popular. [17,29-31]

Most of the integrated microfluidic work has been directed towards the integration of DNA amplification with CE.[1,32,33] Several research groups, including our group, have vigorously pursued integration of sample preparation, polymerase chain reaction (PCR) and DNA microarrays and miniaturization of the whole DNA microarray analysis into a single device. Anderson et al have reported an integrated system that performed RNA purification from a serum lysate, followed by PCR, serial enzymatic reactions and nucleic acid hybridization.[8] Yuen et al reported a microchip module design for blood sample preparation (white blood cell isolation), PCR and DNA microarray analysis.[13] Our group has focused on developing efficient and simple on-chip mixing, valving and pumping techniques and using electrochemical detection based microarray that makes the integrated biochip system desirable for applications valuing portable solutions such as point-of-care diagnostics, in-field environmental testing and on-site forensics. We developed an integrated biochip device that integrated sample preparation with PCR and DNA microarray for sample-to-answer DNA analysis. The on-chip analysis starts with the preparation process of a whole blood sample, which includes magnetic bead-based target cell capture, cell preconcentration and purification and cell lysis, followed by PCR amplification and electrochemical microarray-based detection.

This chapter begins with a description of the design and fabrication of the DNA analysis integrated device, followed by a discussion of the individual microfluidic components essential to this integrated device. Next, a demonstration of the performance of the device for integrated nucleic acid analysis, including pathogenic bacteria detection from ~ mL whole blood samples and single nucleotide polymorphism analysis directly from blood samples, is presented.

Device Design and Fabrication

The biochip device (Fig. 1) consists of a plastic chip, a printed circuit board (PCB) and a Osmetech eSensor® microarray chip.[34] The plastic chip includes a mixing unit for rare cell capture using immunomagnetic separation, a cell preconcentration/purification/lysis/PCR unit and a DNA microarray chamber. The complexity of the chip design is minimized by using some of the chambers for more than one function. For example, the chamber to capture and preconcentrate target cells is also used for subsequent cell lysis and PCR. The PCB consists of embedded resistive heaters and control circuitry. The Osmetech eSensor® (Osmetech Molecular Diagnostics, Pasadena, CA) is a separate PCB substrate with 4×4 gold electrodes on which thiol-terminated DNA oligonucleotides are immobilized via self assembly to detect electrochemical signals of hybridized target DNA.[35,36]

The operation of the biochip device is as followed. A biological sample (such as a blood solution) and a solution containing immunomagnetic capture beads are loaded in the sample storage chamber. Other solutions including a wash buffer, PCR reagents and hybridization buffer are separately loaded in other storage chambers. The biochip device is then inserted into an instrument, which provides electrical power, PCR thermal cycling, DNA electrochemical signal readout and magnetic elements for bead arrest. The PCR chamber of the plastic chip is sandwiched between a Peltier heating element (Melcor Corp., Trenton, NJ) and a permanent magnet. The on-chip sample preparation starts with a mixing and incubation step in the sample storage chamber to ensure target cell capture from the blood using immunomagnetic capture beads. The sample mixture is then pumped through the PCR chamber, where target cell capture and preconcentration occur as the bead-bacteria conjugates are trapped by the magnet. The washing buffer is subsequently pumped through the PCR chamber to purify the captured cells. After the PCR reagents are transferred into the PCR chamber, all the normally open microvalves surrounding the chamber are closed and thermal cell lysis and PCR are performed. Once PCR is completed, the normally closed microvalves are opened, allowing the hybridization buffer and the PCR product to be pumped into the detection chamber, where acoustic mixing of the target and DNA hybridization reaction occurs. The electrochemical hybridization signals corresponding to the redox-reaction of the

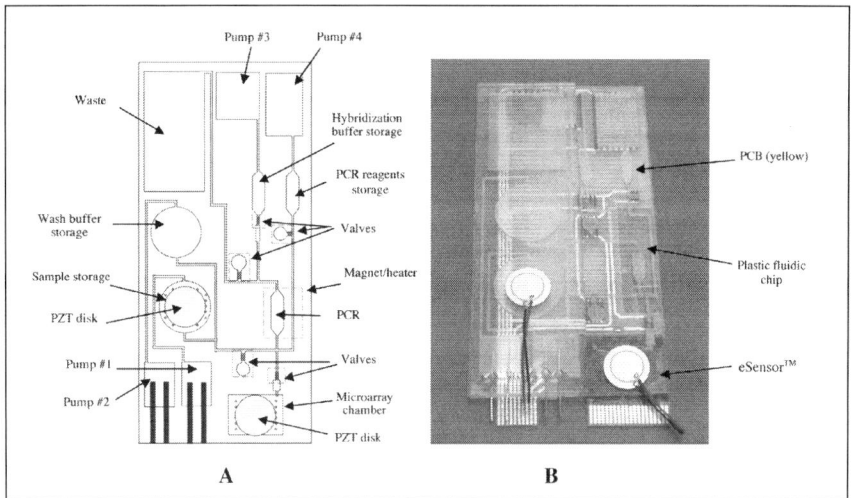

Figure 1. A) Schematic of the plastic fluidic chip. Pumps 1-3 are electrochemical pumps, while Pump 4 is a thermopneumatic pump. B) Photograph of the integrated device that consists of a plastic fluidic chip, a printed circuit board (PCB) and a Osmetech eSensor® microarray chip. Reproduced with permission from ref. 34, copyright 2004 American Chemical Society.

ferrocene-labeled signaling probes that hybridize with the target DNA bound to the immobilized probes are detected on the chip and recorded by the instrument.

The plastic chip measures $60 \times 100 \times 2$ mm and has channels and chambers that range from 300 μm to 1.2 mm in depth and 1 to 5 mm in width. The plastic chip was machined in a polycarbonate (PC) substrate (1.5 mm thick) using conventional computer-controlled machining (Prolight 2,500, Intelitek Inc., Manchester, NH), followed by sealing it with a thin PC cover layer (500 μm thick) using a solvent-assistant thermal bonding technique. During the bonding process, acetone was first applied on one side of the thin cover layer. After 1 min, the cover layer with the gluey surface caused by the chemical reaction with acetone was bonded on the substrate plastic layer followed by a press of 1-ton force at 385°F for 2 min in a hydraulic press (Carver, Inc., Wabash, IN). Platinum wires with 0.5 mm diameter were inserted into the electrochemical pumping chambers, which were then loaded with 20 μL of 5M NaCl solutions to form electrochemical pumps. Two piezoelectric disks (each 15 mm diameter, APC Inc., Mackeyville, PA) that were used to provide acoustic micromixing were bonded onto the external surfaces of the sample storage chamber and the microarray detection chamber, respectively, using a super glue (Duro, Loctite Corp., Avon, Ohio). Fabrication of the paraffin-based microvalves in the biochip began with heating up the plastic chip using a hotplate with a temperature of 90°C that is above the melting temperature of the paraffin. Solid paraffin (~ 10 mg) with a melting temperature T_m of 70°C was then placed into each of the paraffin access holes on the plastic chip. The paraffin was melted instantaneously and capillary force drove the molten paraffin into the channels. The chip was then removed from the hotplate. The paraffin solidified, resulting in an array of microvalves in the device. The paraffin access holes were subsequently sealed using an adhesive tape (Adhesive Research Inc., Glen Rock, PA). The plastic chip was then bonded with the PCB using a double-sided adhesive tape (Adhesive Research Inc., Glen Rock, PA). The eSensor® microarray chip was attached to the detection cavity of the plastic chip using a double-sided adhesive tape.

Fluidic Transport

The fluidic architecture takes advantage of the fluid gravity to remove the air bubbles from the system without the use of porous hydrophobic vents.[8] As in many other microfluidic devices, air plugs or bubbles trapped in the system are of great practical concern, because they often lead to difficulties in controlling the flow and hinder uniform mixing between fluids. Since all the chambers have dimensions of ≥ 500 μm and fluid volumes on the order of μL or mL were handled in the system, the Reynolds number for the flow is on the order of 10 or above.[37] The Reynolds number gives the ratio between inertial forces and viscous forces in a flow. The definition used here is $Re = (Q/A)D/\nu$, where Q is the volumetric flow rate through the channel, A is the cross-sectional area and D_h is the hydraulic diameter of the channel ($4A$/wetted perimeter of the channel) and ν is the kinematic viscosity of the fluid.

When the Reynolds number is on the order of 10 or above, fluid gravity dominated surface forces when the chip was operated in a vertical position. Gas bubbles always migrated to the upper portion of the chamber due to buoyant forces whereas the liquid solution resided in the lower portion. For example, in the PCR chamber where fluids entered at the bottom of the chamber, all air bubbles trapped in the solution escaped towards the top of the chamber and subsequently traveled to the downstream waste chamber, leaving the PCR chamber bubble-free.

Micromixing

Rapid and homogenous mixing is essential in our microfluidic device. First of all, the sample solution in the sample storage chamber needs to be mixed effectively with immunomagnetic capture beads in order to enhance efficient capture and binding of target cells. In the hybridization chamber, the hybridization buffer solution and the amplicon solution need to be mixed in order to obtain homogenous solutions and achieve efficient and rapid DNA hybridization. Mixing in microfluidic systems is generally dominated by diffusion, since turbulence is not practically attainable in micro-scale channel flows or mini systems with small dimensions and thus small Reynolds

(*Re*) numbers. In most of microfluidic systems, *Re* is typically below the critical value for transition to turbulence. Unfortunately, a pure diffusion-based mixing process can be very inefficient and often takes a long time. Mixing is particularly inefficient in solutions containing macromolecules that have diffusion coefficients one or two orders of magnitude lower than that of most liquid molecules. Thus, some mixing techniques must be employed to enhance micromixing.

A few interesting micromixing schemes, including in-line micromixers that enhance mixing between two adjacent flow streams in a microchannel[37-40] and chamber micromixers that utilize stirring mechanisms to mix the fluids in a microchamber,[41,42] have been developed in recent years. One example of in-line micromixers is a multi-stage multi-layer lamination scheme developed by Branebjerg et al.[39] The mixer divides and stacks two flow streams resulting increased contact area and decreased diffusion length. Another in-line micromixer concept was developed by Liu et al[37] using a three-dimensional serpentine microchannel to create rapid stretching and folding of material lines associated with flow-field induced chaotic advection. Electrokinetic instability induced by fluctuating electric fields was also utilized to enhance mixing of electroosmotic channel flows.[38] Chamber micromixing is of particular interest in our applications. Examples of chamber micromixers include those of Moroney et al[41] and Zhu et al.[42] The former used ultrasonic lamb waves (4.7 MHz) traveling in a 4-μm-thick composite membrane of silicon nitride and piezoelectric zinc oxide to induce convectional liquid flow in a chamber. The latter utilized loosely-focused acoustic waves generated by an electrode-patterned piezoelectric film to enhance mixing in an open chamber. Microfluidic motion produced by loosely-focused acoustic waves uses radio frequency (RF) sources with frequencies corresponding to thickness-mode resonance of the thin piezoelectric film. Both devices required a thin chamber membrane (with a thickness of a few microns) between the liquid solution and the piezoelectric film, which was fabricated by silicon (Si) micromachining. The piezoelectric films were driven in ultrasonic frequency range (~ MHz).

We have developed a novel chamber micromixing technique based on the principle of cavitation microstreaming.[43,44] This technique uses air bubbles in a liquid medium as actuators. The bubble surface behaves like a vibrating membrane when the bubble undergoes vibration within a sound field. The behavior of bubbles in sound fields is determined largely by their resonance characteristics. For frequencies in the range considered here (~ kHz), the radius of a bubble at resonant frequency f (Hz) is given by the equation:

$$2\pi a f = \sqrt{3\gamma P_o / \rho} \tag{1}$$

where a is the bubble radius (cm), γ is the ratio of specific heats for the gas, P_o is the hydrostatic pressure (dynes/cm^2) and ρ is the density of the liquid (g/cc). When the bubble undergoes vibration within a sound field, the frictional forces generated at the air/liquid interface induce a bulk fluid flow around the air bubble, called cavitation microstreaming or acoustic microstreaming.[45] It was found that cavitation microstreaming is orderly at low driving amplitudes when the insonation frequency drives the bubbles at their resonance frequency for pulsation and when the bubbles are situated on solid boundaries. The bubble-induced streaming is strongly dependent on frequency for a given bubble radius and on bubble radius for a given frequency. Acoustic microstreaming arising around a single bubble excited close to resonance produces strong liquid circulation flow in the liquid chamber. This liquid circulation flow can be used to effectively enhance mixing. As shown in Figure 2, a set of air bubbles was trapped inside the solution using air pockets (500 μm in diameter and 500 μm in depth) in the cover layer of the chamber.[44] These bubbles were set into vibration by an acoustic field generated using an external piezoelectric transducer (PZT disk, 15 mm diameter, APC Inc., Mackeyville, PA) that was bonded on the external surface of the cover layer using a super glue (DuroTM, Loctite Corp., Avon, Ohio). Fluidic experiments showed that the time taken to achieve a complete mixing in a 50 μL chamber using cavitation microstreaming was significantly reduced from hours (a pure diffusion-based mixing) to only 6 seconds, as shown in Figure 3.[44]

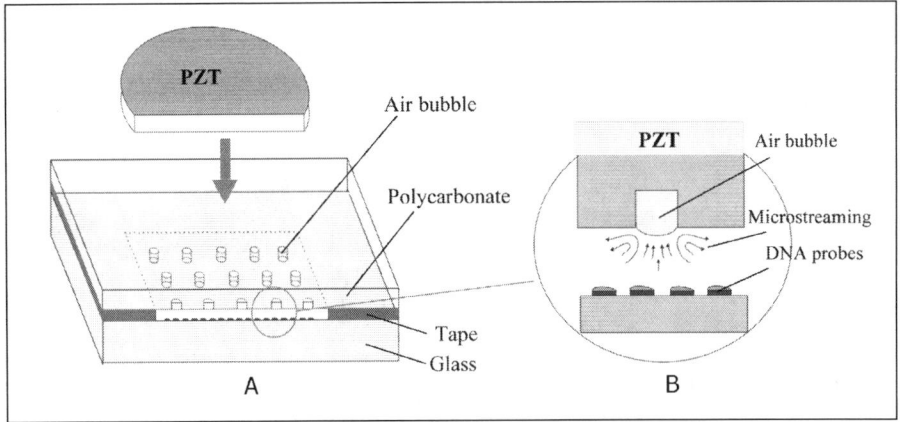

Figure 2. Schematic showing a number of air pockets in the top layer of the DNA biochip chamber. A) Overview; B) sideview. Reproduced with permission from ref. 44, copyright 2003 American Chemical Society.

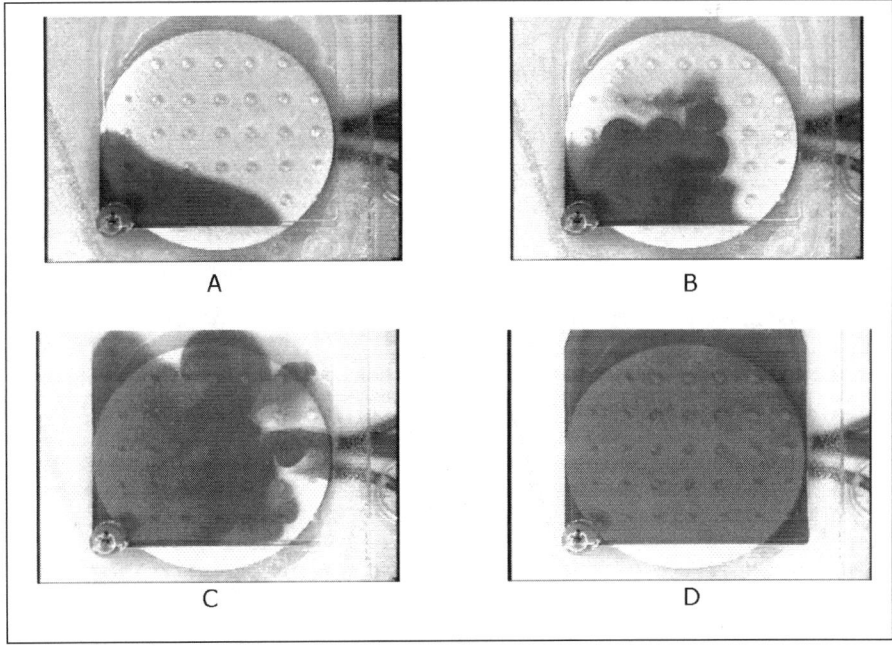

Figure 3. Photographs showing multi-bubble induced (7 × 5 top bubbles) cavitation microstreaming in a 16 × 16 × 0.2 mm chamber at (A) time 0; (B) 2 sec; (C) 4 sec; (D) 6 sec. The PZT disk on the backside of the chamber was driven at 5 kHz and 40 V_{pp}. Reproduced with permission from ref. 44, copyright 2003 American Chemical Society.

Microvalves

Microvalve is an important component in many microfluidic systems. Microvalves that have been realized to date can be divided into two major categories: passive microvalves (without actuation) and active microvalves (with actuation). Since most of passive microvalves are check valves that allow fluid flow only in one direction, they are not as diverse in their use as active valves that can open and close the fluid passage. Most of the conventional active microvalves couple a flexible diaphragm to an electromechanical actuator operating based on thermo-pneumatic,[46,47] bimetallic,[48,49] shape-memory,[50] electrostatic,[51] piezoelectric,[52] or electromagnetic principle.[53] For example, the thermo-pneumatic microvalve developed by Zdeblick et al made use of resistive heating that caused boiling of a trapped liquid and resulted in a high pressure.[46] The pressure pushed a thin silicon (Si) membrane that modulated the "current" of fluid in the regulated channel. Jerman et al demonstrated a normally closed thermal bimorph actuation scheme for a microvalve for gas flow control using an aluminum layer fabricated on a Si membrane.[48] Shoji, et al demonstrated the use of a separately fabricated stacked piezoelectric actuator as a means of closing a Si micromachined valve.[52] Although the active microvalves described above have shown good performance, most of them are fabricated using multi-layer Si process (multi-layer Si stacking and bonding) that is a complicated fabrication process. As a result, the devices often suffer from high cost, poor reliability and high power consumption (note that Si has a high Young's modulus, which makes the Si diaphragm difficult to deform). The displacement of conventional silicon or silicon nitride diaphragms is typically limited to tens of microns or less. Moreover, these active microvalves are made of Si and, thus, are too expensive to be used for single-use bio-microfluidic devices. Other interesting active microvalves include pH-sensitive hydrogel valves[19,20] and a pneumatically actuated PDMS valve.[54] Both devices were made using poly(dimethylsiloxane) (PDMS) microfabrication and used a thin PDMS diaphragm sandwiched between multi-layers of PDMS. PDMS microvalve devices in general suffer from shortcomings of PDMS's physical properties, such as poor thermal conductivity (e.g., it is difficult to use PDMS to construct a DNA polymerase chain reaction device since PDMS expands and tends to degas when heated up) and high liquid/molecule absorption because of its porosity. Since the fabrication of the active microvalves described above involves multi-layer construction, not only is the fabrication process of such microvalves complicated, but the integration into complex microfluidic systems has also proven to be nontrivial.[55]

We have developed a novel microvalving technique that was implemented into the biochip to facilitate a sequential and multi-stage analysis.[34] In this valving mechanism, paraffin is used as an actuator material that undergoes solid-liquid phase transition in response to changes in temperature. Several one-shot valving schemes, including "close-open" and "open-close-open" valves (Fig. 4), are demonstrated. The "close-open" is a normally closed valve that can be opened once (one-shot valve) as shown in Figure 4A,B.[34] The valve uses a bulk of paraffin, which is localized in a heating zone, to close the channel first. The paraffin can be melted using the heat generated by a heater (e.g., a resistive heater) underneath, when the channel needs to be opened. A pressure from the upstream channel section moves the molten paraffin downstream. The downstream channel is incorporated with a wide solidification section ($\sim$ 2 times wider than the upstream channel, 4 $\sim$ 6 mm long), which is located 1 mm away from the heating zone and designed to trap paraffin after solidification and thus prevent the downstream channel from clogging. During the fabrication of the valve, the channel device was first heated up to a temperature (i.e., 90°C), which is above the melting temperature of the paraffin, using a hotplate. A small amount (20 mg) of solid paraffin (Signature Brands LLC, Ocala, FL) with a melting temperature T_m of 70°C was then loaded into each of the paraffin access holes on the channel. The paraffin was melted instantaneously and capillary force drove the molten paraffin into the channel. The channel device was then immediately removed from the hotplate. The paraffin was solidified in the channel, resulting in normally closed microvalves in the channel device. The paraffin access holes were subsequently sealed using an adhesive tape (3M, St. Paul, MN). The fluidic experiments showed that no leakage occurred when the valves were in a "closed" position under normal flow conditions (e.g., pressure less than 20 psi). Maximum hold-up pressure was measured to be about 40 psi, above which we started to

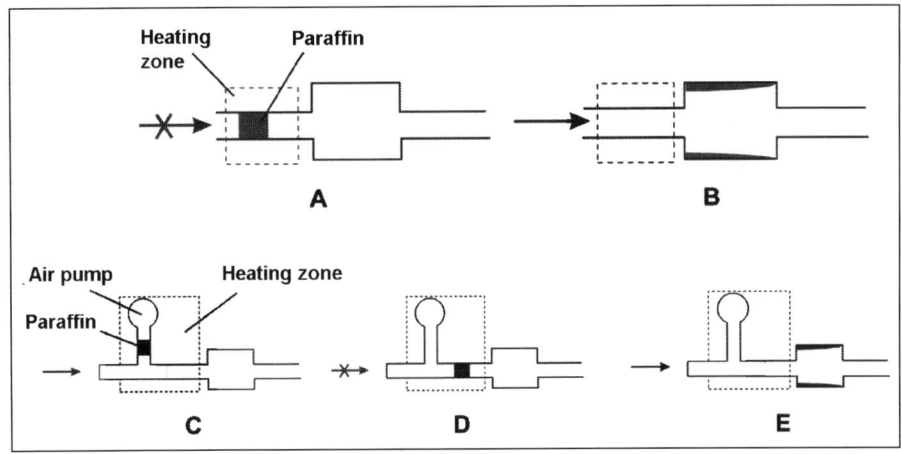

Figure 4. Schematic illustrations of a "close-open" paraffin microvalve design (A and B) and an "open-close-open" microvalve design (C to E). The former has a block of paraffin that initially closes the channel (A). To open the channel, the paraffin is melted using the heater underneath and moved downstream by the pressure from the upstream channel. Once the molten paraffin moves out of the heating zone, it starts to solidify on the wall of a wide channel section resulting in an open channel (B). The latter is a normally open valve with a block of paraffin connected to an air pocket that acts as a thermally actuated air pump (C). When the heater is activated, the air in the pocket expands and pushes the molten paraffin into the regulated channel. If the heater is turned off immediately, the paraffin solidifies in the main flow channel, resulting in a closed channel (D). The channel can be reopened by reactivating the heater (E). Reproduced with permission from ref. 34, copyright 2004 American Chemical Society.

see some leakage at the interface of the paraffin and the channel wall. No dead fluid volume was observed in the valve since the resulted opened channel was a through channel where no fluid element was trapped. The response time of the valves is approximately 20 sec. It was found that the use of a wider solidification channel (i.e., 3 mm wide instead of 2 mm wide) resulted in an improved time response by 10% due to the more space that allowed solidified paraffin to reside in and thus less time for the channel to be opened. It was also found that shortening the distance between the solidification section and the heating zone resulted in a slightly improved time response due to the shortened distance for the molten paraffin to travel prior to solidification. The actuation of a paraffin valve required 100 ~ 200 mW of power.

The "open-close-open" valve, as shown in Figure 4C-E, is a normally open valve that can be closed and reopened, subsequently. The valve is integrated with a thermally actuated air pump that consists of an air pocket (with a diameter of 5 mm and a depth of 0.5 mm) attached with a heater (e.g., Peltier element). As shown in Figure 4C, a small plug of paraffin is first placed in the channel branch connected with the air pump. The main flow channel is initially open. If the heater is turned on to a temperature above the paraffin melting temperature, the paraffin melts. The air trapped in the pocket is also heated up, resulting in increased pressure, which in turn pushes the molten paraffin into the regulated channel. If the heater is turned off immediately, the paraffin solidifies in the main flow channel, resulting in a closed channel (Fig. 4D). The channel can be re-opened by turning on the heater to the temperature above the melting point of the paraffin until the paraffin moves out of the heating zone and solidifies downstream in a solidification section as discussed in previous paragraph (Fig. 4E). Pressure drop measurements in the fluidic experiments showed that the maximum hold-up pressure was about 40 psi when the valve was in a closed mode. No leakage was observed when the channel was closed. The closing operation (including the temperature ramping time of the heater) took approximately 10 sec. The closed channel was re-opened after

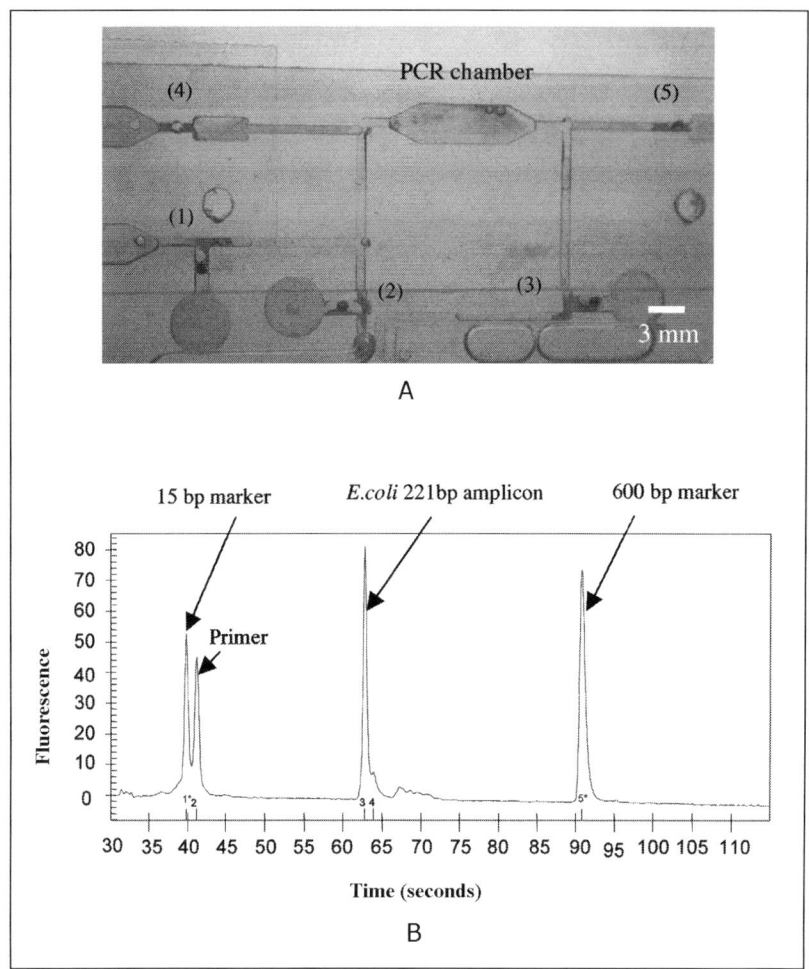

Figure 5. A) A photograph of the polycarbonate micro-PCR device integrated with five paraffin-based microvalves: valves (1-3) are "open-close" valves and (4) and (5) are "close-open" valves (note that the PCB substrate consisting of resistive heaters to actuate the valves is not shown here); (B) PCR results of the micro-PCR device. The *E. coli* K12-specific gene (221 bp fragment) was amplified and analyzed using Agilent BioAnalyzer DNA 500.

the heater was turned back on to heat up the channel and melt the paraffin, which then solidified in the downstream channel section. The reopening operation took about 12 sec.

As shown in Figure 5A, a plastic micro-PCR device that consists of a PCR chamber (5 mm in width, 16 mm in length, 0.5 mm in depth) and five paraffin-based microvalves was fabricated and tested for DNA amplification. A PCB substrate (not shown here) consisting of resistive heaters was attached to the plastic device to provide thermal actuation to the paraffin elements. Following the loading of the PCR reaction mixture into the reaction chamber through the channels regulated by the paraffin microvalves (2) and (3), the chamber was sealed by closing the valves to ensure no leakage. Note that the microvalves (4) and (5) were initially closed. The PCR micro-device was then placed into the Peltier thermal cycler. During PCR thermal cycling, the micro PCR chamber was sandwiched between the two Peltier elements, whereas all the paraffin microvalves were located at a distance of 10 mm away from the heating zone, in order to prevent the bivalve actuation during PCR thermal cycling. After PCR was completed, the microvalves (4) and (5) were opened to retrieve the reaction product from the chamber for off-chip electrophoresis analysis. The electrophoretic results, as shown in Figure 5B, indicate that DNA was successfully amplified in this device. The PCR yield is similar to that of the control PCR performed in a conventional PCR tube (Molecular BioProducts, San Diego, CA) using a conventional DNA Engine™ Thermal Cycler (MJ Research Inc. South San Francisco, CA) under the same conditions, indicating that the paraffin is fully PCR compatible.

The planar designs of the paraffin-based valves do not include a flexible diaphragm and thus are simpler than traditional actuator/diaphragm designs that require multi-layer structures. Although we have only demonstrated one-shot "close-open" and "open-close-open" valving schemes, other configurations can be easily achieved.[56] For example, a toggle valve that consists of a number of "open-close-open" segments operated in sequence can perform a number of "open-close" cycles as designated. Although the time response (~10 sec) of the paraffin-based microvalves is relatively slow as compared to that of many conventional microvalves (~ms), the paraffin-based valves have proven to be practical and useful in our biochip device where rapid response is not critical. It is believed that the time response could be improved by employing a paraffin material with a lower melting temperature. The fabrication process of the paraffin valves is compatible with many other material fabrication processes (such as Si, plastic, etc). The precise loading of paraffin material (melted volume on the order of pL) into the microchannel to form a microvalve can be achieved using a wax injector (Microdrop GmbH, Germany), leading to a simple manufacturing method to implement them into complex microfluidic devices. This process can be much simpler than bulk processes (e.g., bulk etching of Si wafers), surface processes (e.g., thin-film processes), or chemical reactions (e.g., in-situ polymerization[20,57]) used in more traditional valve approaches. Since paraffin is a commonly used and inexpensive material, the paraffin-based microvalves are cost-effective and highly desirable for many single-use and disposable microfluidic applications. It is worth noting that the valving approach is not limited to paraffin and can be extended to many other materials that can undergo a phase transition from solid to liquid in response to changes in temperature.

Micropumps

Micropump is an essential and important component in the integrated microfluidic device. Based on different pump (actuation) mechanisms, conventional micropumps can be classified into two major groups: membrane actuated (mechanical) and nonmembrane actuated pumps.[55] Membrane actuated pumps can be further divided into different types: piezoelectric, electrostatic, thermopneumatic, etc.[54,58,59] Most of these conventional pressure-driven membrane-actuated micropumps suffer from complicated designs, complicated fabrication, or high cost. Nonmembrane pumping principles include electrohydrodynamic, electroosmotic, traveling wave, diffuser, bubble, surface wetting, rotary, etc. Although much progress has been made, micropumps with the appropriate combination of cost, performance and operating requirements are still not available for many applications.

The biochip device reported here requires integrated micropumps for transport of a wide range of sample volumes (μL-mL). In our device, two simple pressure-driven micropumping methods were employed: a thermopneumatic air pump (Pump 4 in Fig. 1A) for pumping of ~μL volumes and electrochemical pumps (Pumps 1-3 in Fig. 1A) for ~mL volumes. The former makes use of the air expansion in an air chamber, which is attached to a resistive heater in the PCB substrate, when heated up. The air expansion is a nearly linear function of temperature. The resulting air expansion pushes the solution from the storage chamber into the downstream channels and chambers. The latter relies on electrolysis of water between two platinum electrodes in a saline solution to generate gases when a DC current is applied.[60] The gas generates a pressure that in turn moves liquid solutions in the biochip. Both pumping mechanisms do not require a membrane and/or check valves in their designs. As a result, their fabrication and operation are much simpler than most conventional micropumps. Flow experiments demonstrated that the thermopneumatic air pumps

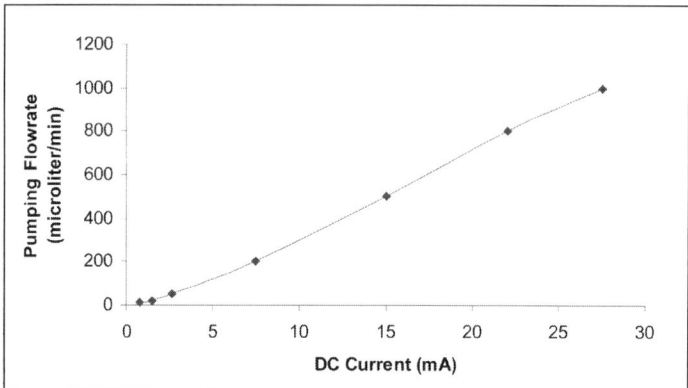

Figure 6. Measurement of the liquid pumping rate as a function of the applied DC current in an electrochemical pump that relies on the electrolysis of water between two platinum electrodes in a saline solution (20 μL 5 M NaCl) (reproduced with permission from ref. 34, copyright 2004 American Chemical Society).

with air pockets of 50 μL internal volume could efficiently move up to 60 μL volume of fluids with a heater power consumption of less than 0.5 W. For pumping of ~mL solution volumes, the electrochemical pump is more efficient and consumes less power. A steady flowrate of up to 0.8 mL/min was achieved with a power consumption of <150 mW (Fig. 6).[34]

Sample Preparation

The integrated biochip device performs automated DNA analysis from complex sample fluids providing a step towards fulfilling the promise of rapid, automated genetic analysis in hand-held devices. Although other attempts have been made in the past to integrate genetic amplification with detection, none of those approaches have addressed the difficult task of sample preparation with rare cell capture capabilities. Sample preparation represents the most time-consuming and labor-intensive procedure in DNA analysis. It also introduces one of the largest variables in subsequent analyses due to its complexity. The developments of back-end detection schemes, resulting in demonstrations of DNA microarray biochips[61,62] and electrophoresis separation chips[63-65] have shifted the bottleneck, impeding further progress in rapid analysis devices, to front-end sample preparation where the 'real' samples of bodily fluids are used.

Target cell isolation and preconcentration from the crude biological sample solution have been studied in the attempt to address front-end stage of sample preparation. Kricka and Wilding[66] have demonstrated physical filters relying on separation of biological cells by size. White blood cells were isolated from whole blood in silicon-glass 4.5 μL microchips containing a series of 3.5 μm feature-sized "weir-type" filters, formed by an etched silicon dam spanning the flow chamber. Genomic DNA targets, e.g., dystrophin gene, were then directly amplified using PCR from cells isolated on the filters. This dual function microchip provides a means to simplify nucleic acid analyses by integrating in a single device two key steps in the analytical procedure, namely cell isolation and PCR.[66-68] Gascoyne et al[69,70] developed dielectrophoretic (DEP) separation techniques which exploit the differential dielectric properties occurring among different biological cells. Becker demonstrated separation of human breast cancer and leukemia cells from blood; Cheng et al[71] showed isolation of cervical carcinoma cells from blood. Ward and Grodzinski[72] introduced immunomagnetic cell separation into microfluidic devices containing gradient producing soft magnetic ridges. These devices were used to separate *E. coli* and mammalian cells from blood sample. Saito et al[73] worked on ultrasonic techniques to arrest euglena and paramecia cells in nodes of acoustic standing wave.

In our integrated diagnostic device, immunomagnetic separation technique has been applied to enrich rare target cells from a large-volume clinical sample such as blood. Considering that the blood sample is a highly heterogeneous cellular and genetic medium (for example, 1 mL of whole blood contains ~10^6 white blood cells, ~10^9 red blood cells and 10^7 platelets, in addition to the serum components), a highly selective and sensitive target preconcentration posts a real challenge for on-chip integration of sample preparation steps. The use of immunomagnetic separation techniques for this purpose has a number of advantages, including high selectivity, ease of implementation and reduced assay complexity, as compared to other approaches.[8,10,66,74] Moreover, the immuno-magnetic approach allows for sample preparation and PCR amplification to be performed in the same chamber, which not only simplifies the device design, fabrication and operation, but also eliminates sample loss due to unnecessary fluid transfer. In the self-contained, integrated biochip system reported here, we have successfully demonstrated capturing 10^6 *E. coli* cells (equivalent to 5 ng of genomic material) from 1 mL whole blood using immunomagnetic beads separation technique. Pathogenic bacteria detection from a whole blood sample was performed using the described device. *Escherichia coli* (*E. coli*) K12 cells inoculated in a whole rabbit blood were used as a model for demonstration. Although this application may have no immediate practical use, it is easy to conceive a system where, for example, the rabbit blood is replaced by a human blood and pathogenic or cancer cells are targeted instead of the *E. coli* K12 cells. The *E. coli*/rabbit blood system was used because of the simplicity of performing the experiments as well as control assays in an ordinary laboratory setting (without BSL-2 or higher laboratory requirements). The input sample is 1 mL of whole citrated rabbit blood (Colorado Serum company, Denvor, CO) containing 10^3~10^6 *E. coli* K12 cells. 10 μL of biotinylated polyclonal rabbit-anti-*E. coli* antibody (ViroStat, Portland, ME) and 20 μL of streptavidin labeled Dynabeads (total 1.3 × 10^7 M-280 beads, Dynal Biotech, Inc., Lake Success, NY) were also added into the sample storage chamber. The bead-antibody-*E. coli* cell complexes were formed during a 20-min cavitation microstreaming based mixing period.

In order to efficiently capture target cells (e.g., *E. coli* K12) from the whole blood sample us-ing immunomagnetic capture beads, cavitation microstreaming was implemented in the sample storage chamber to achieve a homogeneous mixing for the beads-cell complex formation. An experiment was designed and implemented to evaluate cell capture enhancement using cavitation microstreaming technique. A 55 μL sheep blood solution suspended with 100 *E. coli* K12 cells that were already labeled with biotinylated polyclonal rabbit anti *E. coli* antibody (ViroStat, Potland, Maine) was loaded into the mixing chamber followed by a separate load of 5 μL streptavidin-coated colloidal magnetic beads (Miltenyi Biotec, Inc., Auburn, CA). During the mixing process, the beads-antibody-cell complexes would form through specific biotin-streptavidin interaction if the mixing is sufficient. The mixing chamber has a 4 × 4 array of air pockets (0.5 mm in diameter and 0.5 mm in depth) on the chamber wall. Acoustic micromixing was then performed at 5 kHz and 10 V_{pp} (square wave) for 30 min. Following mixing, the mixture solution was retrieved from the mixing chamber and passed through a miniMACS separation column (Miltenyi Biotec, Inc., Auburn, CA) where magnetically labeled bacteria were captured and subsequently plated over-night onto L-Broth Agar plates at 37°C. The *E. coli* colonies formed on the plates were counted. Capture efficiency was calculated by dividing the number of colony forming units found on the elution plate (captured bacteria) by the total number of colony forming units (captured bacteria and escaped bacteria that were not captured by the beads). Capture efficiencies were compared to those using a standard protocol (i.e., vortexing in a microfuge tube) and those using no mixing enhancement except pure diffusion only. Calculated results of capture efficiencies are summarized in Figure 7. Results show that the capture efficiency of *E. coli* cells is comparable between acoustic microstreaming and conventional vortex in tube, indicating mixing efficiency is comparable be-tween them. Both acoustic microstreaming and conventional vortex result in much higher capture efficiency than pure diffusion, suggesting that acoustic microstreaming significantly enhanced cell capture efficiency during the sample preparation step.

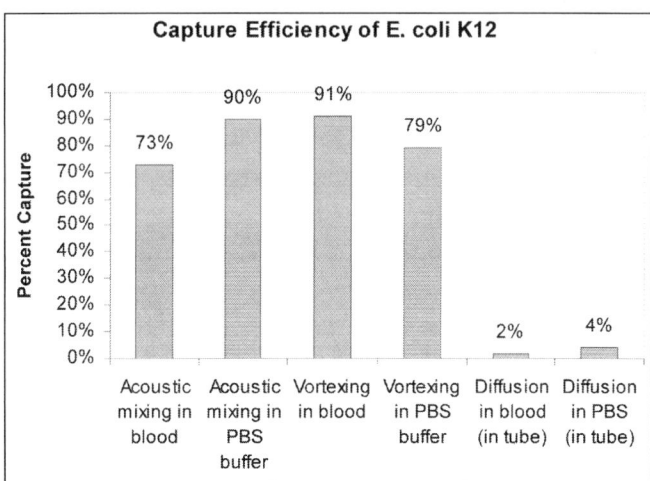

Figure 7. Comparison of normalized capture efficiencies of *E. coli* K12 cells from PBS or blood samples using acoustic, conventional vortexing, or diffusion only.

Following the mixing, the blood sample mixture was pumped using an electrochemical pump (Pump 1 in Fig. 1A) at 0.1 mL/min into the PCR chamber, where the beads-antibody-cells complexes were collected by the magnet. The un-captured samples were flowed into the waste chamber. Subsequently, 1 mL wash buffer was pumped at 0.1 mL/min through the PCR chamber using Pump 2 to wash the captured complexes. As a result of the cell preconcentration and purification steps, the purified bead-antibody-cell complexes were isolated from the blood sample solution and trapped in the PCR chamber. Cell thermal lysis followed by PCR was then performed in the PCR chamber to amplify the DNA of captured *E. coli* K12 cells.

On-Chip DNA Amplification

DNA amplification, such as Polymerase Chain Reaction (PCR), is essential to most genetic analysis applications of integrated biochips. Rapid operation, small sample volume and parallel amplification of different amplicons within the same chip are among the desired features of amplification in the microchip environment. The design and development of such chips is hindered by several challenges including: loss of sample to the chamber walls due to dramatic increase in surface-to-volume ratio, evaporation in small volume regime and effective heat dissipation in order to achieve rapid thermal cycling. Micro PCR devices have been successfully fabricated in glass,[15, 75] silicon[76,77] and plastic.[78-80] Silicon, due to its superior thermal conductivity (~10x of glass and ~700x of polymers), allows for fast temperature ramp time and results in short on-chip protocols.[77] Similarly, successful and fast amplification assays have been demonstrated by many groups in glass.[75,81] Landers' group has used IR heating scheme and glass PCR devices containing 1.7 µl micro chamber; amplification was achieved by 15 thermal cycles in 4 min.[82] In an integrated monolithic silicon-glass device, submicroliter (280 nl) volume was thermally cycled as fast as 30 s/cycle.[83] The low volume limit for micro PCR was also investigated by Nagai and colleagues[84] resulting in a successful amplification in picoliter micro chamber array. Using a real-time PCR device, Belgrader et al reported PCR detection of *Erwinia*, a vegetative bacterium, in 7 min.[77] An integrated rapid PCR-detection system coupled with capillary electrophoresis analysis was presented by Khandurina et al[85] and Lagally et al[81] with amplification times in the 20 min range. Recently with increasing emphasis on the disposable devices, the use of plastic and plastic fabrication methods have become very popular in microreactor development. Plastic chips are inexpensive, optically transparent and biocompatible.[86,87] Despite all the advantages, plastic possesses a major challenge to a designer of PCR micro reactors due to its poor thermal conductivity and resulting difficulty to achieve rapid thermal cycling. Recently, our group has demonstrated a successful DNA amplification in polycarbonate chips; 30 thermal cycles took 30 mins which considering poor thermal conductivity characteristics of polymers is a significant achievement.[88]

Despite all these demonstrations, the sensitivity aspect of micro PCR chip assay has not been well studied. Most of the reports focused on achieving amplification per se, but the systematic evaluation of amplification yield and reaction sensitivity was usually not explored. The initial template concentration used to achieve fast and small volume amplification was usually high, ranging from 0.1 ng phage DNA[82] to hundred ng human genomic DNA.[89] The most sensitive micro PCR assay demonstrated, so far, was by Lagally et al.[83] A single molecule DNA template could be amplified in a glass integrated microfluidic device. Also, in an integrated glass sandwich structure, Mathies and coworkers[81] reported amplification with a starting template concentration as low as 5-6 copies was achieved in their nanoliter-volume glass microchambers. For the "real" sample (containing target cells, rather than purified DNA) analysis, the most sensitive silicon microstructure, which could perform rapid real time PCR analysis from a sample containing low concentration of target (*Erwinia*) cells, was reported by Belgrader and colleagues.[77] A positive amplicon signal was detected in less than 35 cycles (17 s per cycle time) with the starting template concentration as low as 5 cells. In a flat polypropylene tube, Northrup and his group demonstrated a real-time PCR analysis of 1,000 bacillus spores.[90] Recently, our group performed a systematic study on a sensitivity of micro-PCR assay in plastic devices for the first time.[88] We demonstrated a feasibility of amplifying template concentrations as low as 10 *E. coli* cells (50 fg of DNA) in presence of blood. Similarly, we demonstrated PCR multiplexing within the same reactor chamber for four different bacteria species.

As described in the previous section of sample preparation, the purified bead-antibody-cell complexes were isolated from the blood sample solution and trapped in the PCR chamber of the integrated biochip device as a result of the cell preconcentration and purification steps. Following influx of the PCR reagents to the PCR chamber and closing of all the paraffin-based valves, thermal cell lysis and two-primer asymmetric PCR were performed to amplify an *E. coli* K12-specific gene fragment and achieve single-stranded DNA amplicon in the presence of beads. A pair of *E. coli* K12-specific primers were used to amplify a 221 bp *E. coli* K12-specific gene (MG1655) fragment, with forward primer: 5′ AAC GGC CAT CAA CAT CGA ATA CAT 3′ and reverse primer: 5′ GGC GTT ATC CCC AGT TTT TAG TGA 3′. The PCR reaction mixture (20 μL) consists of Tris-HCl (pH 8.3) 10 mM, KCl 50 mM, $MgCl_2$ 2 mM, gelatin 0.001%, dNTPs 0.4 mM each, bovine serum albumin (BSA) 0.1%, forward primer 0.05 μM, reverse primer 5 μM and AmpliTaq DNA polymerase (Applied Biosystems, Foster City, CA) 5 units. Note that during the PCR thermal cycling, the PCR chamber was sandwiched between two Peltier elements, whereas all the paraffin microvalves were located at a distance of 10 mm away from the heating zone, in order to prevent the bivalve actuation during PCR thermal cycling. PCR was performed with an initial denature step at 94°C for 4 min followed by 35 cycles of 94°C for 45 s, 55°C for 45 s and 72°C for 45 s, with a final extension at 72°C for 3 min. After PCR was completed, the normally closed microvalves were opened, allowing the hybridization buffer and the PCR product to be pumped into the microarray detection chamber, where acoustic mixing of the target and DNA hybridization reaction occurred.

DNA Microarray Detection

Following the PCR, two paraffin-based microvalves, one regulating the channel between the PCR chamber and the hybridization buffer storage chamber and the other one between the PCR and the microarray chamber, were opened. The PCR products were then transported to the eSensor® microarray chamber along with hybridization buffer (20 μL) using the Pump 3 as shown in Figure 1A. The hybridization buffer contains 1X hybridization buffer stock, 7% fetal calf serum, 1 mM hexanethiol and 0.5 μM signaling probe.[35,36] The microarray chamber was incubated at 35°C and the electrochemical signals from the eSensor® microarray were measured at 0 min, 15 min, 30 min and 1 hr.

An e-Sensor® chip that is an electrochemical detection based low-density hybridization array[35,36] was used to detect the electrochemical hybridization signals corresponding to the redox-reaction of the ferrocene-labeled signaling probes that hybridized with the target DNA (PCR product) bound to the immobilized probes DNA. The e-Sensor® chip is a printed circuit board (PCB) chip that consisted of an array of gold electrodes modified with a multi-component self-assembled monolayer (SAM) that included presynthesized oligonucleotide (DNA) capture probes that are covalently attached to the electrode through an alkyl thiol linker. When the amplicon solution containing target DNA was introduced into the detection chamber, specific capture probes on an electrode surface encountered complementary DNA from the sample and hybridization occurred. Two ferrocene, electronic labels (capture probe and signaling probe) bind the target in a sandwich configuration. Binding of the target sequence to both the capture probe and the signaling probe connected the electronic labels to the electrode surface through the chains of molecular wires built into SAMs.[36] This added a circuit element to the bioelectronic circuit on that electrode and presence of hybridized (double-stranded) DNA could be detected using alternative current voltammetry (ACV). Since incoming target itself was not labeled, the washing step (to remove excessive, nonbound target) prior to the signal collection was not required and a continuous monitoring of the binding process with a quantitative measurement of the target accumulation is possible. This electrochemical detection (ECD) based microarray platform eliminates the need for an expensive and sensitive laser-based optical system and fluorescent reagents. By replacing the large optical reader with a small ECD reader, the array can be incorporated into a portable or handheld detector instrument for in-field biological detection and clinical diagnostics.

To enhance the rate of DNA hybridization, cavitation microstreaming technique was implemented in the hybridization chamber.[44] Hybridization typically relies on diffusion of DNA target to surface-bound DNA probes and is often a lengthy rate-limiting process (6 to 20 hrs). The bulk of the target solution is at a considerable distance, on the molecular length scale, from the reaction site on the chip surface. For example, the diffusion coefficient of a 250-bp DNA fragment in water at room temperature is approximately 2×10^{-7} cm^2/s and thus its diffusion time along a length of 500 μm is approximately 100 min. This inefficient diffusion greatly limits the throughput of sample analyses and can be overcome by micromixing techniques.

Hybridization kinetics experiments were performed to study the hybridization kinetic enhancement induced by cavitation microstreaming.[44] Figure 8 summarizes the hybridization kinetics results for an acoustic mixing-enhanced device and a diffusion-based device.[44] Each data point is the mean value obtained from four electrodes with identical DNA capture probes in the same

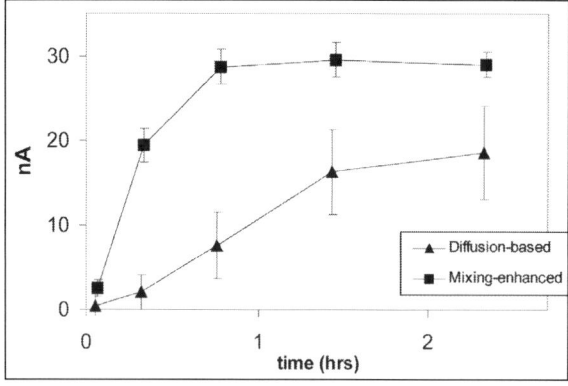

Figure 8. Hybridization kinetics study on static hybridization vs. mixing-enhanced hybridization (5 kHz square wave and 10 V$_{pp}$) performed in eSensor® devices. Each data point is the mean value obtained from four electrodes with identical DNA capture probes in the same device (reproduced with permission from ref. 44, copyright 2003 American Chemical Society).

eSensor® device. The target DNA was obtained by PCR amplification from human genomic DNA. Because of the homogenous nature of the assay, results for the probe coverage were obtained at different time points, giving a current-time curve for the hybridization kinetics. Note that the y-axis in the figure is the measurement of the Faradaic current from the electrodes. The Faradaic current is directly proportional to the number of ferrocene moieties in proximity to the electrode surface that in turn is proportional to the number of target nucleic acid molecules hybridized with the probes.[36] The results show that for static (diffusion-based) hybridization, the hybridization signal evolved slowly and exhibited an initial linear increase. It took approximately 6 hrs for the static sample to reach the saturated level (not shown here). Moreover, the standard deviation of each data point shows that the static hybridization has relatively large electrode-to-electrode variation. For the hybridization assay coupled with cavitation microstreaming, the signal increased rapidly, showing additional acceleration and uniform signal distribution (small standard deviations) compared to the pure diffusion-based device. After 40 minutes of hybridization, the sample in the mixing-enhanced device reached a saturated Faradaic current value. The relative rate of hybridization in the two devices is given by the ratio of the time it takes for the signal to reach one-half of the saturated value. Hybridization in the mixing-enhanced device occurred at approximately 5.3 times the rate of that in the diffusion-based device.

Cavitation microstreaming not only provides rapid lateral mass transport of fluidic elements, but also enhances the vertical mass transport of target DNA in the solution towards the DNA probes on the chip surface. The combination of rapid lateral and vertical fluid movements results in rapid transport of targets in solution to the diffusion boundary layer and thus allows for continuous replenishment of fresh DNA targets around the probes that have been depleted of complementary targets. As a result, the hybridization rate is enhanced. Moreover, the rapid fluid movements can enhance the transport of target within the diffusion boundary layer to the probes on the chip surface by reducing the thickness of the diffusion boundary layer. Cavitation microstreaming in a shallow hybridization chamber reduces the thickness of the diffusion boundary layer by 2.5 fold. Targets are in close proximity to the probes immobilized on the chip surface, resulting in fast hybridization due to short diffusion length. Furthermore, the rapid lateral fluidic movement as observed in the fluidic dye experiments (Fig. 3) ensures a homogenous mixture of targets and sufficient fluid exchange across the large surface area on the chip and thus allows for uniform hybridization signals to be achieved. Uniformity of hybridization signal is critical especially for high-density microarray and/or detection of low-abundance targets. Lack of lateral flow convection could lead to nonhomogeneous array performance and hybridization differences that are independent of differences in target concentration. Although the hybridization kinetics enhancement using acoustic microstreaming is not as significant as in cases of the flow-through approach[91] and the electronic DNA hybridization,[92,93] a distinct advantage that cavitation microstreaming has over the above two methods is its rapid lateral mass transport that significantly enhances uniformity of hybridization. Moreover, cavitation microstreaming requires very simple mixing apparatus compared to most existing chamber micromixers[42,94] and thus can be easily implemented in biochip devices. Other advantages of cavitation microstreaming include low power consumption (~2 mW) and low cost.

Pathogenic Bacteria Detection

Pathogenic bacteria detection from a whole blood sample was performed using the described integrated biochip device (Fig. 1). *E. coli* K12 cells inoculated in a whole rabbit blood were used as a model for demonstration. The electrochemical signal corresponding to the hybridized *E. coli* K12-specific gene as shown in Figure 9A was obtained after 1-hr hybridization.[34] The whole analysis from loading the blood sample and different reagents into the storage chambers of the biochip to obtaining the hybridization results took 3.5 hours. The durations for the different operations are as follows: sample preparation-50 min, PCR amplification-90 min, pumping and valving-10 min, hybridization-60 min. An on-chip assay from 1 mL of whole blood sample containing 10^3 *E. coli*

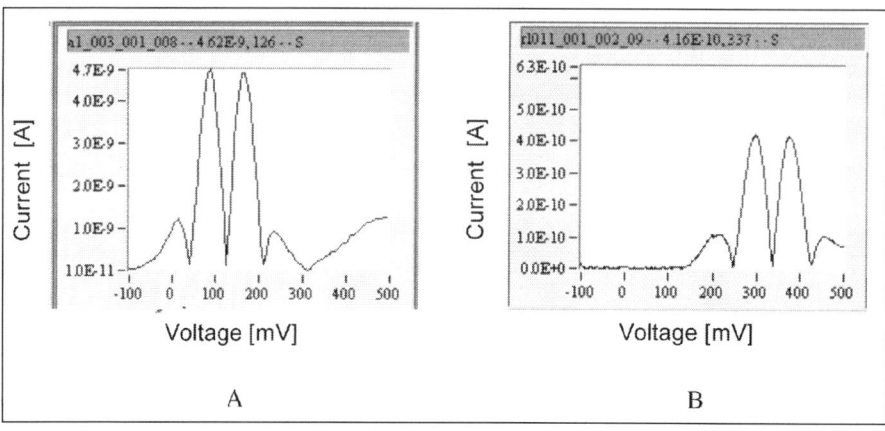

Figure 9. Electrochemical measurement results obtained from the integrated biochips. A) Detection of 10^6 *E. coli* K12 cells from 1 mL of rabbit blood. B) Genotyping identification of HFE-C gene from 1.4 µL of human blood. During the hybridization process, alternating current voltammetry was used to detect the surface-bound redox-molecules attached to the secondary probes. Fourier-transformation was applied to the signal and the 4th harmonic of the transformation, together with the redox-current value, was displayed on screen (reproduced with permission from ref. 34, copyright 2004 American Chemical Society).

K12 cells was also performed demonstrating positive recognition of *E. coli* K12-specific gene but with low signal-to-noise ratio (data not shown here).

The typical cell capture efficiency using Dynabeads in our experiments is about 40%. In the presence of 1.3×10^7 (20 µL) Dynabeads, the amplification efficiency was reduced by 50% as compared to the control PCR reaction performed without Dynabeads present (data not shown). It is believed that the chip assay sensitivity can be improved by assay optimizations, which include: 1) choice of smaller beads (100 nm in diameter as compared to 2.8 µm for the M-280 Dynabeads) that can provide for higher cell capture efficiency due to increased surface-to-volume ratio;[95] and 2) use of paramagnetic beads with lower PCR inhibition rate as compared to the M-280 Dynabeads.

Single Nucleotide Polymorphism Assay

In the hematochromatosis-associated single-nucleotide polymorphism assay performed in our integrated device, the rare target capture and preconcentration steps were omitted since PCR amplification was performed directly from diluted blood samples.[96,97] Thermal cell lysis followed by an asymmetric PCR amplification was performed to amplify DNA sequences containing the sites for HFE-C polymorphism. After PCR, the hybridization buffer and the amplification product were pumped into the microarray chamber, followed by 1-hr hybridization and electrochemical signal scanning. Genotyping result is shown in Figure 9B.[34] The whole analysis from loading the blood sample and different reagents on the chip to obtaining the genotyping results took 2.7 hours: cell thermal lysis/PCR-90 min, pumping and valving-10 min, hybridization-60 min.

Discussion

In this chapter, we have presented a fully integrated microfluidic device that used DNA microarray as back-end detection technology. Many other integrated devices demonstrated in the past used capillary electrophoresis (CE)[1,15,98] or real-time PCR approaches[99] as detection technologies. The CE technique does not provide information on the fragment sequence that is available through the use of PCR and/or hybridization methods. Multiplexing using real-time PCR is challenging since a maximum of four different fluorescent markers have been employed to date in a microchip. This only allows a 4-plex limit for the number of single nucleotide polymorphisms (SNP) that can be identified at a time.[100] Unlike CE and real-time PCR approaches, our electrochemical microarray detection provides a solution for highly multiplexed DNA analysis[101,36] with the capability of processing a complex, heterogeneous sample (i.e., PCR product containing denatured blood). Although the sensitivity of the electrochemical detection described in this chapter is lower than the conventional optical detection techniques, the use of PCR to achieve sufficient concentration of DNA targets compensates for this drawback. The ability of our system to successfully perform genotyping with as little as 1.4 µL of blood (corresponding to about 10,000 white blood cells) demonstrates its utility for clinical diagnostic applications. Moreover, the electrochemical detection system provides excellent specificity because of the usage of a "sandwich" assay.[36] Since our detection system can measure real-time hybridization without the need of a washing step, the assay can be terminated once the genotype is called by the instrument, saving valuable time. Furthermore, the assay performed in our biochip is flexible and has broad applicability due to this system proficiency in detection of both high-abundance and low-abundance DNA from complex biological samples.

The key microfluidic components in the biochip device, including paraffin-based microvalves, cavitation microstreaming mixers and electrochemical pumps, are simple in design, inexpensive and easy to fabricate and integrate into a complex microfluidic system, as compared with most of the existing microvalves, micromixers and micropumps. The use of cavitation microstreaming technique to achieve rapid and homogeneous on-chip microfluidic mixing not only increases the target cell capture efficiency in the sample preparation process but also allows hybridization assays to be performed in less than one third of the time normally needed without mixing enhancement. The use of integrated microfluidic components with low power consumption, along with the employment of the electrochemical microarray detection, suggests that hand-held operation is feasible for the device. The choice of inexpensive, robust microfluidic technologies, coupled with plastic chip fabrication and standard PCB process, facilitates easy commercialization path for this technology. Although technical challenges remain, including increasing cell capture efficiency, detection sensitivity and assay optimization, the platform shown here provides a solution for genetic analysis of complex biological fluidic samples in the fields of point-of-care genetic analysis and disease diagnosis.

Conclusion

We have developed a self-contained disposable biochip for fully integrated genetic assays. The on-chip analysis started with the preparation process of a whole blood sample, which included magnetic bead-based target cell capture, cell preconcentration and purification and cell lysis, followed by PCR amplification and electrochemical DNA microarray-based detection. Crude biological sample and reagent solutions were loaded into the device, while electrochemical signals corresponding to genetic information were the primary outputs. The chip design is capable of handling a large volume (mL) of initial sample to accommodate analysis of rare targets in the sample. The mL volumes were reduced 100-fold when the assay reached the DNA amplification stage. All microfluidic mixers, valves and pumps are integrated on the chip, but use very simple and inexpensive approaches in order to reduce chip complexity. Both low-abundance and high-abundance DNA detections from blood samples were demonstrated using the devices.

Acknowledgements

The authors thank Motorola Labs and Motorola Life Science for all the technical supports. This work has been sponsored in part by NIST ATP contract #1999011104A and DARPA contract #MDA972-01-3-0001.

References

1. Burns MA, Johnson BN, Brahmasandra SN et al. An integrated nanoliter DNA analysis device. Science 1998; 282(5388):484-487.
2. Wilding P, Kricka LJ, Cheng J et al. Integrated cell isolation and polymerase chain reaction analysis using silicon microfilter chambers. Anal Biochem 1998; 257(2):95-100.
3. Kricka L. Miniaturization of analytical systems. Clin Chem 1998; 44(9):2008-2014.
4. Shi YN, Simpson PC, Scherer JR et al. Radial capillary array electrophoresis microplate and scanner for high-performance nucleic acid analysis. Anal Chem 1999; 71(23):5354-5361.
5. Fodor SPA, Read JL, Pirrung MC et al. Light-Directed, Spatially Addressable Parallel Chemical Synthesis. Science 1991; 251:767-773.
6. Harrison DJ, Fluri K, Seiler K et al. Micromachining a Miniaturized Capillary Electrophoresis-Based Chemical-Analysis System on a Chip. Science 1993; 261(5123):895-897.
7. Simpson PC, Roach D, Woolley AT et al. High-throughput genetic analysis using microfabricated 96-sample capillary array electrophoresis microplates. Proc Natl Acad Sci USA 1998; 95(5):2256-2261.
8. Anderson RC, Su X, Bogdan GJ et al. A miniature integrated device for automated multistep genetic assays. Nucleic Acids Res 2000; 28(12):e60.
9. Lenigk R, Liu RH, Athavale M et al. Plastic biochannel hybridization devices: a new concept for microfluidic DNA arrays. Anal Biochem 2002; 311(1):40-49.
10. Taylor MT, Belgrader P, Joshi R et al. Fully Automated Sample Preparation for Pathogen Detection Performed in a Microfluidic Cassette. Micro Total Analysis Systems 2001:670-672.
11. Liu YJ, Rauch CB, Stevens RL et al. DNA amplification and hybridization assays in integrated plastic monolithic devices. Anal Chem 2002; 74(13):3063-3070.
12. Thorsen T, Maerkl SJ, Quake SR. Microfluidic large-scale integration. Science 2002; 298(5593):580-584.
13. Yuen PK, Kricka LJ, Fortina P et al. Microchip module for blood sample preparation and nucleic acid amplification reactions. Genome Res 2001; 11(3):405-412.
14. Khandurina J, McKnight TE, Jacobson SC et al. Integrated system for rapid PCR-based DNA analysis in microfluidic devices. Anal Chem 2000; 72(13):2995-3000.
15. Waters LC, Jacobson SC, Kroutchinina N et al. Microchip device for cell lysis, multiplex PCR amplification and electrophoretic sizing. Anal Chem 1998; 70(1):158-162.
16. Whitesides GM, Stroock AD. Flexible Methods For Microfluidics. Physics Today 2001; 54(6):42-48.
17. Boone T, Fan ZH, Hooper H et al. Plastic advances microfluidic devices. Anal Chem 2002; 74(3):78A-86A.
18. Cheng J, Sheldon EL, Wu L et al. Preparation and hybridization analysis of DNA/RNA from E. coli on microfabricated bioelectronic chips. Nat Biotechnol 1998; 16(6):541-546.
19. Beebe DJ, Moore JS, Bauer JM et al. Functional structures for autonomous flow control inside microfluidic channels. Nature 2000; 404:588-590.
20. Liu RH, Yu Q, Beebe DJ. Fabrication and Characterization of Hydrogel-based Microvalves. J Microelectromechan Syst 2002; 11:45-53.
21. Zeng SL, Chen CH, Santiago JG et al. Electroosmotic flow pumps with polymer frits. Sens Actuators B Chem 2002; 82(2-3):209-212.
22. Dodson JM, Feldstein MJ, Leatzow DM et al. Fluidics cube for biosensor miniaturization. Anal Chem 2001; 73(15):3776-3780.
23. Tsai JH, Lin LW. A thermal-bubble-actuated micronozzle-diffuser pump. J Microelectromechan Syst 2002; 11(6):665-671.
24. Bohm S, Olthuis W, Bergveld P. A plastic micropump constructed with conventional techniques and materials. Sens Actuators A Phys 1999; 77(3):223-228.
25. Harrison DJ, Manz A, Fan Z et al. Capillary Electrophoresis and Sample Injection Systems Integrated on a Planar Glass Chip. Anal Chem 1992; 64:1926-1932.
26. Wilding P, Pfahler J, Bau HH et al. Manipulation and Flow of Biological-Fluids in Straight Channels Micromachined in Silicon. Clin Chem 1994; 40(1):43-47.
27. Xia YN, Whitesides GM. Soft lithography. Annual Review of Materials Science 1998; 28:153-184.
28. Piner RD, Zhu J, Xu F et al. "Dip-pen" nanolithography. Science 1999; 283(5402):661-663.
29. Alonso-Amigo MG, Becker H. Microdevices fabricated by polymer hot embossing. Abstr Pap Am Chem Soc 2000; 219:468-COLL.
30. Becker H, Dietz W, Dannberg P. Microfluidic Manifolds by Polymer Hot Embossing for Micro Total Analysis System Applications. Paper presented at: uTas 98, 1998; Banff, Canada.
31. Grodzinski P, Liu RH, Chen H et al. Development of Plastic Microfluidic Devices for Sample Preparation. Biomed Microdevices 2001; 3(4):275.
32. Waters LC, Jacobson SC, Kroutchinina N et al. Microchip Device for Cell Lysis, Multiplex PCR Amplification and Electrophoretic Sizing. Anal Chem 1998; 70:158-162.

33. Woollery AT, Hadley D, Landre P et al. Functional integration of PCR amplification and capillary electrophresis in a microfabricated DNA analysis device. Anal Chem 1996; 68(23):4081-4086.
34. Liu RH, Yang J, Lenigk R et al. Self-contained, Fully Integrated Biochip for Sample preparation, PCR amplification and DNA Microarray Detection. Anal Chem 2004; 76:1824-1832.
35. Farkas DH. Bioelectronic DNA chips for the clinical laboratory. Clin Chem 2001; 47(10):1871-1872.
36. Umek RM, Lin SW, Vielmetter J et al. Electronic detection of nucleic acids-A versatile platform for molecular diagnostics. J Mol Diagn 2001; 3(2):74-84.
37. Liu RH, Stremler M, Sharp KV et al. A Passive Micromixer: 3-D C-shape Serpentine Microchannel. J Microelectromechan Syst 2000; 9(2):190-197.
38. Oddy MH, Santiago JG, Mikkelsen JC. Electrokinetic Instability Micromixing. Anal Chem 2001; 73(24):5822-5832.
39. Branebjerg J, Gravesen P, Krog JP et al. Fast mixing by lamination. Paper presented at: MEMS '96, 1996; San Diego, CA.
40. Stroock AD, Dertinger SKW, Ajdari A et al. Chaotic mixer for microchannels. Science 2002; 295(5555):647-651.
41. Moroney RM, White RM, Howe RT. Ultrasonically induced microtransport. Paper presented at: MEMS '95, 1995; The Netherlands.
42. Zhu X, Kim ES. Microfluidic Motion Generation with Acoustic Waves. Sens Actuators A Phys 1998; 66(1-3):355-360.
43. Liu RH, Yang J, Pindera MZ et al. Bubble-Induced Acoustic Micromixing. Lab Chip 2002; 2(3):151-157.
44. Liu RH, Lenigk R, Yang J et al. Hybridization Enhancement Using Cavitation Microstreaming. Anal Chem 2003; 75:1911-1917.
45. Nyborg WL. Acoustic streaming near a boundary. J Acoust Soc Am 1958; 30:329-339.
46. Henning AK, Fitch J, Falsken E et al. A thermopneumatically actuated microvalve for liquid expansion and proportional control. Paper presented at: Technical Digest of TRANSDUCERS '97: the 1997 International Conference on Solid-State Sensors and Actuators 1997, 1997; Chicago, IL.
47. Selvaganapathy P, Carlen ET, Mastrangelo CH. Electrothermally actuated inline microfluidic valve. Sens Actuators A Phys 2003; 104:275-282.
48. Jerman H. Electrically-activated, normally-closed diaphragm valves. J Micromech Microeng 1994; 4:210-216.
49. Barth PW, Beatty CC, Field LA et al. A robust normally closed silicon microvalve. Paper presented at: IEEE Solid-State Sensors and Actuator Workshop 1994, 1994; Hilton Head, SC.
50. Ray CA, Sloan CL, Johnson AD et al. A Silicon-based Shape Memory Alloy Microvalve. Mater Res Soc Symp 1992; 276:161-166.
51. Huff MA, Schmidt MA. Fabrication, Packaging and Testing of a Wafer-Bonded Microvalve. Paper presented at: IEEE Solid-State Sensor and Actuator Workshop 1992; Hilton Head Island, SC.
52. Shoji S, Schoot BVd, Rooij Nd et al. Smallest Dead Volume Microvalves for Integrated Chemical Analyzing Systems. Paper presented at: Tech. Digest, Transducers' 91: the 1991 International Conference on Solid-State Sensors and Actuators 1991; San Francisco, CA.
53. Meckes J, Behrens J, Benecke W. Electromagnetically Driven Microvalve Fabricated in Silicon. Paper presented at: Transducers' 97, 1997.
54. Unger MA, Chou H, Thorsen T et al. Monolithic Microfabricated Valves and Pumps by Multilayer Soft Lithography. Science 2000; 288:113-116.
55. Kovacs GTA. Micromachined Transducers Sourcebook. Boston: WCB McGraw-Hill 1998.
56. Liu RH, Bonanno J, Stevens R et al. Thermally Actuated Paraffin Microvalves. Proc Micro Total Analysis Systems 2002, Kluwer Academic Publisher, Dordrecht, The Netherlands 2002:163-165.
57. Yu C, Mutlu S, Selvaganapathy P et al. Flow Control Valves for Analytical Microfluidic Chips without Mechanical Parts Based on Thermally Responsive Monolithic Polymers. Anal Chem 2003; 75:1958-1961.
58. Su YC, Lin LW, Pisano AP. A water-powered osmotic microactuator. J Microelectromechan Syst 2002; 11(6):736-742.
59. Zengerle R, Skluge S, Richter M et al. A Bidirectional Silicon Micropump. Sens Actuators A Phys 1995; 50:81-86.
60. Bohm S, Olthuis W, Bergveld P. An Integrated Micromachined Electrochemical Pump and Dosing System. Biomed Microdevices 1999; 1(2):121-130.
61. Fodor SP, Read JL, Pirrung MC et al. Light-directed, spatially addressable parallel chemical synthesis. Science 1991; 251(4995):767-773.
62. Fortina P, Delgrosso K, Sakazume T et al. Simple two-color array-based approach for mutation detection. EJHG 2000; 8:884-894.

63. Harrison DJ, Fluri K, Seiler K et al. Micromachining a Miniaturized Capillary Electrophoresis-based Chemical Analysis System on a Chip. Science 1993; 261:895-897.

64. Manz A, Harrison DJ, Verpoorte E et al. Planar Chips Technology for Miniaturization and Integration of Separation Techniques into Monitoring Systems: Capillary Electrophoresis on a Chip. J of Chromatogr 1992; 593:253-258.

65. Woolley AT, Mathies RA. Ultra-High-Speed DNA Fragment Separations Using Microfabricated Capillary Array Electrophoresis. Proc Natl Acad Sci USA 1994; 91:11348.

66. Wilding P, Kricka LJ, Cheng J et al. Integrated cell isolation and polymerase chain reaction analysis using silicon microfilter chambers. Anal Biochem 1998; 257(2):95-100.

67. Cheng J, Fortina P, Surrey S et al. Microchip-based Devices for Molecular Diagnosis of Genetic Diseases. J Mol Diagn 1996; 1(3):183-200.

68. Cheng J, Shoffner MA, Hvichia GE et al. II. Investigation of different PCR amplification systems in microbabricated silicon-glass chips. Nucleic Acids Res 1996; 24(2):380-385.

69. Becker FF, Wang X-B, Huang Y et al. Separation of human breast cancer cells from blood by differential dictric affinity. Proc Nat Acad Sci USA 1995; 92:860-864.

70. Gascoyne PRC, Vykoukal J. Particle separation by dielectrophoresis. Electrophoresis 2002; 23:1973.

71. Cheng J, Sheldon EL, Wu L et al. Isolation of cultured cervical carcinoma cells mixed with peripheral blood cells on a bioelectronic chip. Anal Chem 1998; 70(11):2321-2326.

72. Ward MD, Quan J, Grodzinski P. Metal-Polymer Hybrid Microchannels for Microfluidic High Gradient Separations. Eur Cell Mater 2002; 3:123.

73. Saito M, Kitamura F, Terauchi M. Ultrasonic Manipulation of locomotive microorganisms and evaluation of their activity. J Appl Phys 2002; 92(12):7581-7586.

74. Cheng J, Kricka LJ, Sheldon EL et al. Sample preparation in microstructured devices. Microsystem Technology in Chemistry and Life Science Vol 194; 1998:215-231.

75. Kopp M, De Mello A, Manz A. Chemical amplification: Continuous-Flow PCR on a chip. Science 1998; 280:1046-1048.

76. Wilding P, Shoffner MA, Kricka LJ. PCR in a silicon microstructure. Clin Chem 1994; 40(9):1815-1818.

77. Belgrader P, Benett W, Hadley D et al. PCR detection of bacteria in seven minutes. Science 1999; 284(5413):449-450.

78. Boone T, Hooper H, Soane D. Integrated chemical analysis on plastic microfluidic devices. Paper presented at: Solid-State Sensor and Actuator 1998, 1998; Hilton head Island, South Carolina.

79. Kricka L, Fortina P, Panaro N et al. Fabrication of plastic microchips by hot embossing. Lab Chip 2002; 2:1-4.

80. Yu H, Sethu P, Chan T et al. A Miniaturized and Integrated Plastic Thermal Chemical Reactor for Genetic Analysis. Paper presented at: uTAS 2000, 2000; The Netherlands.

81. Lagally ET, Simpson PC, Mathies RA. Monolithic integrated microfluidic DNA amplification and capillary electrophoresisanalysis system. Sens Actuators B Chem 2000; 63:138-146.

82. Giordano B, Ferrance J, Swedberg S et al. Polymerase chain reaction in polymeric microchips: DNA amplification in less than 240 seconds. Anal Biochem 2001; 291:124-132.

83. Lagally ET, Medintz I, Mathies RA. Single-Molecule DNA Amplification and Analysis in an Integrated Microfluidic Device. Anal Chem 2001; 73:565-570.

84. Nagai H, Murakami Y, Morita Y et al. Development of a microchamber array for picoliter PCR. Anal Chem 2001; 73:1043-1047.

85. Khandurina J, Meknight TE, Jacobson SC et al. Integrated System for Rapid PCR-Based DNA Analysis in Microfluidic Devices. Anal Chem 2000; 72:2995-3000.

86. Alonso-Amigo G. Polymer Microfabrication for Microarrays, Microreactors and Microfluidics. J Assoc Lab Aut 2000; 5:96-101.

87. Grodzinski P, Liu RH, Chen H et al. Development of Plastic Microfluidic Devices for Sample Preparation. Biomed Microdevices 2001; 3(4):275.

88. Yang J, Liu Y, Rauch CB et al. High sensitivity PCR assay in plastic micro reactors. Lab Chip 2002; 2:179-187.

89. Cheng J, Waters LC, Fortina P et al. Degenerate oligonucleotide primed-polymerase chain reaction and capillary electrophoretic analysis of human DNA on microchip-based devices. Anal Biochem 1998; 257(2):101-106.

90. Northrup MA, al. e. A Miniature Analytical Instrument for Nucleic Acids Based on Micromachined Silicon Reaction Chambers. Anal Chem 1998; 70:918-922.

91. Cheek BJ, Steel AB, Torres MP et al. Chemiluminescence Detection for Hybridization Assays on the Flow-Thru Chip, a Three-Dimensional Microchannel Biochip. Anal Chem 2001; 73:5777-5783.

92. Sosnowski R, Tu E, Butler W et al. Rapid Determination of Single Base Mismatch Mutations in DNA Hybrids by Direct Electric Field Control. Proc Natl Acad Sci USA 1997; 94:1119-1123.

93. Edman C, Raymond D, Wu D et al. Electric Field Directed Nucleic Acid Hybridization on Microchips. Nucl Acids Res 1998; 25:4907-4914.
94. Moroney RM, White RM, Howe RT. Microtransport Induced by Ultrasonic Lamb Waves. Appl Phys Lett 1991; 59:774-776.
95. Ward MD, Quan J, Grodzinski P. High gradient magnetic separation microchannels for integrated microfluidic devices. Eur Cell Mater 2002; 3:123-125.
96. Burckhardt J. Amplification of DNA from Whole-Blood. PCR-Methods and Applications. 1994; 3(4):239-243.
97. Panaccio M, Georgesz M, Lew AM. FoLT PCR: a simple PCR protocol for amplifying DNA directly from whole blood. Biotechniques. 1993; 14(2):238-243.
98. Woolley AT, Hadley D, Landre P et al. Functional integration of PCR amplification and capillary electrophoresis in a microfabricated DNA analysis device. Anal Chem 1996; 68(23):4081-4086.
99. Ibrahim MS, Lofts RS, Jahrling PB et al. Real-time microchip PCR for detecting single-base differences in viral and human DNA. Anal Chem 1998; 70(9):2013-2017.
100. Klein D. Quantification using real-time PCR technology: applications and limitations. Trends Mol Med 2002; 8(6):257-260.
101. Umek RM, Lin SS, Chen YP et al. Bioelectronic detection of point mutations using discrimination of the H63D polymorphism of the Hfe gene as a model. Mol Diagn 2000; 5(4):321-328.

CHAPTER 5

Integrating Sample Processing and Detection with Microchip Capillary Electrophoresis of DNA

Adam T. Woolley*

Abstract

DNA is the storage medium for inherited information in living systems and thus, techniques for nucleic acid analysis are of great importance. Integrated microchip instrumentation for rapid separation of DNA by capillary electrophoresis (CE) has emerged as an especially promising approach for assaying genetic material. Improvements in separation capabilities and sample capacity have made CE microdevices compatible with the rapid, high-throughput analysis needs in DNA studies. Perhaps the most critical advantage of adopting a microchip format for DNA characterization is the ability to couple sample handling operations, separation and miniaturized detection instrumentation seamlessly on a single device. Indeed, recent work with micromachined, portable PCR-CE systems indicates the exceptional promise of integrated DNA analysis microchips. Lastly, phase-changing sacrificial layer fabrication methods for polymer microdevices should lead to inexpensive, high-performance DNA characterization microchips. Continued progress in integrated microdevices for DNA analysis will enhance capabilities for medical, biological and forensic work.

Introduction

DNA provides the storage mechanism for hereditary information in biological systems and as such, methods for the separation and quantitation of nucleic acids are critical. The development of capillary electrophoresis (CE)[1,2] and its subsequent application to DNA analysis[3-5] have been key advances. Indeed, the size-based fractionation of nucleic acids can now be carried out orders of magnitude more expeditiously, with greater automation and better reliability than was possible 10-20 years ago, largely because of the emergence of capillary array electrophoresis (CAE) instrumentation.[6-8] The downscaling of DNA separations from the slab gel format to CAE provided an early indication of the advantages of miniaturization in DNA characterization.

In this chapter progress in improving DNA analysis capabilities through the implementation of a microchip format will be described. The coupling of DNA sample preparation (amplification, extraction, etc.,) with CE in microdevices will also be detailed. Next, the integration of miniaturized detection systems with CE microchips for DNA analysis will be discussed. Finally, a description will be given of the newly developed phase-changing sacrificial layer microfabrication approach, which shows considerable promise for creating inexpensive polymer microstructures for the characterization of DNA.

*Adam T. Woolley—C100 BNSN, Department of Chemistry and Biochemistry, Brigham Young University, Provo, Utah 84602-5700, U.S.A. Email: wadam@byu.edu

Integrated Biochips for DNA Analysis, edited by Robin Hui Liu and Abraham P. Lee.
©2007 Landes Bioscience and Springer Science+Business Media.

Microchip CE of DNA

In the early 1990s a novel method for miniaturizing CE using planar micromachining techniques developed for the fabrication of integrated circuits was presented by Manz and coworkers.[9] Figure 1 outlines schematically a typical sequence of steps followed in the construction of glass CE microdevices. Since the designs for microchips are generally made using computer programs, the coupling and integration of sample preparation and detection with CE analysis on individual devices can be realized through appropriately designed photolithographic masks for surface patterning. The great potential for these microfluidic systems to speed up CE, miniaturize instrumentation

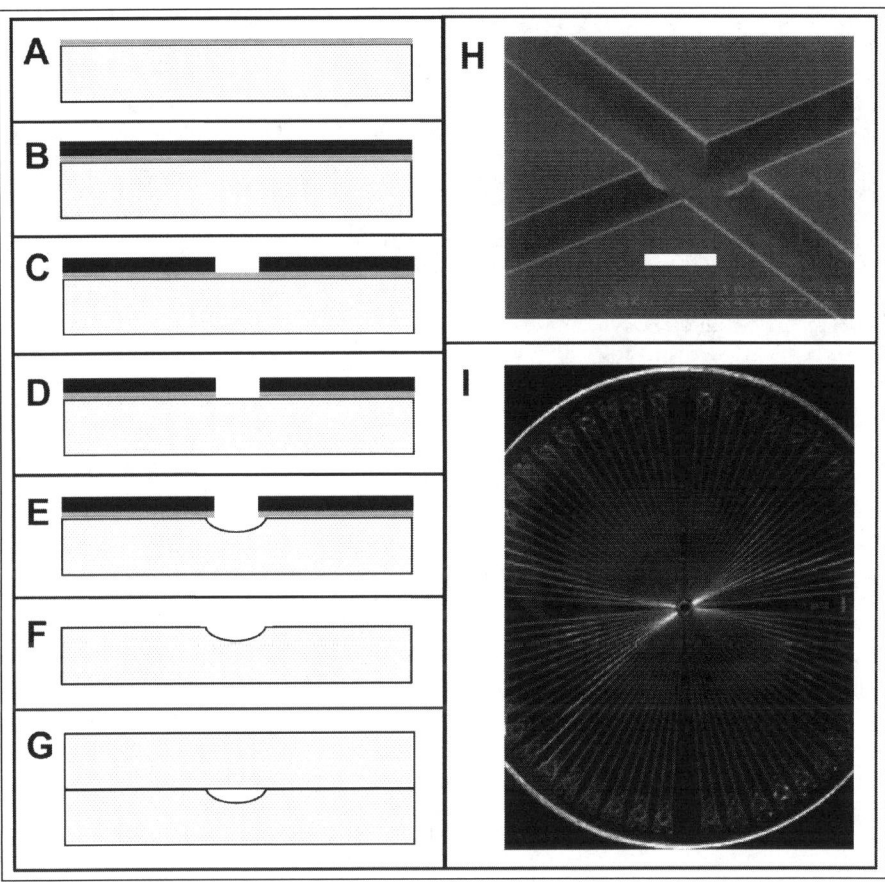

Figure 1. Schematic overview of glass CE microchip fabrication steps. A) A thin protective layer (dark gray) is deposited on a glass substrate (light gray, side view). B) Photoresist (black) is spin coated on top of the protective layer. C) UV exposure through a patterning photomask makes the irradiated regions of photoresist soluble in a developer solution. D) The protective layer that is uncovered following removal of photoresist is etched away. E) The underlying glass exposed in (D) is dissolved in a hydrofluoric acid solution. F) Any remaining photoresist and protective layer are removed from the etched wafer. G) The substrate from (F) is cleaned and bonded in a furnace to a second glass piece. H) Electron micrograph of channels patterned and etched as described in (A-G); the scale bar is 50 μm. Reprinted with permission from ref. 62; copyright 1998, Springer Science and Business Media. I) Top-view photograph of a 96-channel glass microdevice; image courtesy of Professor Richard Mathies from the University of California, Berkeley.

and allow straightforward coupling of sample handling with separation soon became evident and researchers quickly set out to apply CE microchips to enhance DNA analysis.

In 1994 two teams independently demonstrated nucleic acid separations on CE microdevices made similarly to the scheme in Figure 1. Woolley and Mathies separated restriction digests and PCR products in under two minutes,[10] while Effenhauser and coworkers fractionated fluorescently tagged oligonucleotides in less than one minute.[11] Within another year rapid DNA sequencing had also been shown in a CE microchip.[12] In the decade since these early studies, DNA analysis has been improved in terms of both enhanced separation and increased throughput.

Important advances in microchip CE of DNA have been achieved for sequencing fragments. The first publication on DNA sequencing in micromachined channels used a 9% linear polyacrylamide (LPA) gel as the sieving medium and four-color fluorescent labeling; the read length was ~150 bases in 9 minutes with 97% accuracy.[12] Three years later another group carried out a thorough investigation of microchip CE of DNA in LPA and they separated single-color sequencing fragments up to 400 bases long with resolution greater than 0.5 in ~20 minutes.[13] In the next year an optimized procedure for gel polymerization[14] was used to make LPA with a ~1 MDa average molecular weight and this sieving medium was applied in four-color microchip sequencing to 500 bases in 20 minutes with 99.4% accuracy.[15] This threefold improvement in read length made the sequencing performance in microdevices comparable to conventional capillary systems.

However, for microchip CE of DNA to be competitive with either slab gel electrophoresis or CAE, sample throughput needed to be increased through the use of parallel analysis lanes. In 1997 the first CAE microdevice was demonstrated,[16] in which rapid (~160 second) DNA separations for *HFE* typing in a parallel array of twelve independent channels were detected using a linearly rastered confocal laser-induced fluorescence setup. In subsequent work a CAE microplate with 48 lanes was applied in an 8 minute *HFE* genotyping experiment on 96 different specimens.[17] To increase throughput even further, a scanning rotary confocal fluorescence detection instrument was designed for circular CAE microchips whose lanes converged at the center of a device (Fig. 1I).[18] Sequencing with 99% accuracy to an average of ~430 bases per lane in 24 minutes was obtained in a 96-lane radial microdevice.[19] Even higher-capacity, 384-channel microplates were constructed and applied in the rapid typing of 384 *HFE* samples in ~325 seconds.[20] Clearly, the microchip format is very well suited for the speed and throughput requirements of high-volume DNA analysis.

Integrating Sample Processing with Microchip CE of DNA

Perhaps one of the most important benefits of miniaturizing DNA analysis systems is that sample processing steps can be integrated on the same platform as the separation column. In this manner the volume of a specimen loaded on a chip can be scaled down to match more closely the amount actually required for analysis. Such integration can reduce significantly the quantities of reagents consumed and associated reaction expenses.

PCR amplification is one of the most widely utilized DNA sample processing steps. PCR is also an ideal reaction for miniaturization due to the relative ease of mass producing micromachined chambers having volumes less than one microliter and the excellent heat transfer properties of such microreactors. Hence, there has been great interest and important progress made in coupling PCR with microchip CE of DNA, resulting in a number of different device constructs. The initial demonstration of PCR integrated with CE in a miniaturized format was carried out in a hybrid system consisting of: (1) a PCR vessel made from silicon and having patterned thin-film heaters for reaction chamber temperature control and (2) a glass CE microchip.[21] The PCR chamber was inserted into the injection reservoir in the CE microchip to couple the two pieces. During thermal cycling fluid transfer between the PCR vessel and CE microfluidics was prevented by the viscous sieving medium that filled the separation system. Once the DNA sample had been amplified it was transferred electrokinetically through the sieving matrix and into the separation module for CE analysis. This setup was applied in assaying β-globin and *Salmonella* samples, providing the first demonstration that PCR and CE could be integrated functionally in a microchip system. After this initial work,[21] PCR-CE microdevices were also made with the PCR vessel and CE microfluidics all in a single substrate (monolithically integrated PCR-CE).

Whole-device thermal cycling, wherein the entire microchip was taken through the heating and cooling steps in a commercial PCR system, was employed in early experiments; after amplification, CE analysis was performed once the device had been moved manually onto the separation and detection platform.[22] Figure 2A depicts two microchip designs used for whole-device thermal cycling. Whereas this approach benefits from fabrication simplicity and the ability to be interfaced with existing PCR equipment, automation is absent in the manual device transfer step, and repeated exposures of the sieving matrix to temperatures near 100°C can be problematic.

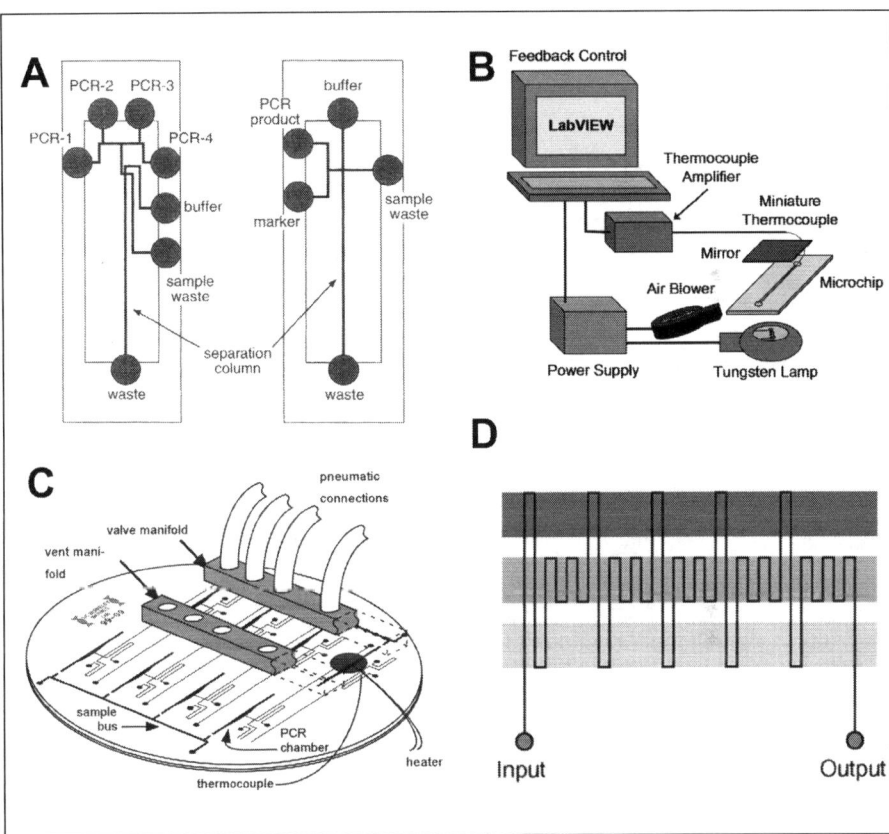

Figure 2. Schematic diagrams of different micromachined setups that integrate PCR amplification with CE separation. A) PCR-CE microchips that are thermally cycled when entire devices are placed in a temperature regulation system. Reprinted with permission from ref. 22; copyright 1998, American Chemical Society (ACS). B) Microchip PCR system with infrared heating for temperature manipulation; adapted with permission from ref. 23; copyright 2001, Elsevier. C) Monolithically integrated PCR-CE instrumentation. A valve manifold controls fluid flow and temperature is regulated by a heater attached to the bottom face of the device. Reprinted with permission from ref. 26; copyright 2001, ACS. D) Microchip flow PCR. Reagents are transported through a microchannel to zones having different temperatures to perform PCR. External heating source temperatures from top to bottom are: 95°C, 77°C, and 60°C; for visualization ease the drawing only shows five amplification cycles. Reprinted with permission from ref. 27; copyright 1998, American Association for the Advancement of Science.

Subsequent monolithically integrated PCR-CE has utilized localized reaction chamber (rather than whole-device) thermal cycling. In one approach focused infrared radiation heated the PCR mixture (Fig. 2B); the rapid temperature transitions yielded cycles as short as 12 seconds.[23] Moreover, in this device design a β-globin target was amplified in ~10 minutes and then identified by CE separation in the same microchip.[24] Integrated PCR-CE microsystems with submicroliter reaction vessels have been evaluated by Mathies' group.[25,26] Heating and temperature monitoring were carried out using micropatterned thin films on the devices and a valve manifold provided control of fluid movement inside the microchannels (Fig. 2C). In these microchips single-template PCR amplification was performed, followed by on-chip assessment in the integrated CE system.[26]

One emerging thermal cycling approach that shows great promise for micromachined devices is continuous flow PCR, wherein the reaction mixture is pumped through a microfluidic manifold that passes the specimen repeatedly through zones of appropriate temperatures to effect amplification (Fig. 2D).[27] New work in microchip flow PCR has evaluated convective[28] and electrokinetic[29] sample transport between different temperature regions. Future flow-based PCR studies should lead to additional improvements in performance and eventual on-chip integration with CE.

Progress has also been made in the integration of DNA sample manipulation procedures other than PCR. Solid-phase extraction of DNA from complex mixtures is an oft-used first step in genetic analysis. Landers' group developed microchips having sol-gel-embedded silica particles and applied these microdevices in extracting DNA from whole blood prior to PCR amplification.[30] Hybridization-based purification of Sanger sequencing products can isolate target DNA fragments from components of the reaction mixture that impair separation performance (high ionic strength buffer, nucleotide triphosphates and DNA template). Hydrogels incorporating oligonucleotides[31] complementary to a common sequence on Sanger fragments have been generated in microchips to purify target DNA prior to separation by CE.[32]

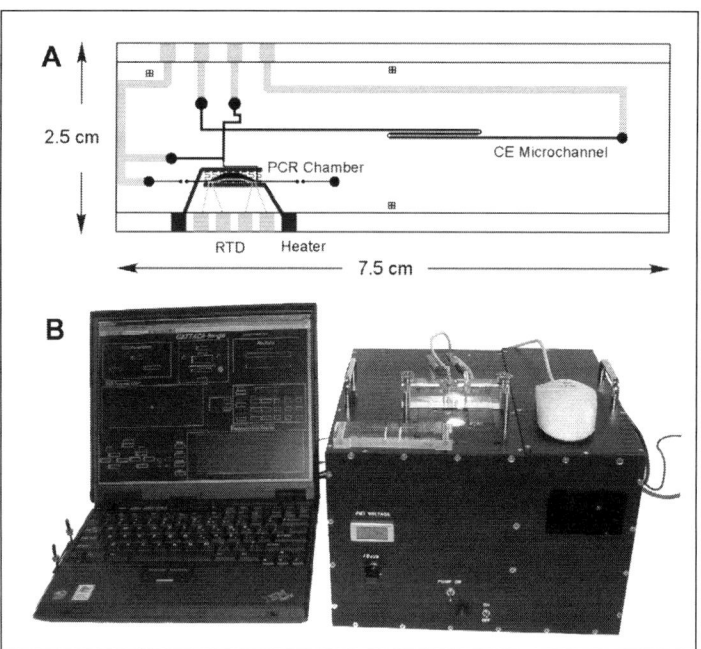

Figure 3. Portable PCR-CE system for cellular analysis. A) Schematic diagram of the miniaturized and integrated PCR-CE microdevice. B) Photograph of the instrument control computer and package for the optical and electronic components in portable PCR-CE analysis. Reprinted with permission from ref. 33; copyright 2004, ACS.

Recently, Mathies' group developed a high-performance, portable PCR-CE system (Fig. 3) that can assay whole bacterial cells.[33] This setup provided excellent analysis capabilities; for example, as few as three *E. coli* cells could be detected in the PCR-CE instrumentation. Moreover, strain-specific *E. coli* identification could be done, even with a four-order-of-magnitude excess of another strain. Finally, the PCR-CE microanalyzer was used in determining whether a methicillin-resistant variant of *S. aureus* was present. This work is a prime example of how integration and miniaturization in DNA analysis offer important new capabilities for addressing critical biological and medical issues.

Integrated DNA Detectors for CE Microchips

The on-chip coupling of processes to extract and amplify DNA from a raw biological sample to facilitate CE separation is critical, as outlined in the previous section. Equally important is the development of miniaturized, integrated systems for the detection of fractionated DNA bands in microchip CE. Unfortunately, laser-induced fluorescence, the most sensitive and widely used detection methodology for CE of DNA, typically requires bulky sources, optics and detectors. Two integrated detection approaches, electrochemical and optical, have now been demonstrated for microchip CE of DNA.

Electrochemical methods were the first to be miniaturized and integrated with microchip CE of DNA, using photolithographically patterned thin-film electrodes for amperometric detection.[34] More recently, portable detector electronic circuitry and high-voltage power supplies have been integrated on CE microdevices,[35] although these systems have not yet been evaluated for DNA separation. A sheath-flow microchip amperometric detector improved the analysis of PCR products tagged with redox intercalators and enabled *HFE* genotyping using ferrocene-labeled primers.[36] Integration of conductivity detection has been pursued in microfluidic systems, with both contact[37-39] and contactless[40,41] designs having been demonstrated; application to DNA separations was shown just recently in the contactless format.[42]

The miniaturization and integration of optical detection of DNA in CE microchips is also developing. A very sophisticated micromachined system on silicon was reported that incorporated an on-column photodiode detector for DNA separations.[43] This initial demonstration hinted at the promise of integration, although an external light emitting diode excitation source was required and the separation performance was suboptimal. Subsequent improvements in microchannel fabrication increased the CE quality somewhat,[44] but the resolution for a restriction digest was still considerably lower than what had been published much earlier.[10] Other advances in optical detection have involved the embedding of miniature light sources[45] or detection systems[46] in polymer CE microchips, although these have not yet been utilized in DNA analysis. The portable PCR-CE setup in the previous section (Fig. 3) demonstrated the downscaling of optical instrumentation for microchip separations of DNA, but the excitation source and detection parts were not fabricated on the same substrate as the microfluidics module.[33] Future work should lead to fully integrated and miniaturized optical and electrochemical detection systems for high-sensitivity microchip analysis of DNA.

Phase-Changing Sacrificial Materials for Polymer Microchip Analysis Systems

While the fabrication of glass microchip substrates dominated microfluidics research initially, interest has grown rapidly in making microstructures from polymeric materials that provide simpler and more flexible device manufacturing. Polymer microchips have been applied in CE of DNA,[47-50] although the performance for sequencing[51,52] still lags that achieved in glass microdevices. For inexpensive polymer microstructures to supplant glass devices in DNA sequencing, two issues will need to be overcome. First, thermally sealed polymer microchips delaminate when ~200 psi are applied,[53] much lower than for glass substrates (~2,000 psi),[54] which makes it difficult to load polymer microchannels with the viscous sieving media often utilized in DNA sequencing. Second, the covalent functionalization of polymer microdevice surfaces to prevent nonspecific adsorption is not established as well as the modification of glass, although my group and others have made important recent progress in this area.[55-57] Hence, easy-to-fabricate and inexpensive polymeric substrates should offer comparable performance to glass microchips, provided device bonding can be made sufficiently robust and the surfaces can be derivatized appropriately.

Recent work from my laboratory has yielded an elegant solution to the issue of polymer microdevice delamination. Phase-changing sacrificial layers (PCSLs) have been developed as a new tool to enable polymer substrates such as poly(methyl methacrylate) (PMMA) to be solvent bonded and thus withstand over tenfold higher internal pressures than thermally enclosed microchips.[53] Figure 4A-C illustrates how a PCSL is deposited as a protective material in polymer microchannels. In brief, a poly(dimethylsiloxane) (PDMS) piece is placed atop imprinted polymer microchannels, liquid PCSL (e.g., melted paraffin wax) is filled into the heated assembly and the substrate is cooled back to room temperature to solidify the PCSL, after which the PDMS is

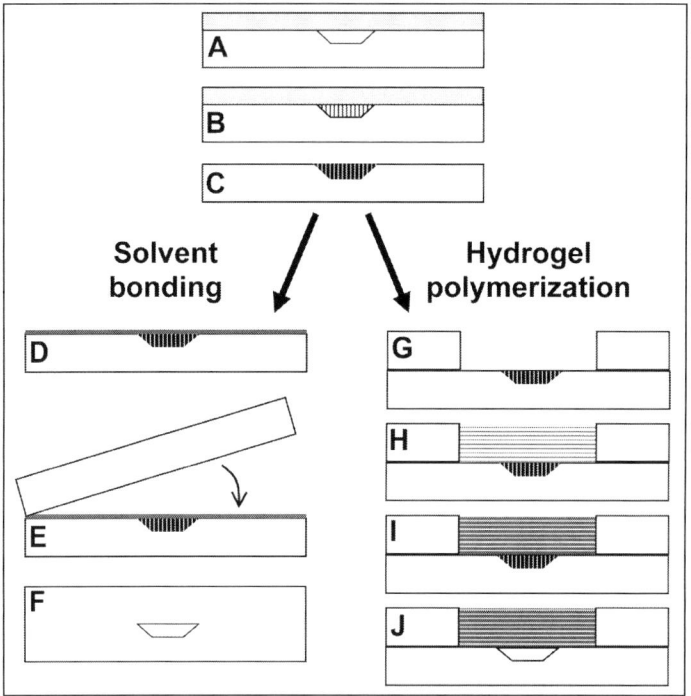

Figure 4. Cross-sectional view of PCSL use for solvent bonding (left) and interfacing polymer hydrogels with microfluidic systems (right). A) PDMS (gray) is placed on a patterned PMMA substrate (white) to form enclosed microchannels. Access holes in the PDMS line up with the PMMA channel ends. B) After the assembly is heated to 70°C, liquid PCSL (white with black vertical lines) is filled into the microchannels. C) The PCSL solidifies (black with white vertical lines) when the device is cooled to room temperature and the PDMS is removed. (D-F) Solvent-bonded microfluidic systems. D) The substrate from (C) is coated with solvent (dark gray). E) A PMMA plate is placed atop the PCSL-protected piece. F) After a bond has formed the microchip is heated to 70°C, the PCSL melts and the channels are emptied using a combination of vacuum and solvent. (G-J) Interfacing an ion-permeable hydrogel with microfluidics. G) A substrate with an opening to contain a hydrogel membrane is placed atop the PMMA piece from (C). H) A monomer solution (white with dark gray horizontal lines) is added to the well. I) An ion-permeable hydrogel (dark gray with white horizontal lines) is formed after polymerization. J) The microdevice is heated to 70°C, and the melted PCSL is removed using applied vacuum.

removed. This provides an inert material to protect microchannels during subsequent processing steps; thus, a solvent (acetonitrile) can weld the protected network to another PMMA plate and the PCSL prevents softened polymer and solvent from blocking the microfluidic array during bonding (Fig. 4D-E). Once enclosure is finished the PCSL can be melted and flushed from the channels readily, leaving a robust polymer microfluidic network (Fig. 4F). These solvent-bonded PMMA microchips offer superb CE performance: hundreds of separations can be carried out in individual devices and very high electric fields (1.5 kV/cm) can be used.[53] Moreover, pressures exceeding 2,200 psi can be applied to these solvent-bonded microstructures without delamination problems,[53] such that it should be feasible to utilize viscous sieving matrixes for DNA sequencing.

PCSLs also enable the creation of ion-permeable hydrogels interfaced with microchannels,[58,59] which may be useful for on-chip DNA sample preconcentration. Previous work in glass microchips that had a porous thin film demonstrated that DNA could be concentrated at the thin film—channel interface;[60,61] however, the device construction procedure was cumbersome. On the other hand, polymer microchannels containing a solid PCSL (Fig. 4C) can be interfaced easily with ionically conductive hydrogels. The fabrication protocol involves first placing atop the PCSL-protected substrate a polymeric piece with an opening to contain the hydrogel (Fig. 4G). Next, the monomer solution is added to the well (Fig. 4H) and UV polymerized to form a rigid, ion-permeable membrane above the PCSL-filled microstructure as depicted in Fig. 4I. Then, an open microchannel beneath the polymer hydrogel is created once the PCSL is liquefied and removed (Fig. 4J). My group initially developed this PCSL fabrication approach for making ionically conductive membranes interfaced with open microcapillaries for electric field gradient focusing of proteins.[58,59] More recently, my group has shown that PCSL-fabricated hydrogels can concentrate proteins by factors of up to 10,000;[59] these devices should allow similar on-chip enrichment of DNA samples. The PCSL fabrication techniques developed in my group have clear potential to facilitate DNA characterization through the making of low-cost, robust polymer microdevices and the integration of sample preconcentration on-chip.

Conclusions

Integrated CE microchips have become powerful tools in the rapid analysis of genetic material. Important improvements in separation capabilities and sample throughput make CE microdevices compatible with the need for rapid, high-volume DNA characterization. The microchip format also enables the seamless coupling of sample handling, separation and miniaturized detection in a single device, such that portable microsystems for DNA-based diagnostics are becoming a reality. Lastly, phase-changing sacrificial materials for polymer microdevice fabrication should lead to inexpensive, versatile and high-performance DNA analysis systems. The further development of microchip instrumentation should lead to continued improvements in low-cost, DNA-based diagnostics in forensics, medicine and biology.

Acknowledgement

I thank all the excellent undergraduate and graduate students who made this work possible. In particular I acknowledge Dr. Ryan T. Kelly for developing PCSLs and Yi Li for applying PCSLs to create sample preconcentration microchips. Finally, I am grateful for financial support from the National Institutes of Health (1 R01 GM064547-01A1).

References

1. Mikkers FEP, Everaerts FM, Verheggen TPEM. High-performance zone electrophoresis. J Chromatogr 1979; 169:11-20.
2. Jorgenson JW, Lukacs KD. Zone electrophoresis in open-tubular glass capillaries. Anal Chem 1981; 53:1298-1302.
3. Kasper TJ, Melera M, Gozel P et al. Separation and detection of DNA by capillary electrophoresis. J Chromatogr 1988; 458:303-312.
4. Cohen AS, Najarian D, Smith JA et al. Rapid separation of DNA restriction fragments using capillary electrophoresis. J Chromatogr 1988; 458:323-333.

5. Drossman H, Luckey JA, Kostichka AJ et al. High-speed separations of DNA sequencing reactions by capillary electrophoresis. Anal Chem 1990; 62:900-903.
6. Huang XC, Quesada MA, Mathies RA. DNA sequencing using capillary array electrophoresis. Anal Chem 1992; 64:2149-2154.
7. Huang XC, Quesada MA, Mathies RA. Capillary array electrophoresis using laser-excited confocal fluorescence detection. Anal Chem 1992; 64:967-972.
8. Mathies RA, Huang XC. Capillary array electrophoresis: An approach to high-speed, high-throughput DNA sequencing. Nature 1992; 359:167-169.
9. Manz A, Harrison DJ, Verpoorte EMJ et al. Planar chips technology for miniaturization and integration of separation techniques into monitoring systems: Capillary electrophoresis on a chip. J Chromatogr 1992; 593:253-258.
10. Woolley AT, Mathies RA. Ultra-high-speed DNA fragment separations using microfabricated capillary array electrophoresis chips. Proc Natl Acad Sci USA 1994; 91:11348-11352.
11. Effenhauser CS, Paulus A, Manz A et al. High-speed separation of antisense oligonucleotides on a micromachined capillary electrophoresis device. Anal Chem 1994; 66:2949-2953.
12. Woolley AT, Mathies RA. Ultra-high-speed DNA sequencing using capillary electrophoresis chips. Anal Chem 1995; 67:3676-3680.
13. Schmalzing D, Adourian A, Koutny L et al. DNA sequencing on microfabricated electrophoretic devices. Anal Chem 1998; 70:2303-2310.
14. Carrilho E, Ruiz-Martinez MC, Berka J et al. Rapid DNA sequencing of more than 1000 bases per run by capillary electrophoresis using replaceable linear polyacrylamide solutions. Anal Chem 1996; 68:3305-3313.
15. Liu S, Shi Y, Ja WW et al. Optimization of high-speed DNA sequencing on microfabricated capillary electrophoresis channels. Anal Chem 1999; 71:566-573.
16. Woolley AT, Sensabaugh GF, Mathies RA. High-speed DNA genotyping using microfabricated capillary array electrophoresis chips. Anal Chem 1997; 69:2181-2186.
17. Simpson PC, Roach D, Woolley AT et al. High throughput genetic analysis using microfabricated 96-sample capillary array electrophoresis microplates. Proc Natl Acad Sci USA 1998; 95:2256-2261.
18. Shi Y, Simpson PC, Scherer JR et al. Radial capillary array electrophoresis microplate and scanner for high-performance nucleic acid analysis. Anal Chem 1999; 71:5354-5361.
19. Paegel BM, Emrich CA, Wedemayer GJ et al. High throughput DNA sequencing with a microfabricated 96-lane capillary array electrophoresis bioprocessor. Proc Natl Acad Sci USA 2002; 99:574-579.
20. Emrich CA, Tian H, Medintz IL et al. Microfabricated 384-lane capillary array electrophoresis bioanalyzer for ultrahigh-throughput genetic analysis. Anal Chem 2002; 74:5076-5083.
21. Woolley AT, Hadley D, Landre P et al. Functional integration of PCR amplification and capillary electrophoresis in a microfabricated DNA analysis device. Anal Chem 1996; 68:4081-4086.
22. Waters LC, Jacobson SC, Kroutchinina N et al. Multiple sample PCR amplification and electrophoretic analysis on a microchip. Anal Chem 1998; 70:5172-5176.
23. Giordano BC, Ferrance J, Swedberg S et al. Polymerase chain reaction in polymeric microchips: DNA amplification in less than 240 seconds. Anal Biochem 2001; 291:124-132.
24. Ferrance JP, Wu QR, Giordano B et al. Developments toward a complete micro-total analysis system for Duchenne muscular dystrophy diagnosis. Anal Chim Acta 2003; 500:223-236.
25. Lagally ET, Emrich CA, Mathies RA. Fully integrated PCR-capillary electrophoresis microsystem for DNA analysis. Lab Chip 2001; 1:102-107.
26. Lagally ET, Medintz I, Mathies RA. Single-molecule DNA amplification and analysis in an integrated microfluidic device. Anal Chem 2001; 73:565-570.
27. Kopp MU, de Mello AJ, Manz A. Chemical amplification: Continuous-flow PCR on a chip. Science 1998; 280:1046-1048.
28. Krishnan N, Agrawal N, Burns MA et al. Reactions and fluidics in miniaturized natural convection systems. Anal Chem 2004; 76:6254-6265.
29. Chen JF, Wabuyele M, Chen HW et al. Electrokinetically synchronized polymerase chain reaction microchip fabricated in polycarbonate. Anal Chem 2005; 77:658-666.
30. Breadmore MC, Wolfe KA, Arcibal IG et al. Microchip-based purification of DNA from biological samples. Anal Chem 2003; 75:1880-1886.
31. Olsen KG, Ross DJ, Tarlov MJ. Immobilization of DNA hydrogel plugs in microfluidic channels. Anal Chem 2002; 74:1436-1441.
32. Paegel BM, Yeung SHI, Mathies RA. Microchip bioprocessor for integrated nanovolume sample purification and DNA sequencing. Anal Chem 2002; 74:5092-5098.
33. Lagally ET, Scherer JR, Blazej RG et al. Integrated portable genetic analysis microsystem for pathogen/infectious disease detection. Anal Chem 2004; 76:3162-3170.

34. Woolley AT, Lao K, Glazer AN et al. Capillary electrophoresis chips with integrated electrochemical detection. Anal Chem 1998; 70:684-688.

35. Jackson DJ, Naber JF, Roussel TJ et al. Portable high-voltage power supply and electrochemical detection circuits for microchip capillary electrophoresis. Anal Chem 2003; 75:3643-3649.

36. Ertl P, Emrich CA, Singhal P et al. Capillary electrophoresis chips with a sheath-flow supported electrochemical detection system. Anal Chem 2004; 76:3749-3755.

37. Guijt RM, Baltussen E, van der Steen G et al. New approaches for fabrication of microfluidic capillary electrophoresis devices with on-chip conductivity detection. Electrophoresis 2001; 22:235-241.

38. Liu Y, Wipf DO, Henry CS. Conductivity detection for monitoring mixing reactions in microfluidic devices. Analyst 2001; 126:1248-1251.

39. Galloway M, Stryjewski W, Henry A et al. Contact conductivity detection in poly(methyl methacylate)-based microfluidic devices for analysis of mono- and polyanionic molecules. Anal Chem 2002; 74:2407-2415.

40. Guijt RM, Baltussen E, van der Steen G et al. Capillary electrophoresis with on-chip four-electrode capacitively coupled conductivity detection for application in bioanalysis. Electrophoresis 2001; 22:2537-2541.

41. Pumera M, Wang J, Opekar F et al. Contactless conductivity detector for microchip capillary electrophoresis. Anal Chem 2002; 74:1968-1971.

42. Abad-Villar EM, Kuban P, Hauser PC. Determination of biochemical species on electrophoresis chips with an external contactless conductivity detector. Electrophoresis 2005; 26:3609-3614.

43. Burns MA, Johnson BN, Brahmasandra SN et al. An integrated nanoliter DNA analysis device. Science 1998; 282:484-487.

44. Webster JR, Burns MA, Burke DT et al. Monolithic capillary electrophoresis device with integrated fluorescence detector. Anal Chem 2001; 73:1622-1626.

45. Uchiyama K, Xu W, Qiu JM et al. Polyester microchannel chip for electrophoresis—incorporation of a blue LED as light source. Fresenius J Anal Chem 2001; 371:209-211.

46. Chabinyc ML, Chiu DT, McDonald JC et al. An integrated fluorescence detection system in poly(dimethylsiloxane) for microfluidic applications. Anal Chem 2001; 73:4491-4498.

47. Xu F, Jabasini M, Baba Y. DNA separation by microchip electrophoresis using low-viscosity hydroxypropylcellulose-50 solutions enhanced by polyhydroxy compounds. Electrophoresis 2002; 23:3608-3614.

48. Wainright A, Nguyen UT, Bjornson T et al. Preconcentration and separation of double-stranded DNA fragments by electrophoresis in plastic microfluidic devices. Electrophoresis 2003; 24:3784-3792.

49. Llopis SD, Stryjewski W, Soper SA. Near-infrared time-resolved fluorescence lifetime determinations in poly(methylmethacrylate) microchip electrophoresis devices. Electrophoresis 2004; 25:3810-3819.

50. Buch JS, Kimball C, Rosenberger F et al. DNA mutation detection in a polymer microfluidic network using temperature gradient gel electrophoresis. Anal Chem 2004; 76:874-881.

51. Shi Y, Anderson RC. High-resolution single-stranded DNA analysis on 4.5 cm plastic electrophoretic microchannels. Electrophoresis 2003; 24:3371-3377.

52. Zhu L, Stryjewski WJ, Soper SA. Multiplexed fluorescence detection in microfabricated devices with both time-resolved and spectral-discrimination capabilities using near-infrared fluorescence. Anal Biochem 2004; 330:206-218.

53. Kelly RT, Pan T, Woolley AT. Phase-changing sacrificial materials for solvent bonding of high-performance polymeric capillary electrophoresis microchips. Anal Chem 2005; 77:3536-3541.

54. Scherer JR, Paegel BM, Wedemayer GJ et al. High-pressure gel loader for capillary array electrophoresis microchannel plates. BioTechniques 2001; 31:1150-1154.

55. Liu J, Pan T, Woolley AT et al. Surface-modified poly(methyl methacrylate) capillary electrophoresis microchips for protein and peptide analysis. Anal Chem 2004; 76:6948-6955.

56. Xiao D, Le TV, Wirth MJ. Surface modification of the channels of poly(dimethylsiloxane) microfluidic chips with polyacrylamide for fast electrophoretic separations of proteins. Anal Chem 2004; 76:2055-2061.

57. Liu J, Sun X, Lee ML. Surface-reactive acrylic copolymer for fabrication of microfluidic devices. Anal Chem 2005; 77:6280-6287.

58. Kelly RT, Woolley AT. Electric field gradient focusing. J Sep Sci 2005; 28:1985-1993.

59. Kelly RT, Li Y, Woolley AT. Phase-changing sacrificial materials for interfacing microfluidics with ion-permeable hydrogels. Anal Chem 2006; 78:2565-2570.

60. Khandurina J, Jacobson SC, Waters LC et al. Microfabricated porous membrane structure for sample concentration and electrophoretic analysis. Anal Chem 1999; 71:1815-1819.

61. Khandurina J, McKnight TE, Jacobson SC et al. Integrated system for rapid PCR-based DNA analysis in microfluidic devices. Anal Chem 2000; 72:2995-3000.

62. Simpson PC, Woolley AT, Mathies RA. Microfabrication technology for the production of capillary array electrophoresis chips. J Biomed Microdevices 1998; 1:7-26.

CHAPTER 6

Integrated Plastic Microfluidic Devices for Bacterial Detection

Z. Hugh Fan* and Antonio J. Ricco

Abstract

This chapter describes integrated plastic microfluidic devices designed and fabricated for bacterial detection and identification. The devices, made from poly(cyclic olefin), contain components for DNA amplification, microfluidic valving, sample injection, on-column labeling and separation. DNA amplification was conducted in a reactor containing a volume ranging from 29 nL to 570 nL; screen-printed graphite ink resistors were used as heaters for thermal cycling. Microfluidic valves were created using in-situ gel polymerization to isolate the DNA amplification region from other components. Electrokinetic injection was used to migrate the amplification products (amplicons) through the gel valves, followed by on-chip electrophoretic separation. On-column labeling of the amplicons was achieved by mixing them with an intercalating dye. The various device functions were demonstrated by detection and genetic identification of two model bacteria, *Escherichia coli* O157 and *Salmonella typhimurium*.

Introduction

Significant advances in microfluidics have occurred since 1990 when Manz et al coined the term "micro total analysis system (μ-TAS)".[1] The μ-TAS concept has been developed from being an alternative to chemical sensors[1] to "lab-on-a-chip" devices and systems.[2,3] The materials used for making microfluidic devices have expanded from silicon[4-6] to glass[7-9] and various plastics.[10-13] Advantages of using microfluidic devices include lower reagent consumption, less sample required, higher separation efficiency, faster reaction kinetics, ease of automation, integration of multiple analytical processes with reduced handling and shorter analysis times and potential for mass production with low cost.[14]

Detection and identification of bacteria are important for diagnosis and therapy of infectious diseases, as well as for countermeasures to potential biological threats. The methods demonstrated for bacterial detection include immunoassay,[15,16] sensor arrays,[17,18] flow cytometry,[19] mass spectrometry,[20] and genetic analysis.[21-25] Among these, genetic analysis has been increasingly used to identify bacteria due to its high sensitivity and unambiguous identification via selective replication of genetic sequences unique to each analytical target. The replication of targets is achieved by DNA amplification using techniques such as the polymerase chain reaction (PCR) and genetic analysis has been used to detect bacteria including antharax,[21] *bacillus subtilis*,[22] *Escherichia coli (E. coli)*,[23,24] and *Salmonella*.[24,25] To take advantage of the miniaturization and integration mentioned above, these analyses have been demonstrated in microfluidic devices.[26-40]

*Corresponding Author: Z. Hugh Fan—Department of Mechanical and Aerospace Engineering and Department of Biomedical Engineering, University of Florida, PO Box 116250, Gainesville, FL 32611-6250, U.S.A. Email: hfan@ufl.edu

Integrated Biochips for DNA Analysis, edited by Robin Hui Liu and Abraham P. Lee.
©2007 Landes Bioscience and Springer Science+Business Media.

Analysis of PCR products (also known as amplicons) could be performed using traditional slab-gel electrophoresis after collecting amplicons from microfabricated devices.[26-31] However, increasing effort has been devoted to the integration of PCR with amplicon analysis on a mono-lithic fluidic platform, using either capillary electrophoresis (CE)[32-38] or DNA hybridization.[39,40] An extensive review of miniaturized PCR devices was recently published in which the reviewers ranked previously published work according to the ratio of [the number of base pairs of the ampli-fied fragment] to [the product of the PCR volume, amplification time and the limit of detection (expressed as DNA copy number)].[41]

Several types of materials, including silicon,[27-29,32,37] glass[30,33-35] and plastics[22,26,38,39] have been used for microfluidics-based PCR amplification. Integration of PCR with analysis in a single disposable plastic device is attractive because of improved reliability, ease of use and elimination of cross contamination.[26] The vast experience and well-developed process in manufacturing low-cost, high-volume plastic parts with micro-scale features (e.g., compact discs) make it possible to manu-facture a device so cost effectively that it can be disposable, which is very important in preventing false-positive results caused by "PCR product carryover".[26] In addition, plastics offer a range of materials and a variety of surface properties to provide biochemical compatibility.[10,11,42,43]

In this chapter, we will discuss integrated plastic microfluidic devices that accomplish DNA amplification, microfluidic valving, sample injection, on-column labeling and separation. PCR was conducted in a channel reactor; thermal cycling utilized screen-printed graphite ink resistors. In-situ gel polymerization was employed to form local microfluidic valves that minimize convective flow of the PCR mixture into other regions. After PCR, amplicons were electrokinetically injected through the gel valve, followed by on-chip electrophoretic separation. An intercalating dye was admixed to label the amplicons; they were detected using laser-induced fluorescence. Two model bacteria, *Escherichia coli* O157 and *Salmonella typhimurium*, were chosen to demonstrate bacterial detection and identification based on amplification of several of their unique DNA sequences.

Device Design and Fabrication

Figure 1 shows the layout of a plastic microfluidic device for bacterial detection and identifica-tion (ID) based on nucleic acid amplification of bacterial genetic signatures. The size of the device is the same as a traditional microplate (8.5 × 12.8 cm). Wells were laid out with a pitch of 4.5 mm or integral multiples thereof, which are the standards defined by the Society for Biomolecular Screening and accepted by the American National Standards Institute. This layout insures compat-ibility of this device with a variety of commercial fluid dispensing systems. This layout includes six device designs, with two copies of each design on one microfluidic device.

Each design in Figure 1 was devised to explore different concepts. ID-1 and -2 were designed for spatially resolved multiplexed PCR reactions. They have long serpentine channels in which PCR primers may be locally immobilized in different zones; this spatial separation of primer sets prevents primers from cross-talking with one other. ID-3 and -4 have a short PCR channel for a single PCR reaction mixture. ID-4 has an extra reservoir to allow the introduction of a DNA sizing ladder as discussed below. ID-5 accomplishes PCR in a well instead of a channel, similar to the work reported by Waters et al;[36] this provides a good comparison with more traditional PCR conducted in tubes. All these designs were tested and found to be functional, but this chapter focuses on ID-6.

Design ID-6 is shown in an exploded view in the inset of Figure 1. The functional regions for PCR and microfluidic valves are indicated. All wells in ID-6 are numbered for reference. Wells 5 and 6 are for introducing a sample into the PCR channel reactor that lies between them. Well 7 is for the introduction of a DNA sizing ladder, which serves as a calibration standard to accurately identify the amplicons according to their sizes. Wells 8 and 4 are for loading a sample plug into the separation channel, while Wells 1 and 3 are for running the separation. Well 2 was not used in this work. Channels are 120 μm wide and 50 μm deep except in the PCR zone where they are 360 μm wide. Channel lengths, measured from intersection "a" in the figure to Wells 1 through 8, are 8.1, 4.1, 48.9, 9.5, 8.4, 13.1, 8.4 and 17.2 mm, respectively. The total volumes of the PCR channels in ID-1, 2, 3, 4 and 6 are 500, 570, 54, 29 and 84 nL, respectively.

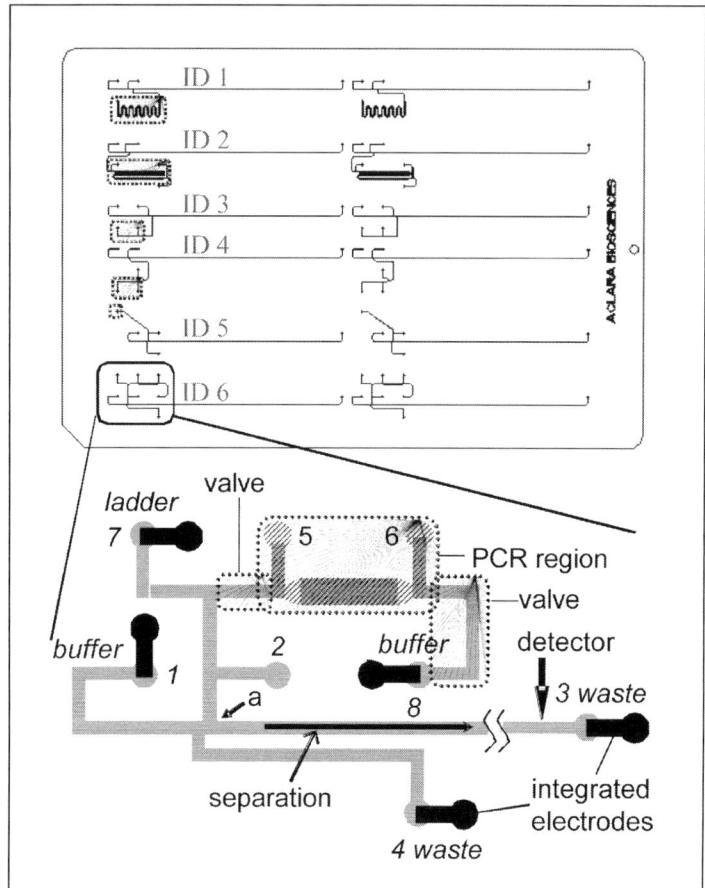

Figure 1. Layout of six integrated plastic microfluidic devices made from poly(cyclic olefin) (*top*). The detailed layout of device ID-6 is depicted in exploded view (*bottom*). The functional regions for PCR and microfluidic valves are indicated. Channels are 120 μm wide x 50 μm deep except in the PCR zone where they are 360 μm wide.

The in-channel PCR volumes of these devices are 2-3 orders of magnitude less than the volumes (5-50 μL) used in other work.[27,29-31] Use of larger PCR volumes was required in earlier efforts because a certain volume of PCR product was needed for slab-gel electrophoresis. In contrast, integrated CE analysis in this work enables the use of nanoliter sample volumes if they contain adequate amounts (copy numbers) of target DNA. The benefit of using nanoliter in-channel PCR includes minimization of reagent consumption (especially for the expensive polymerases) and smaller thermal mass for achieving rapid PCR reactions.

The device was fabricated from poly(cyclic olefin), which was obtained from Zeon Chemicals under the trade name Zeonor. This poly(cyclic olefin) was selected because it possesses low fluorescence background, minimum water absorptivity, good transparency and a high glass transition temperature (Tg, ~ 138°C). We "deselected" two other plastics, polycarbonate and acrylic, which are frequently used for microfluidics. Polycarbonate has a high Tg (~ 145°C) that meets the temperature requirements for PCR reaction,[39] but its high fluorescence background would result in poor detection sensitivity. Acrylic has good optical properties for detection and satisfactory surface properties for DNA separation, but the Tg (~109°C) is too low to allow it to be used for PCR. In addition, acrylic has more than 30 times higher water absorptivity than poly(cyclic olefin), resulting in significant water loss from the channel into the plastic bulk. Several recent efforts describe the use of poly(cyclic olefin) for a variety of applications involving microfluidic chips.[43-47]

The technique used to fabricate devices is a replication method, compression molding, as we reported previously.[11] The method is similar to the approach used in the commercial manufacture of compact discs, which also consist of micro-scale features. The process flow is illustrated in Figure 2. To form the base layer, fluidic features are first microfabricated on a glass plate or silicon wafer using standard photolithographic patterning and etching techniques. Next, a metal mold or "master" is created by electroplating from aqueous solution to deposit anywhere from a few hundred microns to a few millimeters of metal onto the surface of the master, creating a precise replicate "electroform" with inverse topology: a channel in the silicon or glass becomes a ridge in the electroform. Such a metal tool can be mounted as the molding tool on a molding system. Polymer base layers are then formed in volume from melted or softened polymer resin to create smooth and precise channels in the finished fluidic device. Holes at the ends of channels are then created by drilling; they will function as fluid reservoirs. The open microchannels of a molded fluidic base layer are sealed to form closed channels by thermal bonding to a thin polymer film.

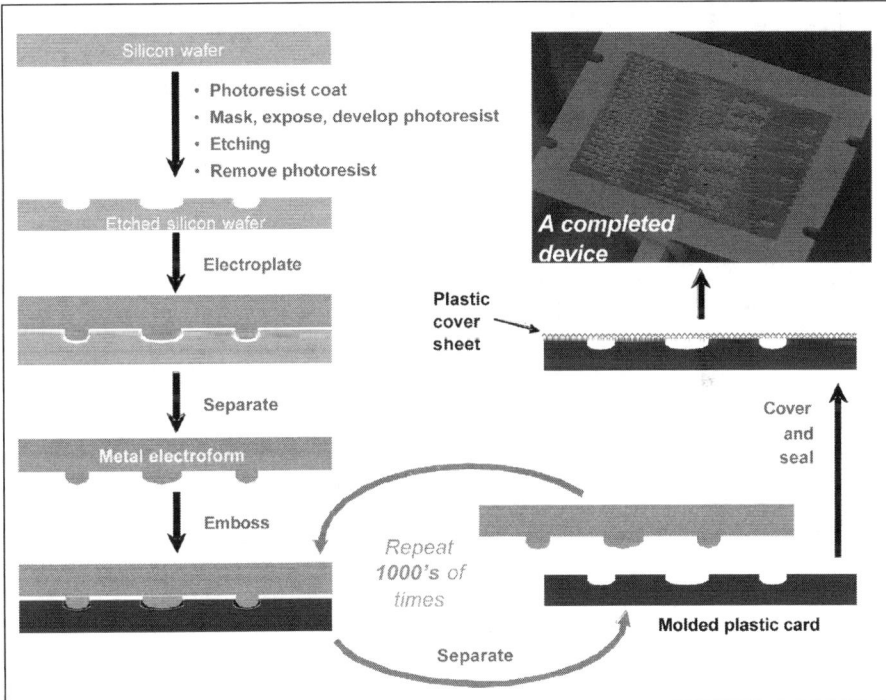

Figure 2. Process overview for mass manufacturing of plastic microfluidic systems. A microfluidic design is patterned and etched on a glass or silicon substrate using standard micromachining techniques to form a positive "master". A metal mold or "electroform" is formed via electroplating onto the surface of the master, creating a precise negative replica of the pattern. Many thousands of plastic parts with channeled structures can be thermally formed against the metal mold tool. Each molded device is then sealed with a plastic layer that encloses the microchannels. (Adapted from Boone et al.[11])

Microfluidic Valves

Of the challenges to integrating PCR with CE separation in low-cost monolithic form, implementation of an appropriate microvalve is the greatest.[33,34] Microfluidic valves are needed to isolate the PCR chamber from other regions of the integrated device.[33,34,39,48] If valves are absent, the PCR mixture flows into the separation medium due to convection, as well as thermal expansion and contraction that cause pressure and volume changes. In addition, diffusion at elevated temperatures during thermal cycling is significantly more rapid than at room temperature.

To address this challenge, we explored the concept of using an in-channel plug of gel as a closed valve during amplification and as an open valve during amplicon injection after PCR, made feasible by the fact that charged amplicons are readily electrophoresed through the gel. Localized gel polymerization has also been explored for other applications including flow control inside microfluidic channels[49] and formation of microreactors.[50] The compatibility of the gel valves with PCR was examined by placing gel in a PCR reaction in a conventional tube, revealing no detectable inhibition.

Photo-initiated gel polymerization can be used to define a gel valve at a precise location without leaving undesirable residue. A UV-curing agent, hydroxycyclohexylphenylketone (HCPK), was used as an initiator for gel polymerization in the device. To make a gel valve at a precise location, 8% acrylamide/bis acrylamide monomer solution containing 5 mM HCPK was first added to all channels and wells. The entire device was filled with the monomer solution using pressure or vacuum. A photomask (or opaque black tape) was placed on top of the device to define the exposed area where microfluidic valves were desired. After an exposure of 5 minutes to collimated light (near 356 nm) from a filtered mercury lamp, the monomer solution in the exposed area polymerized to form a gel plug. The monomer solution in unexposed regions does not polymerize and thus unpolymerized gel can be effectively removed by flushing, then replaced with the desired buffers and reagents. The device must be designed appropriately so that gel valves are not in the way of changing solution in unpolymerized regions. Figure 3 shows a gel valve with a boundary at a T intersection. The gel is dyed for easy visualization.

The gel valve was able to withstand hydrostatic pressures up to 100 psi (689 kPa), at which point the cover film delaminated. That pressure is much higher than the water vapor pressure (12 psi, 83 kPa) at 95°C, the denaturation temperature for PCR. For those applications that require even higher holding pressures, tethering the gel valves to the channel surfaces may be considered.

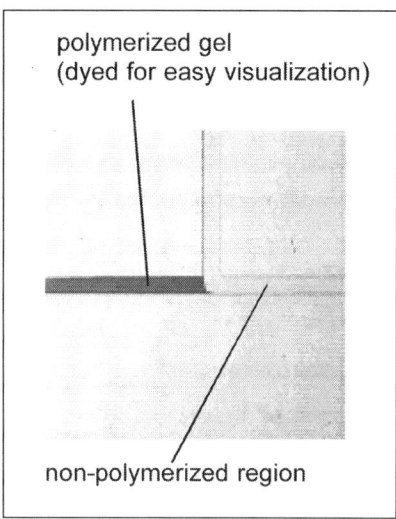

Figure 3. A gel valve precisely located at the edge of a T intersection. (Adapted from Koh et al.[38])

For comparison, the pressure required to move viscous polymer solutions was also determined; such solutions might have been considered as both valve and separation medium. However, various polymer solutions, including 3% poly(ethylene oxide) with a viscosity of 445 cP, were examined and in no case was a pressure differential greater than 5 psi (34 kPa) sustained without appreciable flow.

Screen-Printed Heaters and Electrodes

Integrated heaters offer rapid thermal cycling due to localized heating combined with low thermal mass. Most integrated heaters have been made from patterned thin-film metal resistors[30,32-35,37] or Peltier heaters[27,33,34,39] attached to a device. Other heating elements include non-integrated infrared radiation[51] and heat blocks used in commercial thermal cyclers.[31,36]

Screen-printing was employed to obtain conductive traces as heating resistors for PCR. Screen-printing is a relatively simple, inexpensive and versatile technology that has been used to produce electrodes for electrochemical analysis; screen-printed electrodes have been the transducer of choice for most practical real-life applications (e.g., commercial glucose sensors).[52,53] The screen-printed ink heaters provide intimate thermal contact (no interfacial adhesive layer) and minimum thermal mass. To fabricate heating resistors, silver/graphite inks were screen-printed onto a poly(cyclic olefin) support film (~100 μm thick). Silver/graphite blend ink (sheet resistance ~ 5 Ω/square) was obtained from ERCON (Wareham, MA). The ink-based heating resistor is 2.2 mm wide, with a length adequate to cover the entire PCR channel. After being cured at 95°C for two hours, the ink pattern on its supporting film was aligned and thermally laminated onto the cover film of a device. Figure 4 shows a device with a screen-printed heater.

In addition, screen-printed ink resistors can function as driving electrodes for CE as shown in Figure 4. The CE ink electrodes were fabricated simultaneously with the ink heater. All electrodes are 0.3 mm wide and 4.5 mm long. One end of the ink electrode is in a reagent reservoir while the other end with its contact pad is connected to high voltage via spring-loaded "pogo" contact pins, thus facilitating CE automation. Such an arrangement avoids inserting CE wire electrodes directly into reagents reservoirs as in some commercial instruments, eliminating possible contamination between experiments. Feasibility has been shown for the use of ink traces as driving electrodes for on-chip electrophoresis and it was shown that they have the potential to achieve fast and efficient separations with improved standard deviation for multiple runs relative to noble metal wires.[54]

PCR Thermal Cycling

The setup for PCR thermal cycling was built in house and described previously.[38] It consists of a heater, power supplies, RTD (resistance temperature detector), a fan and a computer. The RTD (5651PDX24A, Minco) was placed in the center of the PCR region using Kapton tape from Omega as shown in Figure 5. The cooling fan directed room-temperature air over the device. The heater and the cooling fan were connected to separate power supplies (GPS-3030D, Instek), each controlled by a signal-conditioning board (National Instruments SC-2042-RTD, Austin,

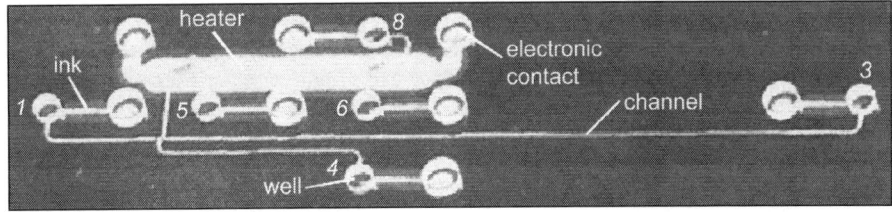

Figure 4. Photograph of a device with a 2.2 mm-wide ink trace as a heater for PCR. For each reservoir numbered, there is a 0.3 mm-wide ink trace connecting the reservoir with an electronic contact well, which allows application of an electric potential via a "Pogo" pin. These ink traces thus function as CE driving electrodes, represented by solid blocks in ID-6 at the bottom of Figure 1. Wells 2 and 7 of ID-6 in Figure 1 were eliminated in this device.

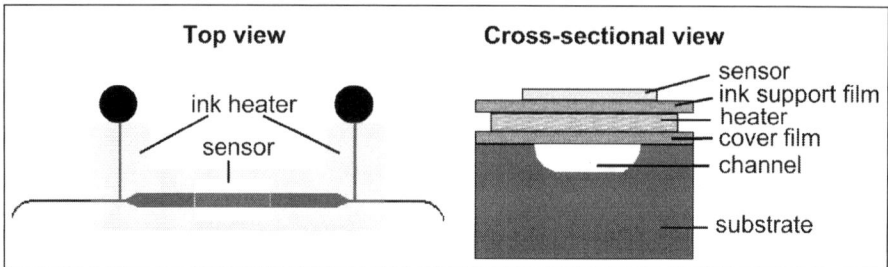

Figure 5. Top view and cross-sectional view of a screen-printed ink heater, a temperature sensor (RTD) and the PCR region of the device designated ID-6.

TX) according to feedback from the RTD. The signal-conditioning board also supplied 1 mA of current to the RTD. A data acquisition board (PCI-MIO-16XE-50, National Instruments) was used to acquire data using a custom LabView (National Instruments) program. A proportional/integral/differential (PID) module within the LabView program was used to control the cycling temperatures. The setup can control multiple PCR reactions simultaneously.

The ink resistor was calibrated against the actual temperatures in the solution inside the PCR channel so that the programmed temperatures for thermal cycling were accurate for successful DNA amplification. To calibrate the temperature measurement system, a micro-thermocouple (K-type, unsheathed, with a diameter of 51 μm, CHAL-002 from Omega) was placed inside the PCR channel; it was connected to a digital thermometer. The channel was filled with the PCR mixture without polymerase for thermal calibration. A linear relationship (R = 1.000) was found between the temperature measured in the PCR solution and that measured by RTD; the calibration slope and intercept were incorporated into the control algorithms for subsequent PCR experiments, allowing PCR without the thermocouple in the channel.

The ink heater was compared with a commercial thermal cycler and several advantages were noted for the microfluidic devices. Figure 6 shows the thermal cycling profiles obtained using a PTC-200 "DNA engine" (thermal cycler; MJ Research, Waltham, MA) and the ink heater. Both temperatures were measured by immersing a thermocouple into a PCR mixture in either a tube or a channel. The data show that the temperature ramp-up speed using the ink heater is ~12°C/s, which is roughly 4 times faster than the MJ thermal cycler. The higher speed allows less time for mismatching that can occur between primers and targets; thus, fewer false positive results should be obtained.[55,56] This quick response is mainly a consequence of the small thermal mass of the heating area. The cooling speed by forced air is ~2°C/s, which is very similar to that in the MJ thermal cycler. Our results, in terms of ramp-up and cool-down speeds, are comparable to micro-scale PCR systems reported by others.[22,27,39,40]

Bacterial Detection

Detection and identification of bacteria are important for diagnosis and therapy of infectious diseases, as well as for countermeasures to potential biological threats. To detect bacteria, all the microfluidic components described above were integrated in one device. Devices of design ID-6 were employed to demonstrate the functionality of the integrated device concept and the interoperability of all microfluidic components.

To implement PCR, a reaction mixture was prepared containing 1X AmpGene PCR Buffer II (Applied Biosystems), 2 mM MgCl$_2$, 0.5 units/μL AmpliTaq Gold DNA polymerase, 200 μM of each deoxynucleotide triphosphate, 2.5 μM of each primer, 0.1 mg/mL bovine serum albumin (BSA) and roughly 4.5 × 10^3 copies of each of the target templates. One half of the reaction mixture was amplified using the commercial thermal cycler; the result served as a positive control. The remaining PCR mix was used for filling the microfluidic device. Prior to loading the PCR mixture, the plastic device was washed with nuclease-free water and 1% BSA solution. The PCR mixture was

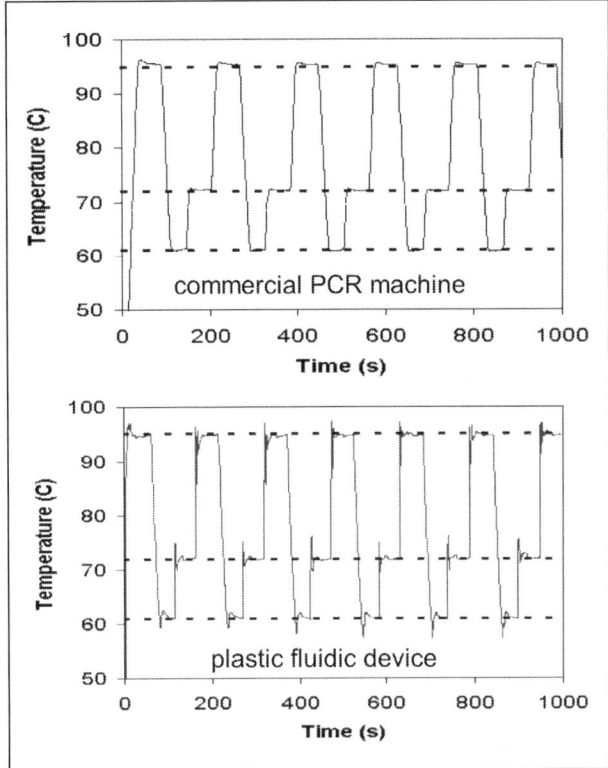

Figure 6. Thermal cycling profiles of a commercial PCR machine (MJ PTC-200 DNA engine) and an ink heater. The temperature ramp-up speed using the ink heater is ~ 4 times faster than that using the MJ thermal cycler, reflected in the number of cycles accomplished within the same period of time.

then pipetted into Well 5; it filled the entire PCR region with the assistance of a "gentle" vacuum at Well 6. After filling the other channels and wells with appropriate reagents as discussed below, all wells were sealed using PCR tape to prevent evaporation. PCR was carried out using temperature cycles including an initial denaturation at 95°C for 5 minutes, followed by 35 cycles of (94°C for 45 s, 56°C for 30 s and 72°C for 45 s) and a final extension step at 72°C for 5 minutes.

E. coli O157 and *Salmonella typhimurium* were chosen as analytes because they are dangerous food-borne pathogens. Thermally-killed *E. coli* O157:H7 and *Salmonella typhimurium* cells were from KPL (Gaithersburg, MD). Target genomic DNA was extracted from these cells using bacterial lysis buffers and "Genomic-Tips" from Qiagen (Valencia, CA). As shown by Legace et al, the primer set designed for *E. coli* O157 and the resultant amplicon of 232 bp is organism-specific,[23] allowing differentiation of *E. coli* from other microorganisms. Primer sets designed for *salmonella* are similarly specific, with an amplicon at 559 bp being specific to *Salmonella typhimurium*, while an amplicon at 429 bp is common to several serotypes of *salmonella*, according to Soumet et al.[25] Oligonucleotide primers were ordered from Integrated DNA Technologies (Coralville, IA). The sequences of all primers and the sizes of the corresponding amplicons are listed in Table 1.

After completion of PCR, the amplicons were electrokinetically migrated through valves and into the separation channel. The channel was filled with a sieving matrix consisting of 1.5% (weight/volume) of hydroxypropylcellulose (HPC, MW 60,000), 0.4% (w/v) of hydroxyethylcellulose (HEC, MW 90,000-105,000), 4 μM of thiazole orange (TO) in 1X TBE buffer at pH 8.4.

Table 1. DNA sequences of primers and sizes of amplicons

Name	Sequence[a]	Amplicon Size
E. coli O157:H7	Forward primer: ATC AAC CGA GAT TCC CCC AGT	
	Reverse primer: TCA CTA TCG GTC AGT CAG GAG	232 bp
Salmonella	Forward primer: GCC AAC CAT TGC TAA ATT GGC GCA	
	Reverse primer: GGT AGA AAT TCC CAG CGG GTA CTG G	429 bp
Salmonella	Forward primer: CGG TGT TGC CCA GGT TGG TAA T	
typhimurium	Reverse primer: ACT CTT GCT GGC GGT GCG ACT T	559bp

Note: (a) The primer sets were designed according to references.[23-25]

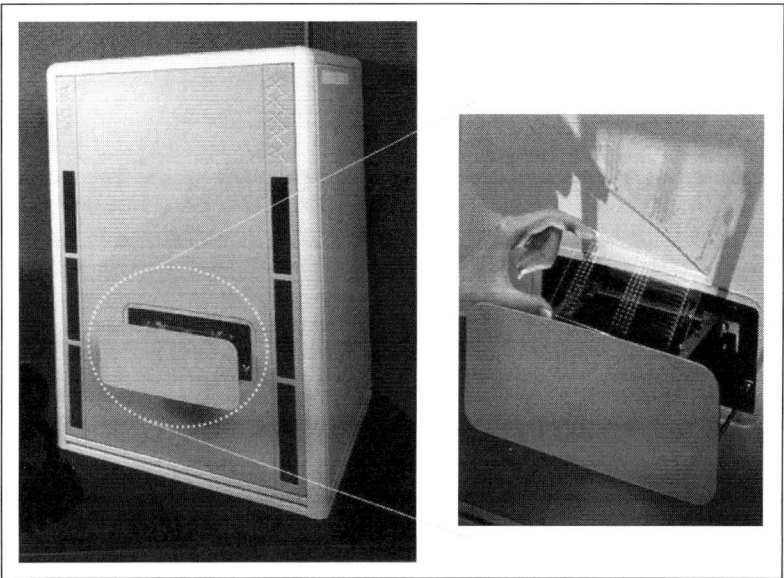

Figure 7. A picture of an instrument developed at ACLARA BioSciences that accepts the device in Figure 1. The inset on the right shows a device being placed in the instrument. The instrument control and data acquisition are provided by a computer and appropriate control software.

The intercalating dye, TO, was purchased from Sigma-Aldrich while HPC and HEC were from Polysciences (Warrington, PA). A buffer consisting of 890 mM Tris, 890 mM boric acid and 20 mM EDTA (10X TBE) at pH 8.4 was from BioRad. The separation medium was loaded into Well 3 and a vacuum was applied to other wells to fill all channels. Wells 1, 2, 3 and 4 were then filled with 2.5μL of the sieving matrix, while Well 7 was loaded with 2.5μL of 10 ng/μL of Precision Molecular Mass Ruler from BioRad (Hercules, CA) for DNA sizing. The injection of amplicons through the gel valve was performed by applying a voltage (300 V) at Well 4 while Wells 7 and 8 were grounded. The injection time was 40 s to ensure that all sizes of amplicon passed through the gel valve and the injection intersection. Separation was achieved by applying a voltage of 1100 V across Wells 1 and 3.

The instrument to implement injections and separations is shown in Figure 7. It accepts a device having a microplate layout. It includes high-voltage power supplies and a laser-induced fluores-

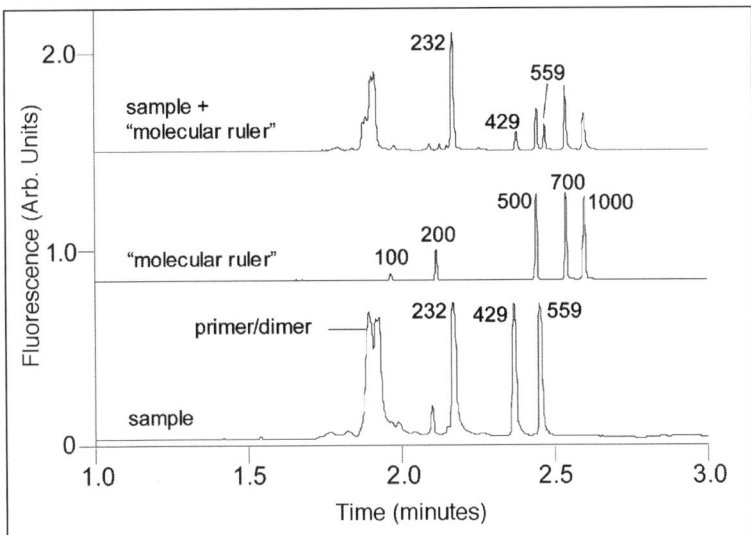

Figure 8. Electropherograms of amplicons (*bottom*), DNA sizing ladder (*middle*) and the mixture formed by simultaneously injecting amplicons and the ladder (*top*). The separation medium consisted of 1.5% HPC, 0.4%HEC, 4 µM TO in 1X TBE buffer, pH 8.4. The field strength was 193 V/cm and the effective separation distance, from the injection point to the detection point, was 4 cm. (Adapted from Koh et al.[38])

cence detector using an argon ion laser (JDS Uniphase, San Jose, CA). The fluorescence signal was detected by a photomultiplier tube (Hamamatsu, Bridgewater, NJ). A computer with a LabView (National Instrument) program was used for instrument control and data acquisition.

Figure 8 shows the electropherograms obtained for amplicons from *E. coli* and *salmonella*; all expected amplicons were produced. The DNA sizing ladder (molecular ruler) was used as a calibration standard to accurately identify the amplicons according to their sizes. Negative PCR results were observed for both amplicons at 429 bp and 559 bp when the sample contained only *E. coli* O157 (data not shown).

Concluding Remarks

Genetic detection and identification of two model bacteria, *Escherichia coli* O157 and *Salmonella typhimurium*, have been demonstrated using an integrated microfluidic device. Integration included the components for PCR, microfluidic valving, sample loading/injection, on-column dye labeling and CE-based separation. Screen-printed heaters and CE driving electrodes were also integrated in the monolithic, molded, poly(cyclic olefin) device. A number of bacteria can be simultaneously detected if unique DNA primer sets are appropriately designed. This work suggests the possible development of this technology for such applications as the diagnosis of bacterial infections, the detection of food-borne pathogens and monitoring public facilities for the presence of biological warfare agents.

Acknowledgements

We thank our former colleagues at ACLARA BioSciences for their contributions to the work described here, particularly W. Tan, C.-G. Koh, Y.-J. Juang and M. Zhao. The work of Tor Björnson in the development of the reader instrument is gratefully acknowledged as well. A part of the work in this chapter was supported by the Defense Advanced Research Projects Agency (DARPA). ZHF thanks Army Research Office (LS-48461), National Science Foundation (CHE-0515711) and National Aeronautics and Space Administration (NASA) for their support for his current research.

Experimental work cited in this chapter was originally conducted at ACLARA BioSciences, Inc.

References

1. Manz A, Graber N, Widmer HM. Miniaturized Total Chemical Analysis System: a Novel Concept for Chemical Sensing. Sensors and Actuators B 1990; 1:244-248.
2. Thorsen T, Maerkl SJ, Quake SR. Microfluidic large-scale integration. Science 2002; 298(5593):580-584.
3. McDonald JC, Whitesides GM. Poly(dimethylsiloxane) as a material for fabricating microfluidic devices. Acc Chem Res 2002; 35(7):491-499.
4. Terry SC, Jerman JH, Angell JB. A gas chromatography air analyzer fabricated on a silicon wafer. IEEE Trans Electron Devices 1979; ED-26:1880-1886.
5. Harrison DJ, Glavina PG, Manz A. Towards Miniaturized Electrophoresis and Chemical-Analysis Systems on Silicon—an Alternative to Chemical Sensors. Sensors and Actuators B-Chemical 1993; 10(2):107-116.
6. Lu CJ, Steinecker WH, Tian WC et al. First-generation hybrid MEMS gas chromatograph. Lab Chip 2005; 5(10):1123-1131.
7. Fan ZH, Harrison DJ. Micromachining of Capillary Electrophoresis Injectors and Separators on Glass Chips and Evaluation of Flow at Capillary Intersections. Anal Chem 1994; 66(1):177-184.
8. Jacobson SC, Hergenroder R, Koutny LB et al. Effects of Injection Schemes and Column Geometry on the Performance of Microchip Electrophoresis Devices. Anal Chem 1994; 66(7):1107-1113.
9. Aborn JH, El-Difrawy SA, Novotny M et al. A 768-lane microfabricated system for high-throughput DNA sequencing. Lab Chip 2005; 5(6):669-674.
10. Soper SA, Ford SM, Qi S et al. Polymeric microelectromechanical systems. Anal Chem 2000; 72(19):642A-651A.
11. Boone TD, Fan ZH, Hooper HH et al. Plastic advances microfluidic devices. Anal Chem 2002; 74(3):78A-86A.
12. Beebe DJ, Moore JS, Yu Q et al. Microfluidic tectonics: A comprehensive construction platform for microfluidic systems. Proc Natl Acad Sci USA 2000; 97(25):13488-13493.
13. Hong JW, Studer V, Hang G et al. A nanoliter-scale nucleic acid processor with parallel architecture. Nature Biotechnology 2004; 22(4):435-439.
14. Manz A, Harrison DJ, Verpoorte E et al. Planar Chips Technology for Miniaturization of Separation Systems—a Developing Perspective in Chemical Monitoring. Advances in Chromatography 1993; 33:1-66.
15. Su XL, Li Y. Quantum dot biolabeling coupled with immunomagnetic separation for detection of Escherichia coli O157:H7. Anal Chem 2004; 76(16):4806-4810.
16. Stokes DL, Griffin GD, Vo-Dinh T. Detection of E coli using a microfluidics-based antibody biochip detection system. Fresenius J Anal Chem 2001; 369(3-4):295-301.
17. Taitt CR, Anderson GP, Lingerfelt BM et al. Nine-analyte detection using an array-based biosensor. Anal Chem 2002; 74(23):6114-6120.
18. Sapsford KE, Shubin YS, Delehanty JB et al. Fluorescence-based array biosensors for detection of biohazards. J Appl Microbiol 2004; 96(1):47-58.
19. McClain MA, Culbertson CT, Jacobson SC et al. Flow cytometry of Escherichia coli on nucrifluidic devices. Anal Chem 2001; 73(21):5334-5338.
20. Tao L, Yu X, Snyder AP et al. Bacterial identification by protein mass mapping combined with an experimentally derived protein mass database. Anal Chem 2004; 76(22):6609-6617.
21. Bradley KA, Mogridge J, Mourez M et al. Identification of the cellular receptor for anthrax toxin. Nature 2001; 414(6860):225-229.
22. Belgrader P, Young S, Yuan B et al. A battery-powered notebook thermal cycler for rapid multiplex real-time PCR analysis. Anal Chem 2001; 73(2):286-289.
23. Riffon R, Sayasith K, Khalil H et al. Development of a rapid and sensitive test for identification of major pathogens in bovine mastitis by PCR. J Clin Microbiol 2001; 39(7):2584-2589.
24. Greisen K, Loeffelholz M, Purohit A et al. PCR primers and probes for the 16S rRNA gene of most species of pathogenic bacteria, including bacteria found in cerebrospinal fluid. J Clin Microbiol 1994; 32(2):335-351.
25. Soumet C, Ermel G, Rose V et al. Identification by a multiplex PCR-based assay of Salmonella typhimurium and Salmonella enteritidis strains from environmental swabs of poultry houses. Lett Appl Microbiol 1999; 29(1):1-6.
26. Findlay JB, Atwood SM, Bergmeyer L et al. Automated closed-vessel system for in vitro diagnostics based on polymerase chain reaction. Clin Chem 1993; 39(9):1927-1933.
27. Wilding P, Shoffner MA, Kricka LJ. PCR in a silicon microstructure. Clin Chem 1994; 40(9):1815-1818.

28. Shoffner MA, Cheng J, Hvichia GE et al. Chip PCR I Surface passivation of microfabricated silicon-glass chips for PCR. Nucleic Acids Res 1996; 24(2):375-379.

29. Northrup MA, Benett B, Hadley D et al. A miniature analytical instrument for nucleic acids based on micromachined silicon reaction chambers. Anal Chem 1998; 70(5):918-922.

30. Kopp MU, de Mello AJ, Manz A. Chemical amplification: Continuous-flow PCR on a chip. Science 1998; 280(5366):1046-1048.

31. Krishnan M, Ugaz VM, Burns MA. PCR in a Rayleigh-Benard convection cell. Science 2002;298(5594):793-793.

32. Woolley AT, Hadley D, Landre P et al. Functional integration of PCR amplification and capillary electrophoresis in a microfabricated DNA analysis device. Anal Chem 1996; 68(23):4081-4086.

33. Lagally ET, Emrich CA, Mathies RA. Fully integrated PCR-capillary electrophoresis microsystem for DNA analysis. Lab Chip 2001; 1(2):102-107.

34. Lagally ET, Medintz I, Mathies RA. Single-molecule DNA amplification and analysis in an integrated microfluidic device. Anal Chem 2001; 73(3):565-570.

35. Khandurina J, McKnight TE, Jacobson SC et al. Integrated system for rapid PCR-based DNA analysis in microfluidic devices. Anal Chem 2000; 72(13):2995-3000.

36. Waters LC, Jacobson SC, Kroutchinina N et al. Multiple sample PCR amplification and electrophoretic analysis on a microchip. Anal Chem 1998; 70(24):5172-5176.

37. Burns MA, Johnson BN, Brahmasandra SN et al. An integrated nanoliter DNA analysis device. Science 1998; 282(5388):484-487.

38. Koh CG, Tan W, Zhao MQ et al. Integrating polymerase chain reaction, valving and electrophoresis in a plastic device for bacterial detection. Anal Chem 2003; 75(17):4591-4598.

39. Liu Y, Rauch CB, Stevens RL et al. DNA amplification and hybridization assays in integrated plastic monolithic devices. Anal Chem 2002; 74(13):3063-3070.

40. Trau D, Lee TM, Lao AI et al. Genotyping on a complementary metal oxide semiconductor silicon polymerase chain reaction chip with integrated DNA microarray. Anal Chem 2002; 74(13):3168-3173.

41. Roper MG, Easley CJ, Landers JP. Advances in polymerase chain reaction on microfluidic chips. Anal Chem 2005; 77(12):3887-3893.

42. Johnson TJ, Ross D, Gaitan M et al. Laser modification of preformed polymer microchannels: Application to reduce band broadening around turns subject to electrokinetic flow. Anal Chem 2001; 73(15):3656-3661.

43. Kameoka J, Craighead HG, Zhang HW et al. A polymeric microfluidic chip for CE/MS determination of small molecules. Anal Chem 2001; 73(9):1935-1941.

44. Lee DS, Yang H, Chung KH et al. Wafer-scale fabrication of polymer-based microdevices via injection molding and photolithographic micropatterning protocols. Anal Chem 2005; 77(16):5414-5420.

45. Mela P, van den Berg A, Fintschenko Y et al. The zeta potential of cyclo-olefin polymer microchannels and its effects on insulative (electrodeless) dielectrophoresis particle trapping devices. Electrophoresis 2005; 26(9):1792-1799.

46. Yang Y, Li C, Kameoka J et al. A polymeric microchip with integrated tips and in situ polymerized monolith for electrospray mass spectrometry. Lab Chip 2005; 5(8):869-876.

47. Kim DS, Lee SH, Kwon TH et al. A serpentine laminating micromixer combining splitting/recombination and advection. Lab Chip. 2005; 5(7):739-747.

48. Anderson RC, Bogdan GJ, Puski A et al. Genetic Analysis Systems: Improvements and Methods. Solid-State Sensor and Actuator Workshop. Hilton Head Island, SC: Transducer Research Foundation, 1998:7-10.

49. Beebe DJ, Moore JS, Bauer JM et al. Functional hydrogel structures for autonomous flow control inside microfluidic channels. Nature 2000; 404(6778):588-590.

50. Zhan W, Seong GH, Crooks RM. Hydrogel-based microreactors as a functional component of microfluidic systems. Anal Chem 2002; 74(18):4647-4652.

51. Oda RP, Strausbauch MA, Huhmer AFR et al. Infrared-mediated thermocycling for ultrafast polymerase chain reaction amplification of DNA. Anal Chem 1998; 70(20):4361-4368.

52. Ball JC, Scott DL, Lumpp JK et al. Electrochemistry in nanovials fabricated by combining screen printing and laser micromachining. Anal Chem 2000; 72(3):497-501.

53. Wang J, Tian BM, Nascimento VB et al. Performance of screen-printed carbon electrodes fabricated from different carbon inks. Electrochimica Acta. 1998; 43(23):3459-3465.

54. Zhao M, Ricco AJ, Nguyen U et al. Functional and Efficient Electrode-integrated Microfluidic Plastic Devices. In:Ramsey JM, van den Berg A, eds. Micro Total Analysis Systems. Netherland: Kluwer Academic Publishers, 2001:193-194.

55. Wittwer CT, Garling DJ. Rapid cycle DNA amplification: time and temperature optimization. Biotechniques 1991; 10(1):76-83.

56. Wittwer CT, Herrmann MG, Gundry CN et al. Real-time multiplex PCR assays. Methods 2001; 25(4):430-442.

CHAPTER 7

PCR in Integrated Microfluidic Systems

Victor M. Ugaz*

Abstract

Miniaturized integrated DNA analysis systems offer the potential to provide unprecedented advances in cost and speed relative to current benchtop-scale instrumentation by allowing rapid bioanalysis assays to be performed in a portable self contained device format that can be inexpensively mass-produced. The polymerase chain reaction (PCR) has been a natural focus of many of these miniaturization efforts, owing to its capability to efficiently replicate target regions of interest from small quantities template DNA. Scale-down of PCR has proven to be particularly challenging, however, due to an unfavorable combination of relatively severe temperature extremes (resulting in the need to repeatedly heat minute aqueous sample volumes to temperatures in the vicinity of 95°C with minimal evaporation) and high surface area to volume conditions imposed by nanoliter reactor geometries (often leading to inhibition of the reaction by nonspecific adsorption of reagents at the reactor walls). Despite these daunting challenges, considerable progress has been made in the development of microfluidic devices capable of performing increasingly sophisticated PCR-based bioassays. This chapter reviews the progress that has been made to date and assesses the outlook for future advances.

Introduction

Advances in genomic analysis technology continue to be made at a rapid pace and have contributed to the development of new instrumentation that is paving the way for high-throughput low-cost DNA assays to become commonplace. These technologies have the potential to impact an unprecedented array of fields including medical diagnostics, forensics, biosensing and genome-wide analysis.[1-4] Many of these methodologies rely on the ability to replicate selected sub-regions within a larger DNA template. The polymerase chain reaction (PCR) offers a straightforward and highly efficient means to perform this replication, thereby making it one of molecular biology's key enabling technologies.

The PCR process involves repeatedly cycling a reagent mixture containing template DNA, primers, dNTPs, a thermostable polymerase enzyme and other buffering additives, through thermal conditions corresponding to (1) denaturation of the double-stranded template ($\sim 95°C$), (2) annealing of single-stranded oligonucleotide primers at complementary locations flanking the target region (~ 50-$65°C$) and (3) enzyme directed synthesis of the complementary strand ($\sim 72°C$). The number of target DNA copies increases exponentially as this cycling process is repeated, doubling with each cycle under ideal conditions. Although the kinetics associated with each step in the PCR process are rapid when considered individually,[5,6] a typical 30-40 cycle replication still requires timescales of order 1-2 hours to complete. These prolonged reaction times are largely a reflection of the highly inefficient design of many conventional thermocycling instruments, where thermally massive hardware components (e.g., metal thermal blocks) and low-conductivity plastic reaction vessels (e.g., polypropylene tubes) combine to produce an unfavorable coupling between

*Victor M. Ugaz—Artie McFerrin Department of Chemical Engineering, Texas A&M University, College Station, TX, 77843-3122, U.S.A. Email: ugaz@tamu.edu

Integrated Biochips for DNA Analysis, edited by Robin Hui Liu and Abraham P. Lee.
©2007 Landes Bioscience and Springer Science+Business Media.

increased thermal energy requirements associated with rapid heating and cooling and the necessity to hold the temperature constant at each step for a sufficiently long time to ensure that the entire reagent volume attains thermal equilibrium. Progress has been made in the development of faster benchtop-scale thermocyclers, most notably by approaches that involve replacing plastic reaction tubes with thin glass micro-capillaries and using techniques ranging from forced air circulation to infrared heating for temperature control.[7-14] These configurations allow the reaction mixture to be distributed over a greater surface area for more efficient heat transfer, although overall throughput can still be limited by the capillary loading, sealing and unloading processes.

PCR in Microfluidic Systems

In addition to these benchtop-scale approaches, increasing interest has been focused on performing PCR in miniaturized systems where ultra-small reaction volumes (typically in the nL range) not only greatly reduce reagent consumption, but can also be rapidly heated and cooled while simultaneously lowering costs due to greatly reduced reagent consumption. Moreover, the use of photolithographic microfabrication offers the potential to produce hundreds or thousands of devices at once, bringing hardware costs to a level of $1 or less. These highly desirable characteristics have stimulated considerable interest in the area of microfabricated thermocycling systems. However the use of such small reaction volumes can pose significant challenges including adverse effects associated with evaporation and nonspecific absorption to the microchannel walls under high surface to volume conditions. Progress continues to be made toward addressing the challenges facing development of microfluidic thermocycling systems and the interested reader is referred to several recent reviews for more details.[6,15-24]

Despite its importance as a key enabling technology, PCR is only one step in a broader sequence of processes that comprise a complete molecular biology assay. Some combination of sample isolation and collection, pre and postreaction purification, subsequent biochemical reactions and product detection and analysis must also be performed. On the macroscale, these steps are typically carried out in a conventional laboratory setting, often requiring dedicated instruments and personnel to be employed at each stage along with the need to repeatedly prepare and dispense precise sample and reagent mixtures. These processes can not only be tedious and time consuming, they also introduce multiple opportunities for measurement errors and sample contamination. The ability to adapt a series of these operations to a miniaturized 'lab-on-a-chip' format has the potential to address these limitations, yielding significant reductions in analysis time and cost. Moreover, a fully self-contained portable design offers the potential to minimize tedious manual fluidic manipulations and sample contamination issues.

This enhanced functionality, however, is accompanied by a corresponding increase in the level of design and operational complexity. A myriad of challenges must be addressed, including ensuring biochemical compatibility among successive steps in the analysis process, careful thermal design to ensure that the relatively high temperatures achieved during PCR thermocycling do not negatively impact activity on other parts of the chip, the ability to accurately dispense and transport nanoliter liquid volumes within a microchannel network and ensuring that reagents can be properly sealed to prevent evaporation during thermocycling. In this chapter, we review some of the remarkable progress that has been made toward developing sophisticated miniaturized devices that incorporate PCR as part of an integrated molecular biological assay system. The review focuses on developments reported in refereed journals, with the understanding that additional studies may be documented in conference proceedings and patent literature.

Integrated PCR and Gel Electrophoresis

Considerable progress has been made in the development of microdevices capable of combining PCR with postreaction product analysis performed by electrophoretic separation of the amplified DNA fragments (Table 1). The Mathies group was the first to demonstrate this approach using a hybrid design consisting of a 20 μL microfabricated silicon PCR reactor bonded on top of a glass electrophoresis microchip.[25] Unfavorable surface interactions at the reactor walls that inhibited PCR amplification were avoided through the use of disposable polypropylene liners, allowing targets

Table 1. Summary of microfluidic systems demonstrating integration of PCR with gel electrophoresis

Reference	Integrated Functions	Device Fabrication	Fluid Handling	Temperature Control	Templates	Targets
Wooley et al (1996) [25]	PCR, gel electrophoresis	Microfabricated Si reactor with polypropylene liner bonded to glass	Electrokinetic	Integrated polysilicon heaters	Plasmid DNA	268 bp
					Bacterial genomic (*Salmonella*)	159 bp
Lagally et al (2000) [26]	PCR, gel electrophoresis	Microfabricated chip (glass/glass)	Pneumatic valve/ vent manifold	Thin film heaters and temperature sensors affixed to chip	136 bp fragment	136 bp
Lagally et al (2001) [27]	PCR, gel electrophoresis	Microfabricated chip (glass/glass)	Pneumatic valve/ vent manifold	Thin film heaters and temperature sensors affixed to chip	136, 231 bp fragments, 2686 bp cloning vector	136, 231 bp
Lagally et al (2001) [28]	PCR, gel electrophoresis	Microfabricated chip (glass/glass)	Pneumatic valve/ vent manifold	Integrated micro-fabricated heaters and temperature sensors	Human genomic	157, 200 bp
Legally et al (2004) [29]	PCR, gel electrophoresis	Microfabricated chip (glass/glass)	Pneumatic valve/ vent manifold	Integrated micro-fabricated heaters and temperature sensors	Bacterial genomic (*E. coli*)	348, 625 bp
					Bacterial genomic (*S. aureus*)	219, 310 bp
Waters et al (1998) [32]	PCR, gel electrophoresis	Microfabricated chip (glass/glass)	Electrokinetic	Benchtop thermocycler	Lambda DNA	500 bp
					Bacterial genomic (*E. coli*)	154, 264, 346 bp
					Plasmid DNA	410 bp

Table 1. Continued

Reference	Integrated Functions	Device Fabrication	Fluid Handling	Temperature Control	Templates	Targets
Waters et al (1998) [31]	PCR, gel electrophoresis	Microfabricated chip (glass/glass)	Electrokinetic	Benchtop thermocycler	Lambda DNA	199, 500 bp
					Bacterial genomic (E. coli)	346 bp
Dunn et al (2000) [30]	PCR, gel electrophoresis	Microfabricated chip (glass/glass)	Electrokinetic	Benchtop thermocycler	Plasmid DNA Mouse genomic	410 bp 107,114, 121, 149 bp
Khandurina et al (2000) [33]	PCR, gel electrophoresis	Microfabricated chip (glass/glass)	Electrokinetic	Thermoelectric elements affixed to chip	Lambda DNA	199 bp
Zhou et al (2004) [34]	PCR, gel electrophoresis	Microfabricated chip (glass/glass)	Electrokinetic	Thermoelectric elements affixed to chip	Reverse transcribed cDNA from SARS-CoV	240, 438 bp
Hong et al (2001) [35]	PCR, gel electrophoresis	Microfabricated chip (PDMS/glass)	Electrokinetic	Thermoelectric elements affixed to chip	Lambda DNA	500 bp
Rodriguez et al (2003) [36]	PCR, gel electrophoresis	Microfabricated chip (Si/glass)	Electrokinetic	Integrated micro-fabricated heaters and temperature sensors	Plasmid DNA	350 bp
Koh et al (2003) [37]	PCR, gel electrophoresis	Molded poly(cyclic olefin) chip	Electrokinetic, with polyacrylamide gel sealing valves	Integrated micro-fabricated heaters and temperature sensors	Bacterial genomic (E. coli, Salmonella)	232, 429, 559 bp
Ferrance et al (2003) [38]	PCR, gel electrophoresis	Microfabricated chip (glass/glass)	Electrokinetic	Non-contact infrared heating	Human genomic	380 bp
Easley et al (2000) [39]	PCR, gel electrophoresis	Multilayer microfabricated chip (PDMS/glass)	On-chip pneumatic PDMS valves	Non-contact infrared heating	Bacterial genomic (Salmonella)	278 bp

from both plasmid and bacterial genomic DNA templates to be successfully amplified. Following PCR, the products were electrokinetically injected into a separation channel filled with a sieving gel matrix containing a fluorescent intercalating dye and detected using a laser excited confocal imaging system. Total analysis times ranging from 20 to 45 minutes were reported.

Subsequent refinements to this basic design involved integrating both the PCR reactor and electrophoresis column into a single continuous glass microchannel network. In one adaptation reagents were loaded and sealed inside a 280 nL reaction chamber using a valve/vent manifold mounted on top of the chip and temperature control was provided by thin film heater and thermocouple elements affixed to the back side.[26] Upon completion of the reaction, the manifold was removed and platinum electrode wires were inserted into the access holes so that the products could be electrokinetically transported to the electrophoresis channel for analysis. This design was used to demonstrate amplification of targets from cloning vector and genomic control templates with sufficient sensitivity to permit single molecule detection.[27] A further improvement to the design involved integration of microfabricated heaters and temperature sensors to actuate thermocycling within a 200 nL reaction chamber.[28] This device was used to perform multiplex sex determination from human genomic DNA in under 15 minutes. A self-contained portable version was also constructed and used to perform multiplex PCR directly from whole bacterial cells in 30 minutes, with product detection accomplished using fluorescently labeled primers (Fig. 1A).[29]

Another notable family of glass microchips for performing integrated PCR and gel electrophoresis has been developed in the Ramsey group. Here, the basic design consisted of a 10-20 μL reaction reservoir fabricated by drilling a hole in one of the glass substrates. After loading, the PCR reagents were covered with mineral oil or wax to prevent evaporation and thermocycling was performed by placing the chip inside a conventional thermocycler.[30-32] After the reaction was completed, an intercalating dye was added to the sample reservoir and electrophoretic product analysis was performed with laser-induced fluorescence detection. Subsequent improvements included the use of a specially designed thermoelectric fixture with dual thermoelectric heating elements and incorporation of a porous membrane structure for injection of reaction products into the electrophoresis gel.[33] Successful amplification of targets from a variety of templates ranging from lambda DNA to bacterial and mouse genomic DNA have been reported using these devices, with total analysis times of 20 minutes or less.

Zhou and coworkers also demonstrated a glass microchip design integrating PCR and gel electrophoresis with an external thermoelectric thermocycling apparatus to perform a duplex PCR analysis of 240 and 438 base pair (bp) targets associated with the SARS coronavirus.[34] Other examples of integration include designs employing hybrid PDMS/glass microchips[35] and designs interfacing a silicon PCR microchip with integrated temperature control to a glass electrophoresis microchip.[36] A particularly novel design reported by Koh et al consisted of a plastic microfluidic chip with integrated temperature control that incorporated on-chip photopolymerized gel valves that not only sealed reagents inside the PCR reactor but also allowed the products to be electrokinetically extracted through the valve material and directed into an electrophoresis channel for separation and detection.[37] The device was used for successful analysis of targets from two different bacterial genomic templates with a detection limit on the order of six DNA copies.

Recent work by the Landers group has yielded further reductions in analysis time through the use of an innovative noncontact infrared heating technique to actuate thermocycling. This concept was successfully employed in an integrated device capable of amplifying a 380 bp β-globin target in a 600 nL reactor followed by electrokinetic injection and electrophoretic separation.[38] In addition, a separate microchip was used to perform solid phase extraction of DNA from whole blood prior to PCR. Subsequent work has resulted in further optimizations to this design by a reduction in reactor volume to 280 nL and through development of a fluid handling system based on a multilayer hybrid glass/PDMS assembly that provides an addressable array of pneumatic valves capable of performing sealing, pumping and injection into the electrophoresis gel matrix (Fig. 1B).[39] This design offers an impressive capacity for analysis speed, as demonstrated by the ability to perform integrated PCR and electrophoresis of bacterial targets with total analysis times of about 12 minutes.

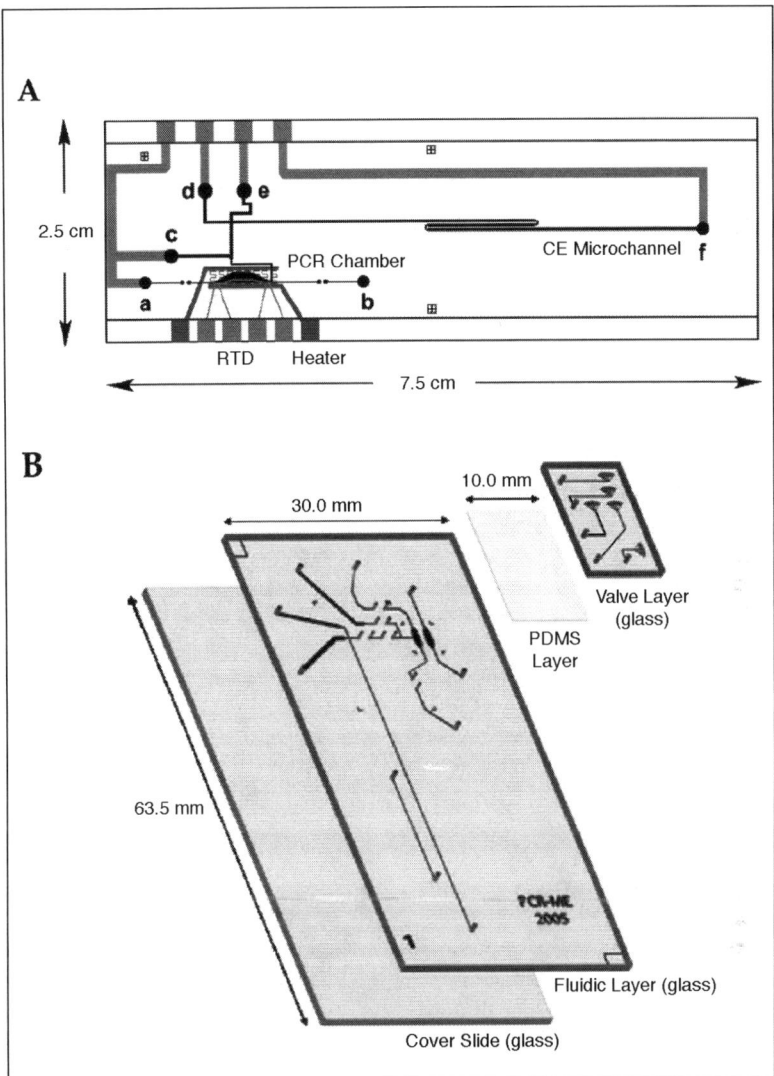

Figure 1. Microdevice designs capable of performing integrating sequential PCR and gel electrophoresis. A) Schematic layout of a device incorporating an etched glass PCR reactor and electrophoresis channel network (indicated in black) bonded to a second glass wafer patterned with electrodes and temperature sensors (indicated in green) along with a heating element (indicated in red)[29] (reproduced with permission, copyright 2004 American Chemical Society). B) Four-layer microdevice design incorporating integrated pneumatically actuated PDMS valves for liquid handling.[39] Thermocycling is actuated using a noncontact infrared heating process (reproduced with permission, copyright 2006 Royal Society of Chemistry).

Integrated PCR and Sample Purification

Advances have also been made in the design of microdevices integrating PCR with other upstream and downstream sample handling and analysis steps (Table 2). The Wilding group has developed a series of microdevices incorporating micromachined 'weir-type' filtration structures

Table 2. Summary of microfluidic systems demonstrating integration of PCR with sample purification

Reference	Integrated Functions	Device Fabrication	Fluid Handling	Temperature Control	Templates	Targets
Wilding et al (1998) [40]	Cell filtration, PCR	Microfabricated chip (Si/glass)	External syringe	Thermoelectric elements affixed to chip	Human genomic	202 bp
Yuen et al (2001) [41]	Cell filtration, PCR	CNC machined Plexiglas chip	External syringe	Thermoelectric elements affixed to chip	Human genomic	226 bp
Panaro et al (2005) [42]	Cell filtration, PCR	Microfabricated chip (Si/glass)	External manifold syringe pump	Thermoelectric elements affixed to chip	Human genomic	379 bp
Lee et al (2005) [43]	Cell lysis, PCR	Microfabricated chip (PDMS/glass)	Electroosmotic pumping	Integrated micro-fabricated heaters and temperature sensors	Bacterial genomic (*Streptococcus pneumoniae*)	273 bp
Cady et al (2004) [44]	Purification, real-time PCR	Microfabricated chip (PDMS/Si)	Syringe pump	Thermoelectric elements affixed to chip	Bacterial genomic (*Listeria mono-cytogenes*)	544 bp

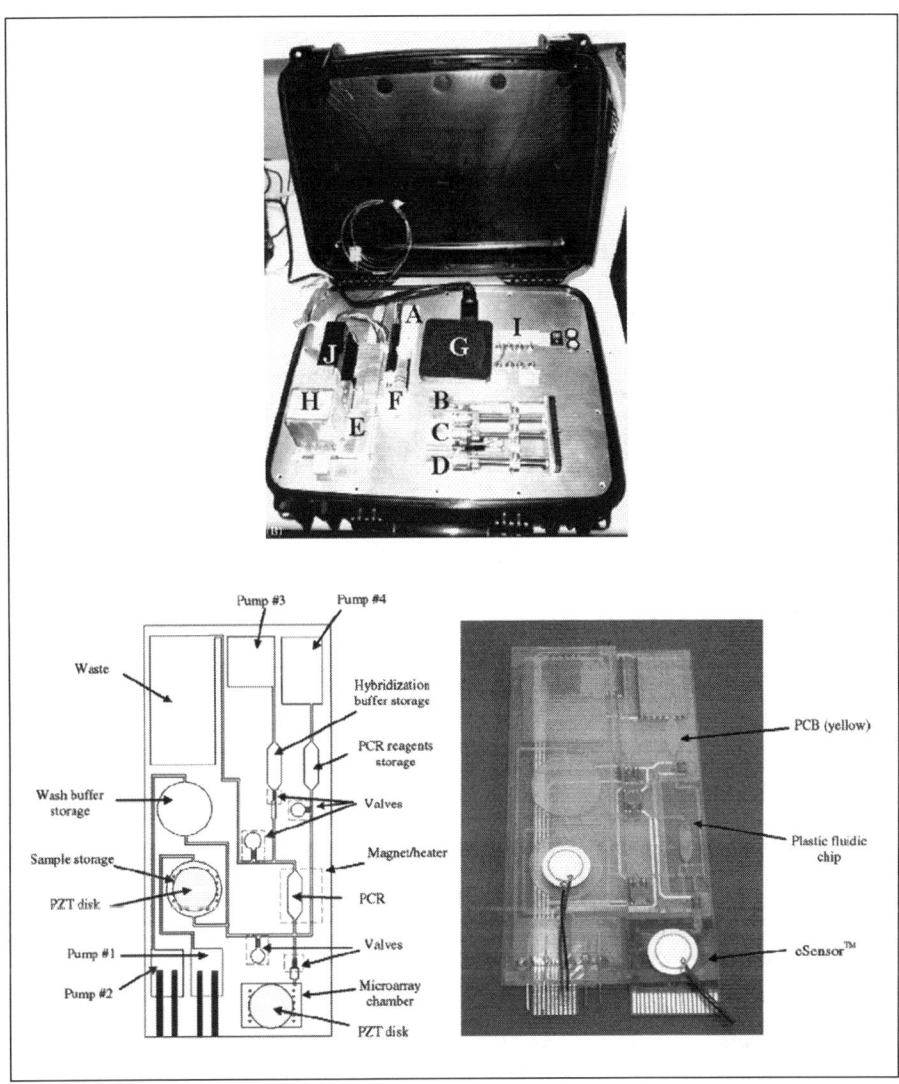

Figure 2. A) Microchip-based bacterial detection system integrating sample purification and real time PCR.[44] The system includes integrated syringe pumps (A-D), Moog micro valve (F), cooling fan (G), LED-based fluorescence excitation/detection system (H—dotted outline) with PMT detector (J), power toggle switches (I). The microfluidic purification/detection chip (E) is inserted into the unit directly above the thermoelectric heater cooler. The entire unit measures 36 cm × 28 cm × 15 cm (reproduced with permission, copyright 2004 Elsevier B.V). b) Self-contained bioanalysis device incorporating the ability to perform sequential sample preparation, PCR amplification and DNA microarray hybridization[46] (reproduced with permission, copyright 2004 American Chemical Society).

capable of isolating white blood cells from whole blood samples in the 1-10 µL range. Filtration occurred as samples were pumped through the chamber where the larger white blood cells were selectively retained by the filter, after which a PCR reagent mixture was injected and thermocycling was performed using a thermoelectric apparatus.[40,41] Both hybrid glass/silicon and Plexiglas substrate materials have been used to construct the devices, and the potential for further integration has been demonstrated by performing product sizing analysis using separate electrophoresis microchips.[42] An integrated microdevice capable of performing cell lysis followed by PCR was reported by Lee et al.[43] The design consisted of a hybrid glass/PDMS configuration with PDMS thermal lysis and PCR chambers interconnected by a glass microchannel network in which the lysis products were mixed with PCR reagents and electroosmotically transported to the PCR reactor. The device was used to successfully amplify a 273 bp target from whole bacterial cells. Finally, a portable pathogen detection system integrating DNA purification and real-time PCR detection has been developed by Cady and coworkers (Fig. 2A).[44] Here, samples subjected to off-chip chemical lysis were pumped through a flow network containing arrays of etched silicon pillars that allowed purification to be performed via sequential binding and washing steps. The purified samples were then pumped into a PDMS PCR reactor for real-time amplification using a SYBR green fluorescence chemistry. The ability to detect between 10^4 and 10^7 bacterial cells in about 45 minutes was demonstrated in characterization studies involving amplification of a 544 bp target.

Integrated PCR and Hybridization

Several groups have also explored integration of microchip PCR with DNA hybridization analysis (Table 3). Anderson and coworkers demonstrated this concept using a device containing a credit card sized polycarbonate fluidic cartridge interfaced with an Affymetrix GeneChip microarray.[45] The integrated system was capable of performing sequential reverse transcription, PCR, enzymatic reactions and hybridization. Further progress in microarray integration was reported by Liu et al who presented a fully self-contained biochip device that incorporated the ability to perform sample preparation (cell capture, concentration, purification and lysis), PCR and hybridization using an integrated Motorola eSensor microarray (Fig. 2B).[46] This device was used to perform pathogenic bacteria detection from whole rabbit blood and single-nucleotide polymorphism analysis from diluted human blood samples. Another example of integration with hybridization involved employing an array of hybrid glass/silicon microreactors with integrated on-chip heaters and temperature sensors.[47] The bottom surface of each microreactor was patterned with hybridization oligonucleotides allowing reaction products to be detected with a confocal fluorescence scanner. Finally, Liu et al demonstrated a polycarbonate microdevice where PCR and hybridization are performed in separate fluidically interconnected reactors.[48] Here, reagents were sealed in the PCR chamber during thermocycling by employing phase change valves based on Pluronic F127, a block copolymer that liquefies at low temperature ($\sim 5\,^{\circ}$C) but becomes a solidified gel at higher temperatures. Operation of the device was demonstrated by using it to perform a bacterial detection assay.

Further Advancements in Integration

Increases in device complexity that accompanies simultaneous miniaturization and integration of multiple sample processing and analysis steps can pose daunting challenges, however progress is steadily being made toward addressing many of these issues resulting in the development of increasingly sophisticated designs (Table 4). An early compelling example illustrating the power of miniaturized genomic analysis systems was the hybrid glass-silicon design developed in the Burns group,[49] capable of performing a series of liquid metering, thermal reaction and analysis operations. This pioneering device was used to amplify a 106 bp DNA target from a bacterial genomic template via an isothermal strand displacement amplification (SDA) process followed by gel electrophoresis with integrated photodetection of the fluorescently labeled reaction products. A greatly improved version of this design was recently reported that is capable of performing a complete genotyping assay involving two sequential reactions followed by an electrophoretic separation in an ultra-compact 1.5 × 1.6 cm hybrid glass/silicon microfluidic chip (Fig. 3A).[50] Sample DNA and PCR reagents were loaded into the chip and pneumatically propelled into a reaction zone where they were sealed using paraffin phase change valves. Thermocycling was actuated using an array of integrated resistive heaters and temperature sensors, with the reaction zone thermally

Table 3. Summary of microfluidic systems demonstrating integration of PCR with DNA hybridization

Reference	Integrated Functions	Device Fabrication	Fluid Handling	Temperature Control	Templates	Targets
Anderson et al (2000) [45]	Purification, PCR, microarray hybridization	CNC machined polycarbonate cartridge	On-chip pneumatic silicone valves	Thermoelectric elements affixed to chip	Serum samples with HIV virus	1.6 kb
Liu et al (2004) [46]	Purification, cell lysis, PCR, microarray hybridization	CNC machined polycarbonate cartridge	On chip electro-chemical pumping, acoustic micro-mixing, paraffin phase change microvalves, and magnetic bead capture	Thermoelectric elements affixed to chip	Bacterial genomic (*E. coli*) in rabbit blood sample	221 bp
Trau et al (2002) [47]	PCR, hybridization	Microfabricated chip (Si/glass)	N/A	Integrated microfabricated heaters and temperature sensors	Plant genomic	Various
Liu et al (2002) [48]	PCR, hybridization	Laser machined polycarbonate	Syringe pump, Pluronic phase change microvalves	Thermoelectric elements affixed to chip	Bacterial genomic (*E. coli, E. faecalis*)	195, 221 bp

Table 4. Summary of highly integrated microfluidic DNA analysis systems

Reference	Integrated Functions	Device Fabrication	Fluid Handling	Temperature Control	Templates	Targets
Burns et al (1998) [49]	SDA, electrophoresis, integrated photodetection	Microfabricated chip (Si/glass)	Pneumatic pumping, drop placement at hydrophobically patterned zones	Integrated micro-fabricated heaters and temperature sensors	Bacterial genomic (Mycobacterium tuberculosis)	106 bp
Pal et al (2005) [50]	PCR, restriction enzyme digestion, electrophoresis	Microfabricated chip (Si/glass)	Pneumatic pumping, drop placement at hydrophobically patterned zones, paraffin phase change microvalves	Integrated micro-fabricated heaters and temperature sensors	Reverse transcribed viral genomic (Influenza)	690 bp
Blazej et al (2006) [51]	Sanger sequencing, product purification, electrophoresis	Multilayer micro-fabricated chip (PDMS/glass)	Electrokinetic with on-chip pneumatic PDMS valves	Integrated micro-fabricated heaters and temperature sensors	Plasmid DNA	N/A

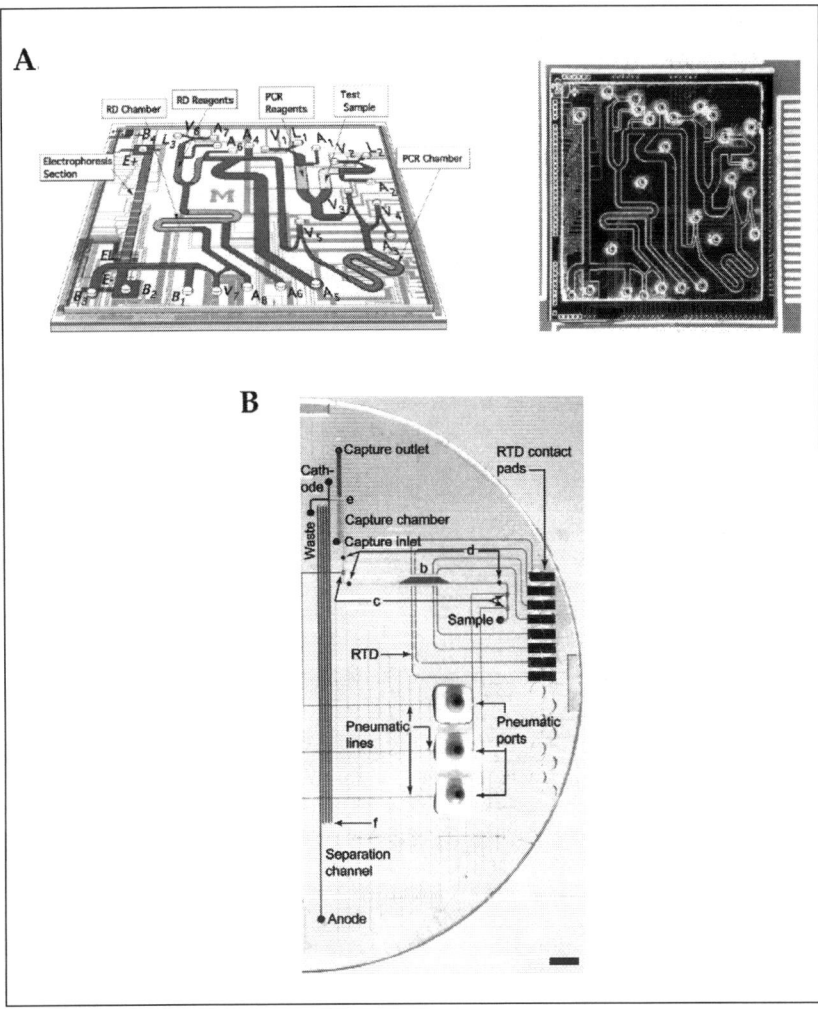

Figure 3. A) Schematic (left) and photograph (right) of a microfluidic device capable of performing a sequential PCR, restriction enzyme digestion and gel electrophoresis.[50] An etched glass microchannel network is bonded to a silicon chip containing integrated electrodes, heaters and temperature sensors. Fluid handling is accomplished using an array of addressable on-chip phase change valves. The chip measures 1.5 × 1.6 cm (reproduced with permission, copyright 2006 Royal Society of Chemistry). B) Layout of a microfluidic bioprocessor for performing Sanger cycle sequencing of DNA (scale bar = 5 mm).[51] Thermocycling reactors are interconnected with post reaction purification and gel electrophoresis components. A multilayer glass/PDMS design is used to incorporate an array of pneumatically actuated microvalves (reproduced with permission, copyright 2006 National Academy of Sciences, USA).

isolated from the rest of the chip. Upon completion of the PCR reaction, the sealing valves were opened and the sample was propelled forward, mixed with additional reagents and incubated to perform a restriction enzyme digestion reaction. Upon completion of the second reaction, the products were directed into a gel electrophoresis column for analysis. This system was used to perform a PCR-RFLP assay for influenza A virus detection, illustrating tremendous potential as a generic platform suitable for use in a wide range of genotyping applications. Another notable recent advancement has been reported by the Mathies group, where a multilayer glass/PDMS device was used to perform integrated Sanger cycle sequencing of DNA—a reaction requiring similar thermal cycling parameters as PCR (Fig. 3B).[51] Sequencing reactions were performed in 250 nL reactors, after which the products were electrokinetically transported into a novel capture gel for purification. Here, oligonucleotides complementary to the region immediately adjacent to one of the sequencing primers were immobilized in a sparsely crosslinked polyacrylamide gel plug, such that the sequencing products could be retained inside the gel while unincorporated primers and other reagents passed through. Upon heating the gel to a temperature above the primer melting point, the hybridized products were released and electrokinetically injected into the separation channel. Read lengths of 556 bases were achieved with a 99% base call accuracy, demonstrating tremendous potential for significant reductions in cost and time scales associated with genome sequencing.

Novel Micro-PCR Methods

Buoyancy driven natural convection phenomena have also been investigated as a means of accelerating the thermocycling process. By designing reaction chambers that harness an imposed static temperature gradient to generate a circulatory convective flow field, PCR reagents can be automatically transported through temperature zones associated with denaturing, annealing and extension conditions. Convective flow thermocyclers may be broadly classified as *cavity-based* or *loop-based* systems, depending on the nature of the flow field generated. Cavity-based designs typically consist of reactor geometries in which the PCR reagents are enclosed between upper and lower surfaces maintained at annealing and denaturing temperatures, respectively.[52-54] When the aqueous PCR reagent mixture is heated from below, an unstable "top heavy" arrangement is created which can provide sufficient driving force to establish a continuous circulatory flow in much the same fashion as in an ordinary lava lamp. Here, thermocycling parameters (e.g., cycling time, residence time within each temperature zone) are controlled by an interplay between reactor geometry (height to diameter aspect ratio) and the magnitude of the imposed temperature gradient. Krishnan et al successfully demonstrated this concept by amplifying a 295 bp β-actin target from a human genomic DNA template in a 35 μL cylindrical reactor (Fig. 4A).[52] In subsequent work, this simple design has been adapted to perform amplification of a 191 bp target associated with membrane channel proteins M1 and M2 of the influenza-A virus with cycling times ranging from 15 to 40 minutes in a multiwell cartridge format that offers potential for use in high throughput settings.[53,55] Convectively driven PCR of a 96 bp target from a bacterial genomic template has also been demonstrated by Braun et al in low aspect ratio cylindrical cavities using either focused infrared heating or a micro immersion heater inserted at the center of a low aspect ratio cavity to drive the flow (Fig. 4B).[56-58]

Successful PCR amplification has also been demonstrated in closed-loop convective flow systems. These designs are attractive because of their ability to generate unidirectional flows along a closed path enabling cycling parameters to be precisely controlled. Wheeler and coworkers designed a novel disposable polypropylene reactor in which opposite sides of a racetrack-shaped flow loop were maintained at 94 and 55°C respectively to amplify targets ranging from 58 to 160 bp from a bacterial genomic template (Fig. 4C),[59] while Chen et al employed a triangular arrangement with three independently controlled temperature zones maintained at 94, 55 and 72°C to amplify 305 and 700 bp targets from a bacterial genomic template (Fig. 4D).[60] Krishnan et al employed both triangular and racetrack shaped designs to amplify 191 and 297 bp targets from control and human genomic DNA targets respectively.[55] Most recently, Agrawal et al have demonstrated a triangular design in which flow loops are constructed using disposable plastic tubing (Fig. 4E).[61] Here, two opposing sides of the loop are maintained at denaturing and extension temperatures using independently controlled thermoelectric heaters while the third side passively attains annealing conditions. This system is capable of performing single and multiplex PCR for targets ranging from 191 bp to 1.3 kb within 10 to 50 minutes using 10 to 25 μL reaction volumes, highlighting

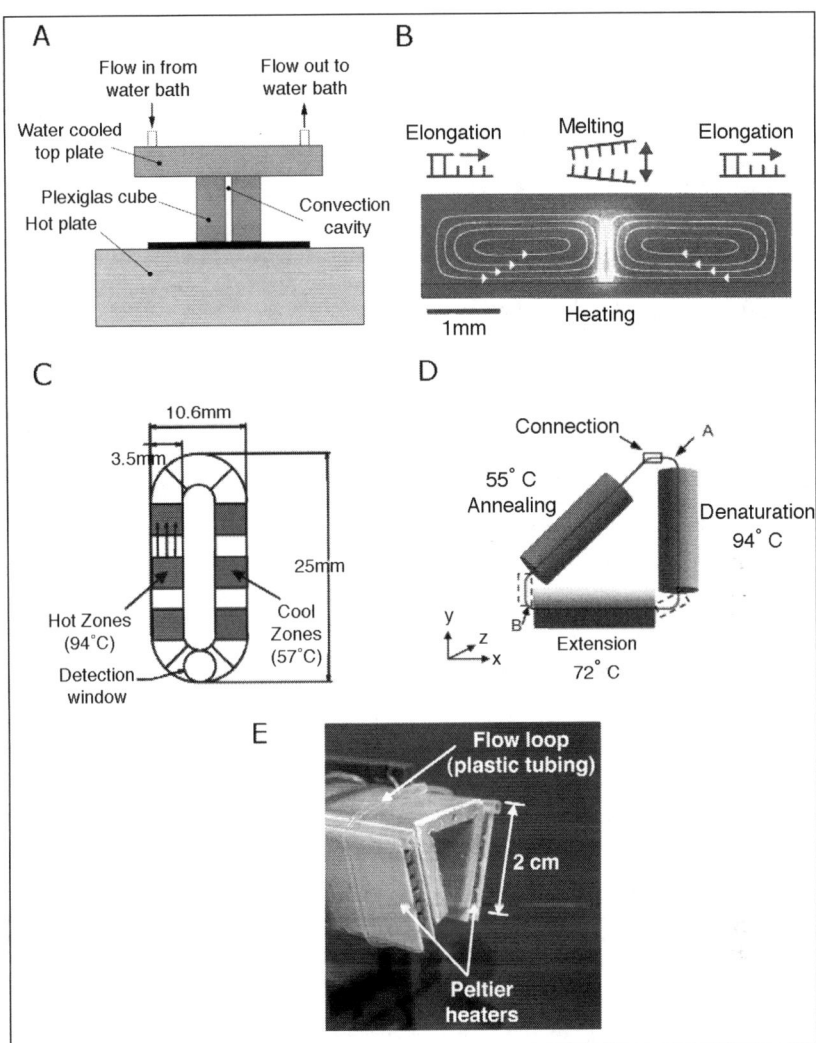

Figure 4. Examples of convective flow-based PCR thermocycler designs. Cavity-based systems may consist of (A) high aspect ratio reactor geometries subjected to a vertical temperature gradient,[52] or (B) low aspect ratio reactor geometries heated at the center[57] (reproduced with permission, copyright 2003 The American Physical Society). Examples of loop-based systems include (C) racetrack shaped designs[59] (reproduced with permission, copyright 2004 American Chemical Society) and triangular loop designs heated along (D) three[60] (reproduced with permission, copyright 2004 American Chemical Society) or (E) two sides of the flow path.[61]

the attractive combination of simplified hardware design and rapid cycling performance possible in convective flow-based systems.

Conclusions and Future Outlook

Impressive progress continues to be made in the development of integrated microfluidic systems capable of performing PCR-based bioanalysis assays. These advances are largely driven by the compelling benefits offered by miniaturization, including lower hardware costs arising from mass production using microfabrication technology, extremely rapid analysis times and greatly reduced reagent requirements. A number of challenges remain, however, including development of improved miniaturized fluid handling technology capable of precisely metering, transporting and sealing nanoliter liquid volumes to prevent evaporation without relying on bulky off-chip mechanical hardware. Potential inhibitory effects that arise under ultra-high surface to volume geometric conditions are also likely to remain a serious consideration as reactor volumes continue to shrink. Finally, improvements are still needed in the area of product detection, where the necessity to employ benchtop-scale fluorescence imaging systems often limits portability. Developments in these areas are ongoing and ultimately have the potential to greatly expand the use of genomic analysis technology by making the capability to perform an increasingly sophisticated array of assays accessible for use in a wide range of settings by those who need it most.

References

1. Collins FS, Green ED, Guttmacher AE et al. A vision for the future of genomics research. Nature 2003; 422:835-847.
2. Sauer S, Lange BMH, Gobom J et al. Miniaturization in functional genomics and proteomics. Nat Rev Gene 2005; 6:465-476.
3. Syvänen A-C. Toward genome-wide SNP genotyping. Nat Gene 2005; 37:55-59.
4. Yang S, Rothman RE. PCR-based diagnostics for infectious diseases: uses, limitations and future applications in acute-care settings. Lancet Infect Dis 2004; 4:337-348.
5. Cantor CR, Smith CL. Genomics: The science and technology behind the human genome project. New York: Wiley Interscience; 1999.
6. Spitzack KD, Ugaz VM. Polymerase chain reaction in miniaturized systems: big progress in little devices. In: Minteer SD, ed. Microfluidic Techniques: Reviews and Protocols. Vol 321. Totowa, New Jersey, USA: Humana Press; 2005:Chapter 10.
7. Belgrader P, Elkin CJ, Brown SB et al. A reusable flow-through polymerase chain reaction instrument for the continuous monitoring of infectious biological agents. Anal Chem 2003; 75:3446-3450.
8. Ebmeier RJ, Whitney SE, Sarkar A et al. Ranque–Hilsch vortex tube thermocycler for fast DNA amplification and real-time optical detection. Rev Sci Instrum 2004; 75:5356-5359.
9. Friedman NA, Meldrum DR. Capillary tube resistive thermocycling. Anal Chem 1998; 70:2997-3002.
10. Soper SA, Ford SM, Xu YC et al. Nanoliter-scale sample preparation methods directly coupled to polymethylmethacrylate-based microchips and gel-filled capillaries for the analysis of oligonucleotides. J Chromatogr A 1999; 853:107-120.
11. Swerdlow H, Jones BJ, Wittwer CT. Fully automated DNA reaction and analysis in a fluidic capillary instrument. Anal Chem 1997; 69:848-855.
12. Wittwer CT, Fillmore GC, Garling DJ. Minimizing the time required for DNA amplification by efficient heat transfer to small samples. Anal Biochem 1990; 186:328-331.
13. Wittwer CT, Fillmore GC, Hillyard DR. Automated polymerase chain-reaction in capillary tubes with hot air. Nucleic Acids Res 1989; 17:4353-4357.
14. Zhang NY, Tan HD, Yeung ES. Automated and integrated system for high-throughput DNA genotyping directly from blood. Anal Chem 1999; 71:1138-1145.
15. Erickson D, Li DQ. Integrated microfluidic devices. Anal Chim Acta 2004; 507:11-26.
16. Handal MI, Ugaz VM. DNA mutation detection and analysis using miniaturized microfluidic systems. Expert Rev Mol Diagn 2006; 6:29-38.
17. Kelly RT, Woolley AT. Microfluidic systems for integrated, high-throughput DNA analysis. Anal Chem 2005; 77:96A-102A.
18. Kricka LJ, Wilding P. Microchip PCR. Anal Bioanal Chem 2003; 377:820-825.
19. Lagally ET, Mathies RA. Integrated genetic analysis microsystems. J Phys D Appl Phys 2004; 37: R245-R261.
20. Lagally ET, Soh HT. Integrated genetic analysis microsystems. Critical Reviews in Solid State and Materials Sciences 2005; 30:207-233.
21. Roper MG, Easley CJ, Landers JP. Advances in polymerase chain reaction on microfluidic chips. Anal Chem 2005; 77:3887-3894.
22. Soper SA, Brown K, Ellington A et al. Point-of-care biosensor systems for cancer diagnostics/prognostics. Biosens Bioelectron 2006; 21:1932-1942.
23. Wilding P. Nucleic acid amplification in microchips. In: Cheng J, Kricka LJ, eds. Biochip Technology. New York: Taylor & Francis Books, Inc.; 2003:173-184.

24. Zhang CS, Xu JL, Ma WL et al. PCR microfluidic devices for DNA amplification. Biotechnol Adv 2006; 24:243-284.
25. Woolley AT, Hadley D, Landre P et al. Functional integration of PCR amplification and capillary electrophoresis in a microfabricated DNA analysis device. Anal Chem 1996; 68:4081-4086.
26. Lagally ET, Simpson PC, Mathies RA. Monolithic integrated microfluidic DNA amplification and capillary electrophoresis analysis system. Sens Actuators B Chem 2000; 63:138-146.
27. Lagally ET, Medintz I, Mathies RA. Single-molecule DNA amplification and analysis in an integrated microfluidic device. Anal Chem 2001; 73:565-570.
28. Lagally ET, Emrich CA, Mathies RA. Fully integrated PCR-capillary electrophoresis microsystem for DNA analysis. Lab Chip 2001; 1:102-107.
29. Lagally ET, Scherer JR, Blazej RG et al. Integrated portable genetic analysis microsystem for pathogen/infectious disease detection. Anal Chem 2004; 76:3162-3170.
30. Dunn WC, Jacobson SC, Waters LC et al. PCR amplification and analysis of simple sequence length polymorphisms in mouse DNA using a single microchip device. Anal Biochem 2000; 277:157-160.
31. Waters LC, Jacobson SC, Kroutchinina N et al. Multiple sample PCR amplification and electrophoretic analysis on a microchip. Anal Chem 1998; 70:5172-5176.
32. Waters LC, Jacobson SC, Kroutchinina N et al. Microchip device for cell lysis, multiplex PCR amplification and electrophoretic sizing. Anal Chem 1998; 70:158-162.
33. Khandurina J, McKnight TE, Jacobson SC et al. Integrated system for rapid PCR-based DNA analysis in microfluidic devices. Anal Chem 2000; 72:2995-3000.
34. Zhou ZM, Liu DY, Zhong RT et al. Determination of SARS-coronavirus by a microfluidic chip system. Electrophoresis 2004; 25:3032-3039.
35. Hong JW, Fujii T, Seki M et al. Integration of gene amplification and capillary gel electrophoresis on a polydimethylsiloxane-glass hybrid microchip. Electrophoresis 2001; 22:328-333.
36. Rodriguez I, Lesaicherre M, Tie Y et al. Practical integration of polymerase chain reaction amplification and electrophoretic analysis in microfluidic devices for genetic analysis. Electrophoresis 2003; 24:172-178.
37. Koh CG, Tan W, Zhao M et al. Integrating polymerase chain reaction, valving and electrophoresis in a plastic device for bacterial detection. Anal Chem 2003; 75:4591-4598.
38. Ferrance JP, Wu QR, Giordano B et al. Developments toward a complete micro-total analysis system for Duchenne muscular dystrophy diagnosis. Anal Chim Acta 2003; 500:223-236.
39. Easley CJ, Karlinsey JM, Landers JP. On-chip pressure injection for integration of infrared-mediated DNA amplification with electrophoretic separation. Lab Chip 2006; 6:601-610.
40. Wilding P, Kricka LJ, Cheng J et al. Integrated cell isolation and polymerase chain reaction analysis using silicon microfilter chambers. Anal Biochem 1998; 257:95-100.
41. Yuen PK, Kricka LJ, Fortina P et al. Microchip module for blood sample preparation and nucleic acid amplification reactions. Genome Res 2001; 11:405-412.
42. Panaro NJ, Lou XJ, Fortina P et al. Micropillar array chip for integrated white blood cell isolation and PCR. Biomol Eng 2005; 21:157-162.
43. Lee CY, Lee GB, Lin JL et al. Integrated microfluidic systems for cell lysis, mixing/pumping and DNA amplification. J Micromech Microeng 2005; 15:1215-1223.
44. Cady NC, Stelick S, Kunnavakkam MV et al. Real-time PCR detection of Listeria monocytogenes using an integrated microfluidics platform. Sens Actuators B Chem 2005; 107:332-341.
45. Anderson RC, Su X, Bogdan GJ et al. A miniature integrated device for automated multistep genetic analysis. Nucleic Acids Res 2000; 28:E60.
46. Liu RH, Yang J, Lenigk R et al. Self-contained, fully integrated biochip for sample preparation, polymerase chain reaction amplification and DNA microarray detection. Anal Chem 2004; 76:1824-1831.
47. Trau D, Lee TMH, Lao AIK et al. Genotyping on a complementary metal oxide semiconductor silicon polymerase chain reaction chip with integrated DNA microarray. Anal Chem 2002; 74:3168-3173.
48. Liu YJ, Rauch CB, Stevens RL et al. DNA amplification and hybridization assays in integrated plastic monolithic devices. Anal Chem 2002; 74:3063-3070.
49. Burns MA, Johnson BN, Brahmasandra SN et al. An Integrated Nanoliter DNA Analysis Device. Science 1998; 282:484-487.
50. Pal R, Yang M, Lin R et al. An integrated microfluidic device for influenza and other genetic analyses. Lab Chip 2005; 5:1024-1032.
51. Blazej RG, Kumaresan P, Mathies RA. Microfabricated bioprocessor for integrated nanoliter-scale Sanger DNA sequencing. Proc Natl Acad Sci USA 2006; 103:7240-7245.
52. Krishnan M, Ugaz VM, Burns MA. PCR in a Rayleigh-Bénard convection cell. Science 2002; 298:793.
53. Ugaz VM, M. K. Novel convective flow based approaches for high-throughput PCR thermocycling. JALA 2004; 9: 318-323.

54. Yariv E, Ben-Dov G, Dorfman K. Polymerase chain reaction in natural convection systems: A convection-diffusion-reaction model. Europhys Lett 2005; 71:1008-1014.
55. Krishnan M, Agrawal N, Burns MA et al. Reactions and fluidics in miniaturized natural convection systems. Anal Chem 2004; 76:6254-6265.
56. Braun D. PCR by thermal convection. Modern Physics Letters B 2004; 18:775-784.
57. Braun D, Goddard NL, Libchaber A. Exponential DNA replication by laminar convection. Phys Rev Lett 2003; 91:158103.
58. Hennig M, Braun D. Convective polymerase chain reaction around micro immersion heater. Appl Phys Lett 2005; 87:183901.
59. Wheeler EK, Benett W, Stratton P et al. Convectively driven polymerase chain reaction thermal cycler. Anal Chem 2004; 76:4011-4016.
60. Chen Z, Qian S, Abrams WR et al. Thermosiphon-based PCR reactor: experiment and modeling. Anal Chem 2004; 76:3707-3715.
61. Agrawal N, Hassan YA, Ugaz VM. A pocket-sized convective PCR thermocycler. Angew Chem Int Ed 2007; in press.

CHAPTER 8

Integrated Nucleic Acid Analysis in Parallel Matrix Architecture

Jong Wook Hong*

Abstract

With the advent of the postgenomic era, accurate and fast nucleic analyses are becoming more important with reasonable running costs. Nucleic acids carry precious information important for the understanding of many complex biological processes critical for diagnosis and therapy of a wide range of diseases. This information is also instrumental in the classification of organisms. Recent developments in microfluidics provide a new methodology with unprecedented sensitivity for mRNA isolation from single mammalian cells and genomic DNA processing of multiple bacterial cells. In this chapter, nucleic acid analyses on microfluidic chips with particular emphasis on parallel matrix architecture, which will soon play an important role in applications of microfluidics on the development of proteomics and functional genomics, will be discussed.

Introduction

Nucleic Acid Analyses

To understand fundamental blueprints of cells or organs, nucleic acid analyses are one of the most important steps over the status quo analyses, i.e., morphological or physiological analyses. In most cases, with conventional bench-top tube methods, nucleic acid analyses are restricted by sample amounts as well as time and costs allowed for the analyses. With conventional methods, typically $10^3 \sim 10^4$ identical cells are required to start nucleic acid analysis. However, the preparation of identical cells is limited only to laboratory culturable bacterial cells or established mammalian cell lines. It is estimated that more than 99% of bacterial cells on earth have not been identified and one of the main reasons is because of the difficulties of hampering laboratory culture steps. Indeed, when it comes to rare cells such as omnipotent stem cells, indispensable steps are to chart individual cell behavior and analyze individual cell characteristics in single cell sensitivity. Unfortunately, there has not been a tool to handle single cell level samples and nucleic acid analysis. Therefore, a new methodology of nucleic acid analyses from single cell or a few cells is required. The purpose of this chapter is to summarize and understand the current technology status for nucleic acid analyses on integrated biochips,[1-4] specifically in a parallel matrix architecture with microfluidic sample handling capabilities.

Pneumatic Control of Microfluid

There are two basic approaches to microscale fluid control: electrokinetic and pneumatic controls. Although there have been many impressive examples of electrokinetic microfluidic

*Jong Wook Hong—Materials Research and Education Center, Department of Mechanical Engineering, Auburn University, Auburn, AL 36849, U.S.A. Email: jwhong@eng.auburn.edu

Integrated Biochips for DNA Analysis, edited by Robin Hui Liu and Abraham P. Lee.
©2007 Landes Bioscience and Springer Science+Business Media.

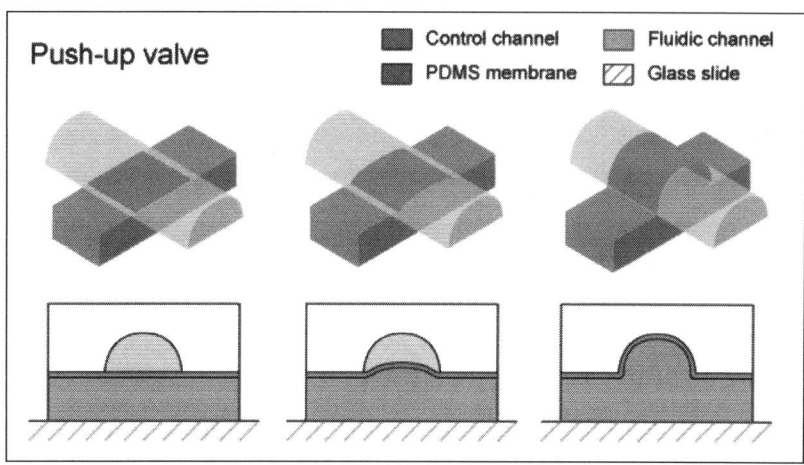

Figure 1. Schematic explanation of a single micromechanical valve geometry and operation. Thin membrane between control and fluid channels is inflated by the outside pressure through the control channel. A color version of this figure can be found at www.eurekah.com.

devices for parallel gene analysis,[5] the electokinetic method has limited applications in parallel and matrix architecture because of intrinsic diffusion occurring at the virtual valve junctions. Pneumatic control of microfluid with robust micromechanical valves has provided an alternative with the advantage of easy construction of multiple experimental conditions in parallel matrix methods. The micromechanical valves are fabricated through multilayer soft lithography with a silicone rubber, polydimethysiloxane (PDMS).[6] The simple principle of this useful invention is based on the deflection of silicone thin film, which lies between a fluidic channel and a control channel in different layers. When pressurized air or nitrogen is introduced into the control line, a thin membrane between the control line and the fluidic line is deflected upward and seals off the fluidic line as depicted in Figure 1. The simplicity and flexibility in multilayer soft lithography operations on the same chip allow for the demonstration of many different applications.

Integration and Parallelization on a Microfluidic Chip

During the first decade of microfluidic developments in the 1990s, most chip nucleic acid analyses were focused on capillary electrophoresis. The chips were fabricated out of silicon, glass or polymers and housed as a single process for nucleic acid analysis on a chip. At that time, the devices were deficient of integrity and parallelization because achieving a high-degree integrity for a fluidic device, without increasing the control complexity, was a challenging problem in microfluidics for both electrokinetic and pneumatic controls. This problem was solved by using a pneumatic control method with two innovative ways: crossing-over of the control lines over fluidic lines and microfluidic multiplexing.[7] By narrowing the width of a control channel smaller than that of the fluidic channel, the control channel is bridged over the fluidic channels without stopping the flows in the fluid channel. Only with the desired fluid channel part having control channel width equal or wider than the flow channel, can a working pneumatic valve be achieved. The fluidic multiplexor has also contributed greatly to achieve complex fluid manipulations with a minimal number of valve operations. The multiplexor is a combinatorial array of binary valve patters which can make a flow in single channels allowing for 'N' flow channels controlled with only $2\log_2 N$ control channels. For instance, only 20 control channels are required to specifically address 1024 flow channels. One of the first examples of integrated parallel microfluidic processing was demonstrated with an example of enzymatic cell screening on a microfluidic chip, which holds 256 picoliter microfluidic chambers. The chip works as a microfluidic comparator to test for the expression of a particular enzyme. An electronic comparator circuit is designed to provide a large output signal when the input signal exceeds a reference threshold value. A population of *Escherichia coli* expressing recombinant cytochrome c peroxidase (CCP) is loaded into the device and a fluorogenic substrate, Apmlex Red, is loaded into separate input channels with the central mixing barriers. Opening the barrier and allowing for a certain period of incubation time, the presence of one or more CCP-expressing cells in an individual chamber produces a strong amplified output signal. Recovery from the chip can be performed by selecting a single chamber and then purging the contents to a collection output. This system could be applied to replace conventional agar plate based screening method, allowing automation of biotechnology on a chip.

Further advancement of the microfluidic integration and parallelization has been achieved by introducing matrix architecture. Through the matrix architecture, not only integrating several identical chambers and conducting experiments with the same condition, but also changing detailed experimental conditions, i.e., the ratios of sample and reagent on the same experiments, can be achieved. Hence, advanced experimental flexibility of handling hundreds of samples with different conditions could be realized through the parallel matrix architecture. This is useful for many combinatorial experiments.

Nucleic Acid Analysis

DNA Analyses with Complex Parallelization

Although with relatively simple microfluidic parallelization, the first parallel microfluidic device for complicated microfluidic reactions was used for DNA isolation from multiple bacterial cells (Fig. 2).[8,9] The microfluidic chip capable of three parallel sets of processes on a 20 × 20 mm space is operated using the same number of micromechanical valves for a single set. In addition, by changing the mixing ratio of two solutions, reaction conditions of the DNA processing for each processor can be changed in a parallel architecture (referred to the three different microfluidic mixer colors in Fig. 2). The chip has 26 access holes, 1 waste hole and 54 valves within a 20 × 20 mm space. To operate the microscaled pneumatic valves inside the microfluidic chips, the control channels are connected to polyethylene tubing through stainless steel pins. The microvalves are controlled individually by an outside pressure source utilizing a user-friendly visualized programming language, LabView. The chip is made up of two different layers, the actuation layer and the fluidic layer, which are bonded together through multilayer soft lithography.

Using this nanoliter fluidic processor, complex biochemical processes, such as bacterial cell capture, mixing different ratios of solutions, cell lysis, DNA purification from the cell lysates and the purified DNA recovery from the microfluidic chip, were carried out in a vertical, left to right, direction. Also, parallel processing of the complicated biochemical reaction was achieved in a horizontal, top to down, direction on a single microfluidic chip (Fig. 3). To demonstrate automatic DNA purification from a limited number of cells without any pre or postsample treatments, *E. coli* (BL21-2+, expressing eGFP) culture were directly introduced into the microfluidic device after 12 hours of cultivation at 37°C. The introduced bacterial cells were lysed with a commercial lysis buffer (Xtra Amp®, lysis buffer series 1, Xtrana, Inc., Ventura, CA). *E. coli* samples, lysis buffer and dilution buffer were accurately metered to sub-nanoliter accuracy and introduced into the microfluidic mixers, with a volume of 5 nL each. Then, the microfluidic mixers were operated to achieve instant mixing of all three different solutions. By integrating and operating a microfluidic mixer, diffusion-limited microfluidic mixing can be enhanced.

Another remarkable aspect of this parallel nanoliter fluidic processor is that solid extraction methods have been incorporated in the microfluidic regime. By selecting beads having specific affinity functionality on the surfaces, specific target molecules could be purified on a microfluidic chip. This bead-dam method is attributed to the soft micromechanical valve, which can control the opening degree of the valve. The opening of the valve can be controlled accurately with the silicone membrane due to the material's low Young's Modulus. The microfluidic bead dam method has been applied in RNA isolation and synthesized chemical purification on chip.[10] After the parallel cell

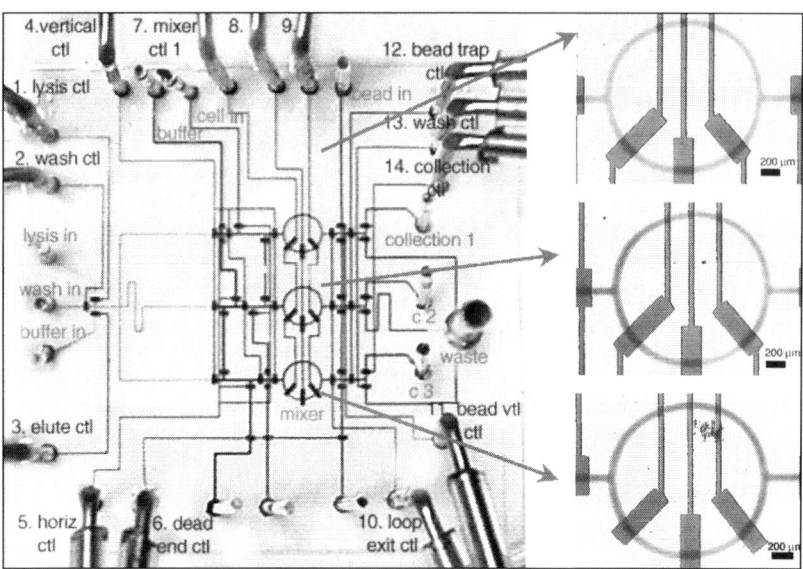

Figure 2. Integrated bioprocessor chip with parallel architecture. This chip is made of two different layers, the actuation layer and the fluidic layer, which are bonded together. The actuation channels are filled with a green food coloring and the fluid channels are filled with yellow, blue and red food coloring according to their functionalities. The width of the fluid channels is 100 μm and that of valve actuation channels is 200 μm. Input and output ports and vias have been annotated. Multiple parallel processes of DNA recovery from living bacterial cells are possible in three processors with sample volumes of 1.6 nL, 1.0 nL and 0.4 nL. The chip has 26 access holes, 1 waste hole and 54 valves within 20 x 20 mm. The control channels are connected to polyethylene tubing through stainless steel pins. (Right) The volumes of bacterial cell solution used in the top, middle and bottom processors are 1.6, 1.0 and 0.4 nL, respectively. The total volume of each rotary reactor is the same (5 nL), so 0.4, 1.0 and 1.6 nL of dilution buffer were metered on a chip and added to the bacterial cell volumes, respectively, along with 3.0 nL of lysis buffer. This metering process is illustrated with the use of food dyes.

lysis, the lysates containing genomic bacterial DNA were introduced over the purification beads, which were then introduced and stacked before the bacterial cell sample introduction. The beads used for the DNA purification were 2.8 μm (Dynabeads DNA Direct Universal). A valve at the location on the chip where the column is to be constructed was partially open and the beads were allowed to flow in. The beads are larger than the valve opening, so they cannot pass through and are quickly concentrated into a small column. The beads loaded on the chip had a total capacity of 100 pg of DNA, enough to recover genomic DNA from around 20,000 bacteria. After the purification, an elution buffer (Tris-EDTA buffer, pH 7.8) was introduced into the chip and the purified DNA solutions were collected from the three collection ports. To verify DNA purification from the chip, gene amplifications were carried out with conventional PCR methods. The target gene was 461 bp fragment of pppD (prepilin peptidase-dependent protein D precursor) gene and PCR reactions were successfully carried out with 0.4 nL, 1.0 nL and 1.6 nL of bacterial cell samples. The numbers of target bacterial cells used for the genomic DNA isolation from the cell culture were 280, 700 and 1120. Further sensitive experiments were carried out with a 1/10 times diluted bacterial cell culture and successful gene isolation from less than 28 *E. coli* cells was confirmed. The genomic DNA chip device has increased the sensitivity of genetic DNA isolation five orders of magnitude as compared to the conventional methods which requires at least $10^5 \sim 10^6$

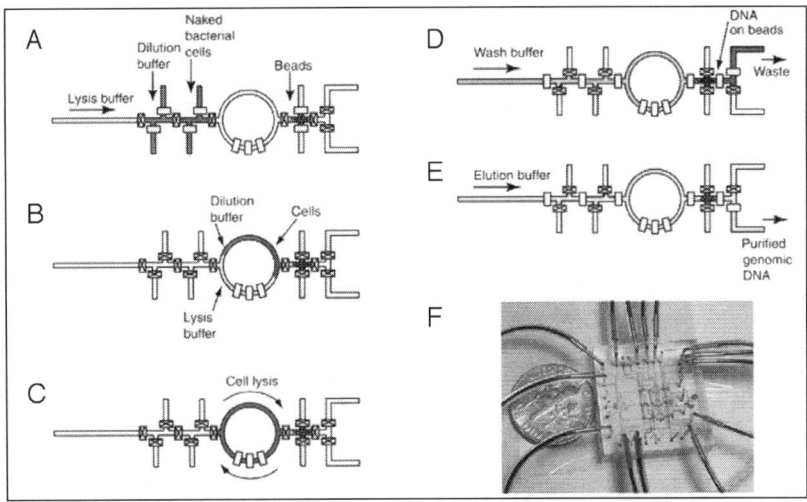

Figure 3. Process flow detailing how an integrated processor for DNA purification can be implemented in a nanofluidic system. (Open valve, rectangle; closed valve, x in rectangle.) A) Bacterial cell culture (indicated in red) is introduced into the chip through the 'cell in' port located in the uppermost part of the chip. Buffer (in green) for dilution of the cell sample is introduced through the 'buffer in' port located next to the 'cell in' port. The amount of dilution is determined by the ratio of channel lengths for cells and buffer; in this case it is 1:1. Lysis buffer (yellow) is introduced from the left side of the chip. B) The cell sample, dilution buffer and lysis buffer slugs are introduced into the rotary mixer, which has a total volume of 5 nL. C) The three liquids are circulated inside the reactor, resulting in efficient mixing and consequent lysis of the bacterial cells. D) The lysate is flushed over a DNA affinity column and drained to the 'waste port'. E) Purified DNA is recovered from the chip by introducing elution buffer from the left side of the chip. The recovered DNA can either be recovered from the chip or sent to another part of the chip for further analysis or manipulation. F) Photograph of an actual nanofluidic system that implements three simultaneous parallel processes of the DNA recovery scheme illustrated above. The three parallel processes use distinct sample volumes of 1.6 nL, 1.0 nL and 0.4 nL, respectively. The parallel architecture allows complex processes to be implemented in parallel without increasing the control complexity of the system.

identical bacterial cells. This is an example of how microfluidic chips can be used in new applications that were not accessible before (Fig. 4).

RNA Analyses

Analysis of RNA, such as microRNA (miRNA) and messenger RNA (mRNA), is becoming more important to profile the expression levels of genes and understand gene regulation mechanisms. In Figure 4 an example of mRNA recovery from single mammalian cells, NIH 3T3 (mouse embryonic fibroblast cell line), on an integrated microfluidic chip is depicted. This RNA chip was also fabricated by multilayer soft lithography. The functionalities of the chip are similar to the genomic DNA chip explained in the previous section with the exception of the parallelization and the micromixer parts. All the processes from single mammalian cell capturing, cell lysis to messenger RNA recovery were carried out automatically with computerized pneumatic control. For the purification of mRNA, 2.8 μm polymer magnetic beads having oligo-dT (Dynabeads Oligo(dT)25) were stacked to form an affinity column inside the microfluidic chip and used for the mRNA recovery. 3T3 cells were grown to near confluency, growth of cells throughout the culture medium, on 100 mm tissue culture plates. The cells were rinsed with 1X phosphate buffered saline (PBS) and then trypsinized to remove the proteins that cause cells to stick to each other or

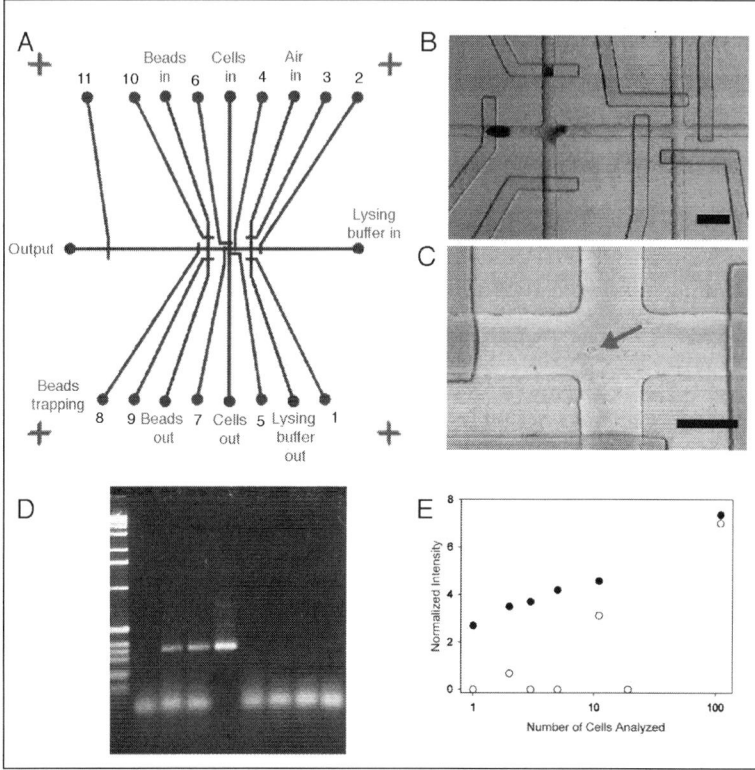

Figure 4. mRNA purification chip and RT-PCR analysis of isolated mRNA. A) Layout of the microfluidic chip. The blue lines represent the 100 μm wide fluidic channels and the red lines are the 100 μm wide valve actuation (control) channels. The fluidic ports are named; the actuation ports are numbered 1 to 11. The lysis buffer chamber is composed of the channel space delineated by valves 1, 2, 3 and 4. The 'cell chamber' is composed of the channel space delineated by valves 4, 5, 6 and 7 (see also b,c). The 'bead chamber' is composed of the channel space delineated by valves 7, 8, 9 and 10 (see also b). B) Photograph of the in situ affinity column construction. A column of 2.8 μm diameter paramagnetic beads covered with oligo dT is being built against a partially closed microfluidic valve. Scale bar, 200 μm. C) A single mammalian cell is loaded in the 'cell chamber' before the lysis step. Scale bar, 100 μm. D) A series of experiments, together with negative controls, were done to demonstrate single-cell sensitivity of the microfluidic chip. The reverse transcriptase (RT)-PCR products were analyzed on a 2% agarose gel; the amplified gene is β-actin. Lane 1 (from left), 1 kb ladder; lane 2, PBS, no cells (on the chip experiment); lane 3, one cell (on the chip experiment); lane 4, nine cells (on the chip experiment); lane 5, 200 μL supernatant from 1 day old cells + 5 μL of beads (in a test tube, not on chip experiment); lane 6, 200 μL lysis buffer + 5 μL of beads (in a test tube, not on chip experiment); lane 7, cells load on the chip but none are trapped in the chamber (on the chip); lane 8, 200 μL deionized water + 5 μL beads (in a test tube); lane 9, PCR reagents only. E) A second series of experiments were done to test the chips with variable numbers of cells and both β-actin and OZF transcripts. The RT-PCR products for both transcripts were analyzed on a 2% agarose gel, whose bands were quantified, normalized and plotted on a graph. Zero values indicate the absence of a detectable band in the gel; neither transcript was detected in the experiment with 19 cells, possibly due to RNAase contamination or chip failure. For high-abundance β-actin mRNA (closed circles), detection is down to the single-cell level; for the moderately abundant OZF mRNA (open circles), detection is at the level of 2-10 cells. (After ref. 9)

stick to the surface. The tripsinized cells were resuspended in 1 x PBS and the final concentration was adjusted to 10^6-10^7 cells/mL. The first step of the RNA purification is to load the cells on the chip by introducing the cells into the chamber through the 'cell in' channel (Fig. 4A), followed by introducing a lysis buffer through the 'lysing buffer in' part. The lysis buffer used for the mammalian cell lysis was a commercially available Dynal buffer. The cell lysate was allowed to flow through the stack of beads, which were introduced to the chip prior to the experiment. The typical speed of the lysate passing over the stacked bead was 100 μm/s that could be adjusted by changing the pressure on the air in channel (Fig. 4A) to trap the mRNA. After the isolation of mRNA through the affinity chromatographic separation, a short tubing was connected to the 'output' port of the chip and was placed in a 0.1 mL PCR tube to retrieve the beads from the chip. The beads containing mRNA were flushed with a lysis buffer into the PCR tube and a magnet was used to draw out any beads remaining in the outflow channel. The beads were centrifuged and resuspended in 100 μL of fresh lysis buffer containing 1 μL of RNase inhibitor. The tube was vortexed and the mRNA was either analyzed immediately by amplifying the target gene from mRNA through conventional RT-PCR, cDNA synthesis by reserve transcription followed by PCR amplification, or stored in the tube frozen at $-80\,^{\circ}$C. Tiny amounts of messenger RNA from single mammalian cells were recovered successfully with enough sensitivity of target gene amplification through conventional RT-PCR. The targets used for the confirmation of mRNA isolation were the high-abundance β-actin transcript or the moderate-abundance OZF transcript. Successful purification of mRNA from single cells with the microfluidic chip was verified by gel electrophoresis as shown in Figure 4D. Indeed, Marcus et al. recently demonstrated a microfluidic chip to perform 72 parallel 450-pL RT-PCRs.[11] The device and method can be used for highly parallel single cell gene expression analysis. Advanced and integrated study of mRNA was conducted on mRNA isolation and cDNA synthesis on a single microfluidic chip by utilizing the power of microfluidics. All the steps of cell capture, cell lysis, mRNA purification, cDNA synthesis and cDNA purification were demonstrated providing the first quantitative calibrations for microfluidic mRNA isolation and cDNA synthesis.[12]

Matrix PCR on a Microfluidic Chip

An advanced study on how microfluidics can reduce the complexity of pipetting operations was demonstrated by Liu et al[13] with the matrix PCR method. The chip was fabricated by multilayer soft lithography and holds 400 nanoliter PCR reactions. The device consists of three layers, one matrix fluidic layer between two layers with hydraulic valves and pneumatic pumps. With the conventional fluid handling method, 1200 pipetting steps are required to handle 400 PCR reactions. The same number of reactions can be achieved on the matrix PCR chip with only 41 pipetting steps. This is a simultaneous achievement with two important goals: dealing with the macro/micro interface problem that has hampered the impact of microfluidic technology and the development of a matrix device which has immediate applications in medical diagnosis and gene testing.

Pico Titer Plate-Based PCR

In the microfluidic format, most of gene amplification methods are carried out in solution. A unique approach of solid-phase amplification through immobilizing (attaching) the amplified PCR product to a bead that captures DNA in picoliter volumes was demonstrated by Leamon et al.[14] The picoliter chambers were made by anisotropic etching of fiber optic face plates with different chamber size ranging from 26~76 μm in depth and 39~44 μm in diameter yielding about 40~50 picoliter volumes of 300,000 discrete individual reactors (Fig. 5). 25~35 μm beads from commercially available affinity chromatography column (Amersham Bioscience) were used to capture the amplified DNA. A 20 bp capture primer is attached to the beads prior to the bead loading process which is intended to put a single bead in the individual picoliter chambers on the optical plate. Both a loading cartridge for complete introduction of a sample solution to the plate and an amplification chamber to prevent evaporation of the reaction liquid are required to run this small volume parallel device. After loading PCR mix to the plate pneumatically with a syringe, the plate is released from the loading cartridge and placed in a flat thermal cycler for PCR. Gene amplification results on the plate were demonstrated. Advanced research of the picotiter plate

Figure 5. Picoliter titer plates. Highly integrated picoliter titer plates are filled with beads. The beads are carrying immobilized enzymes required for pyrophosphate sequencing. Picture courtesy of Dr. Jonathan M Rothberg of 454 Life Science Corp., CT.

has been published by Margulies et al[15] showing bacterial (Mycloplasma genitalium) genome sequencing with a readlength of 113 bp. It is a highly parallel emulsion-based method to isolate and amplify DNA fragments in vitro to perform pyrophosphate-based sequencing in picoliter-sized wells. Although the matrix approach was not tried, they showed feasibility of high throughput with picoliter wells.

Microfluidic Digital PCR

An example of the microfluidic systems for microbial gene analysis through digital PCR has been demonstrated by Ottesen et al.[16] The chip is made by multilayer soft lithography and 1,176 nanoliter volume chambers are partitioned by pneumatically controlled microvalves. Digital PCR is a way to transform the exponential, analog nature of the PCR into a digital signal from single templates. They showed that the integrated microfluidic system was able to determine the genetic capacities of not-yet-cultivated species resident in complex microbial ecosystems. Microfluidic digital PCR yields a fluorescent signal upon amplification of a gene regardless of the number of template copies present in a cell localized in a reaction chamber. Thus, this approach can be employed to estimate, independent of gene copy number, the fraction of a target gene when the given species within the samples are provided.

Microfluidic Sanger DNA Sequencing

Integrated microfluidic nucleic acid processor for Sanger DNA sequencing has been demonstrated by the Mathies' group at Berkeley.[17] In addition to the integration of microfluidic PCR reactors to amplify a gene, affinity-capture purification chambers for purification and capillary

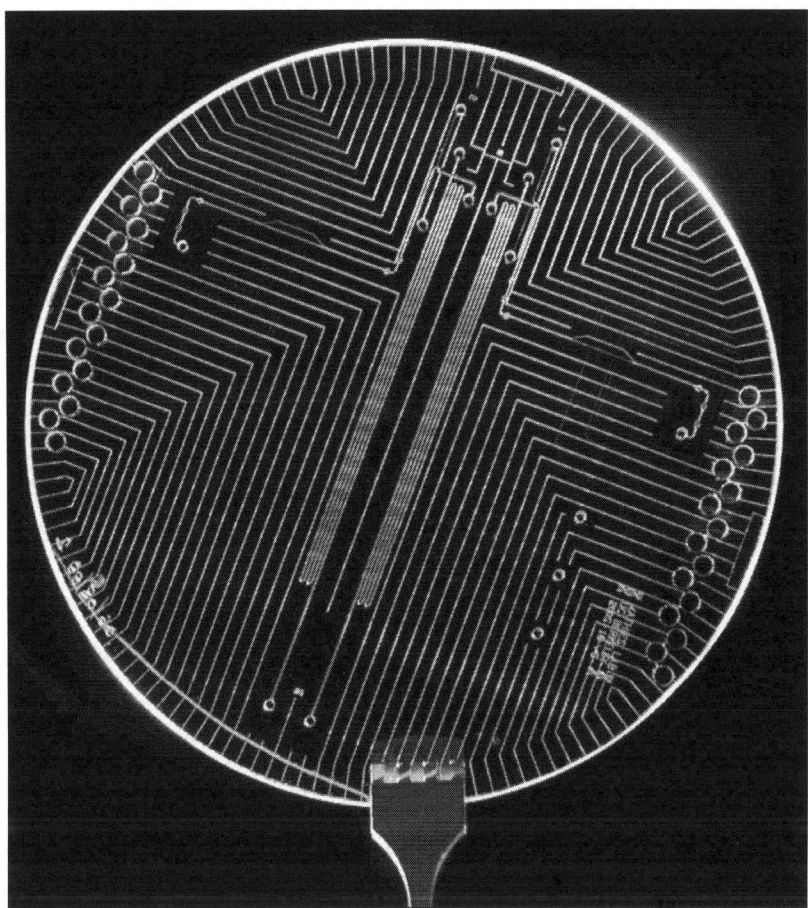

Figure 6. Integrated microfluidic Sanger sequencing chip. Multiple processes of DNA sequencing are carried out on a wafer size hybrid microchip. Picture courtesy of Dr. Richard Mathies of University of California at Berkeley.

electrophoresis channels for sequencing, integrating pneumatic valves and pumps onto a microfabricated device reduced the amount of the sequencing samples down to 250 nL from microliter levels of their previous on-chip study. Sequencing of up to 556 bases with 1 fmol of a 750-bp pUC18 amplicon was demonstrated on the chip (Fig. 6). All the processes of gene amplification, sample handling and sequencing though 30 cm of gel-filled microfluidic channels were carried out on the hybrid glass-PDMS wafer scale microfluidic device with pneumatic and electrokinetic fluid controls. The microfluidic system consists of three glass layers hosting microfluidic channels, temperature detectors, valves and manifold for the valve control, respectively and one membrane layer for pneumatic valves.

Microfluidic DNA Sequencing-by-Synthesis

The first fully integrated microfluidic system for DNA sequencing-by-synthesis was demonstrated by Emil Kartalov et al.[18] The device is capable of sequencing up to 4 consecutive base pairs using fluorescence detection. For sequencing-by-synthesis (SBS), surface chemistry to attach the DNA to the microchannel to avoid DNA loss during washing stage has been incorporated. The chip can be employed to sequence only a few base pairs, e.g., up to 4 base pairs to date. It consumes 500 nL of reagents per feed, a considerably small amount as compared to the conventional methods and <1 fmol of DNA, much smaller amounts than using other methods such as pyrosequencing.[19] However, the chip could incorporate all the advantages of microfluidics, i.e., cost effectiveness, speed, control, high level of integration and parallelization, in DNA sequencing application.

Summary and Perspective

Microfluidics has great potentials of improving nucleic acids analyses through parallelization in matrix. Throughout this chapter, unique microfluidic approaches for nucleic acid analyses, including handling of single or multiple nucleic acids on the integrated parallel microfluidic devices, have been discussed. Further development in this "buzz" area will produce high-throughput microfluidic devices for genetic analyses including DNA isolation, messenger RNA isolation and cDNA synthesis as well as gene sequencing in microfluidic reactor arrays.

Acknowledgements

The author would like to thank Sachin Jambovane at Auburn University for his help during the preparation of the manuscript.

References

1. Erickson D, Li D. Integrated microfluidic devices. Anal Chim Acta 2004; 507:11-26.
2. Lagally ET, Soh HT. Integrated genetic analysis microsystems. Critical Reviews in Solid State and Materials Sciences 2005; 30:207-233.
3. Lagally ET, Mathies RA. Integrated genetic analysis microsystems. J Phys D: Appl Phys 2004; 37: R245-R261.
4. Sun Y, Kwok YC. Polymeric microfluidic system for DNA analysis. Anal Chim Acta 2006; 556:80-96.
5. Emrich CA, Tian H, Medintz IL et al. Microfabricated 384-lane capillary array electrophoresis bioanalyzer for ultrahigh-throughput genetic analysis. Anal Chem 2002; 74:5076-5083.
6. Unger MA, Chou HP, Thorsen T et al. Monolithic microfabricated valves and pumps by multilayer soft lithography. Science 2000; 288:113-116.
7. Thorsen T, Maerkl SJ, Quake SR. Microfluidic large-scale integration. Science 2002; 293:580-584.
8. Hong JW, Quake SR. Integrated nanoliter systems. Nat Biotechnol 2003; 21:1179-1183.
9. Hong JW, Studer V, Hang G et al. A nanoliter-scale nucleic acid processor with parallel architecture. Nat Biotechnol 2004; 22:435-439.
10. Lee CC, Sui G, Elizarov A et al. Multistep synthesis of a radiolabeled imaging probe using integrated microfluidics. Science 2005; 16:1793-1796.
11. Marcus JS, Anderson WF, Quake SR. Parallel picoliter RT-PCR assays using microfluidics. Anal Chem 2006; 78:956-958.
12. Marcus JS, Anderson WF, Quake SR. Microfluidic single cell mRNA isolation and analysis. Anal Chem 2006; 78:3084-3089.
13. Liu J, Hansen C, Quake SR. Solving the "world-to-chip" interface problem with a microfluidic matrix. Anal Chem 2003; 75:4718-4723.
14. Leamon JH, Lee WL, Tartaro KR et al. A massively parallel PicoTiterPlate based platform for discrete picoliter-scale polymerase chain reactions. Electrophoresis 2003; 24:3769-3777.
15. Margulies M, Egholm E, Altman WE et al. Genome sequencing in microfabricated high-density picoliter reactors. Nature 2005; 437:376-380.
16. Ottesen EA, Hong JW, Quake SR et al. Microfluidic digital PCR enables multigene analysis of individual environmental bacteria. Science 2006; 314:1464-1467.
17. Blazej RG, Kumaresan P, Mathies RA. Microfabricated bioprocessor for integrated nanoliter-scale Sanger DNA sequencing. Proc Natl Acad Sci USA 2006; 103(19):7240-7245.
18. Kartalov EP, Quake SR. Microfluidic device reads up to four consecutive base pairs in DNA sequencing-by-synthesis. Nucleic Acids Res 2004; 32:2873-2879.
19. Ronaghi M, Karamohamed S, Petterson B et al. Real-time DNA sequencing using detection of pyrophosphate release. Anal Biochem 1996; 242:84-89.

CHAPTER 9

Chip-Based Genotyping by Mass Spectrometry

Kai Tang*

Abstract

DNA analysis by mass spectrometry has the advantage of being the most accurate method for genotyping. Since the mass of molecular ions is measured and each allele has its own intrinsic mass, this method provides direct readout of allele types without the need for any labeling. Coupled with different chip-based platforms, it has also become one of the highest throughput and most accurate quantitative methods for determining allele frequency. Matrix-assisted laser desorption/ionization (MALDI) time-of-flight mass spectrometry (TOFMS) is the most commonly used genotyping method. Silicon chips have been used as a sample positioning and concentration device for MALDI-MS. Alternatively, it can also be functionalized and used as a capturing device for DNA template. Subsequent enzymatic reactions and MALDI-MS can be performed on the chip surface. Electrospray ionization (ESI) has also been used for genotyping, mainly for the analysis of microsatellites. It could also be coupled with chip-based nano-electrospray nozzles to increase sensitivity and throughput.

Introduction

Long before the human genome was sequenced, short tandem repeat (STRs, also known as microsatellites) polymorphisms[1-4] had been used, mainly in linkage analysis and association studies for gene mapping.[5] STRs are DNA sequences consist of 4-20 repeated units with each unit ranging from two to several bases. They are abundant in the human genome (1 in every 10-20 kb)[3] and highly polymorphic (i.e., have many possible alleles).[1,6] Genotyping STRs usually involves amplification of the individual loci by polymerase chain reaction (PCR) and sizing the PCR products by electrophoresis.[7-11] The analysis is usually time-consuming, not easily automated or multiplexed.

The second type of genetic markers is single nucleotide polymorphisms (SNPs). SNPs are stable sequence variations that occur at a single base position and usually bialleleic in humans. Although less informative compared to STRs, they are more abundant in human genome[12,13] (1 in every 300-1000 bases) and can be analyzed in highly multiplexed and automated formats. Linkage studies with thousands of markers and association studies with large number of individual samples are more readily carried out with SNP assays in high-throughput mode.

There are numerous technology platforms for SNP genotyping. They differ in throughput, sensitivity, specificity, quantitative ability and cost. Reviews of these techniques are available.[14-16] Among the different methodologies, genotyping by matrix-assisted laser desorption/ionization (MALDI) time-of-flight mass spectrometry (TOFMS) offers the most versatility and is one of the most powerful and reliable methods developed so far. MALDI is an innovative technique for ionizing macromolecules in gas phase.[17,18] In MALDI, large molecules of interest are cocrystallized

*Kai Tang—School of Biological Sciences, Nanyang Technological University, 60 Nanyang Drive, Singapore 637551. Email: tkai@ntu.edu.sg

Integrated Biochips for DNA Analysis, edited by Robin Hui Liu and Abraham P. Lee.
©2007 Landes Bioscience and Springer Science+Business Media.

in excess of selected matrix molecules. Those matrix molecules are usually small organic acids that have strong absorption at the wavelength of the laser pulse, usually 337 nm or 355 nm. When the crystals are hit by a laser pulse, desorption of the matrix and analyte molecules occurs. Collision of the matrix and analyte molecules in gas phase results in proton transfer, creating molecular ions. In TOF mass spectrometers, those protonated molecular ions are accelerated in a small region with strong electric field to different velocities depending on their masses. Ions with larger mass acquire lower velocity and vise versa. Ions with different masses are further separated in a long field-free region due to their different velocities and arrive at the detector at different time. The flight time of different ions are recorded at the detector and converted to masses. A schematic diagram is shown in Figure 1. When applying MALDI-TOFMS to SNP genotyping, samples are usually prepared by primer extension reactions to generate allele specific products of different masses, then desalted and analyzed in the mass spectrometer. The enzymatic primer extension reaction adds specificity to genotyping and also increases sensitivity when compared to assays based only on hybridization. Specificity and accuracy are further enhanced by using mass spectrometry, since mass, the intrinsic value of products, can be accurately measured for identification, not tagged reporters used in other platforms.

The power of mass spectrometry is also demonstrated when it is applied to genotyping of STRs. After PCR amplification of the repeat region, electrospray ionization (ESI)[19] mass spectrometry is used to analyze the mass of the PCR product. In ESI, the analyte solution is sprayed out from a charged needle. Evaporation of solvent from charged droplets eventually produces multiply charged gas phase ions. For analyzing large intact biomolecules, ESI is best coupled with a Fourier transform ion cyclotron resonance (FTICR) mass spectrometer.[20,21] In such a device, ions are trapped and circulate in an ICR cell at the center of a high magnetic field. FTICR mass spectrometer measures the frequency of the circular motion, which is inversely proportional to mass, through Fourier

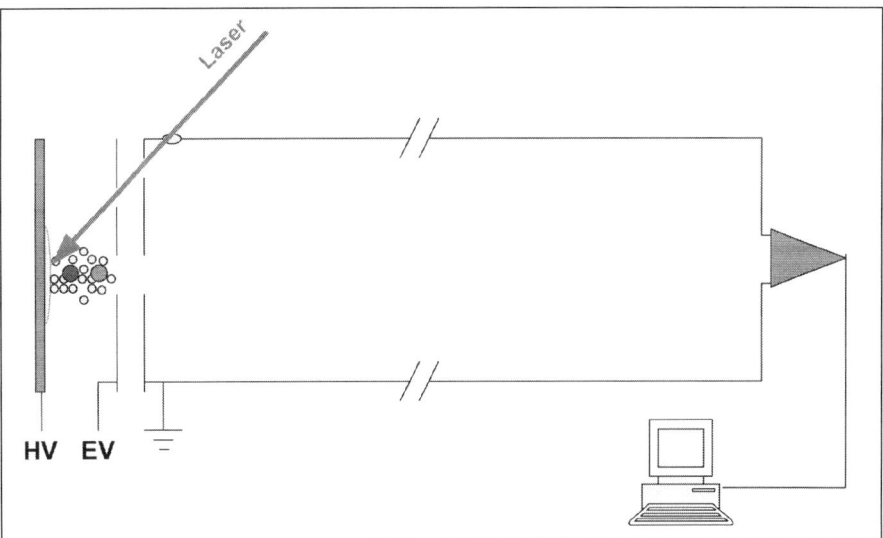

Figure 1. Linear time-of-flight (TOF) mass spectrometer used for SNP genotyping. Ions are generated by MALDI and extracted in the mild electric field between the sample plate (biased at high voltage HV) and extraction plate (at slightly lower voltage EV) and further accelerated in the strong electric field between the extraction plate and the entrance of the flight tube (held at ground potential). The speed of ions reaches its maximum when entering into the flight tube and remains constant until they reach the detector. This maximum speed is dependent on the mass to charge ratio of an ion. Therefore, the mass value can be obtained from the ion flight time, which can be measured accurately using a digitizer.

transformation of the time-domain image current signal generated on the ICR cell.[22] Ultra high resolution has been demonstrated for FTICR mass spectrometry.[23] By direct measurement of PCR products, repeat polymorphisms can be characterized in detail.

Many review papers on SNP genotyping by MALDI mass spectrometry have been published.[24-34] Reviews on STR genotyping by ESI MS are relatively rare, however, an excellent work was published by Null and Muddiman,[35] and recently by Hofstadler et al.[36] The aim of this chapter is not to repeat these reviews but to provide some insight on how chips can be used to improve the performance of genotyping by mass spectrometry.

MALDI Based Genotyping Methods

Although MALDI-TOFMS has been used for STR genotyping under various experimental designs,[37-43] its application has been limited to simple STRs that contain only one repeat sequence, or that with a second site mutation. The main reason is due to the limited resolution and mass accuracy offered by MALDI-TOFMS at m/z ≥ 20 kDa. Those STRs with nonconsensus alleles, compound STRs and complex repeats are difficult to identify with this approach. However, MALDI-TOFMS is best suited for SNP genotyping since allele specific products can be generated within the optimal detection range (1 kDa < m/z < 10 kDa).

The most commonly used SNP genotyping assay for MALDI-TOFMS detection is the primer extension assay.[44] After PCR amplification of the region of interest from genomic DNA, a genotyping primer is annealed adjacent to the polymorphic site in the presence of DNA polymerase and a mixture of dNTPs and ddNTPs. The mixture of dNTPs and ddNTPs is carefully chosen so that one ddNTP terminates at the polymorphic site and the same type of dNTP is not present in the mixture. An example is illustrated in Figure 2. Such reaction mixture generates allele specific products of different lengths, which allows easy interrogation by MALDI-TOFMS.

A few critical sample preparation steps are prerequisite for a successful MALDI analysis of DNA products. One is to desalt the genotyping reaction products, as cations are known to interfere with MALDI.[45] This can be achieved by simply adding cation exchange resins after primer

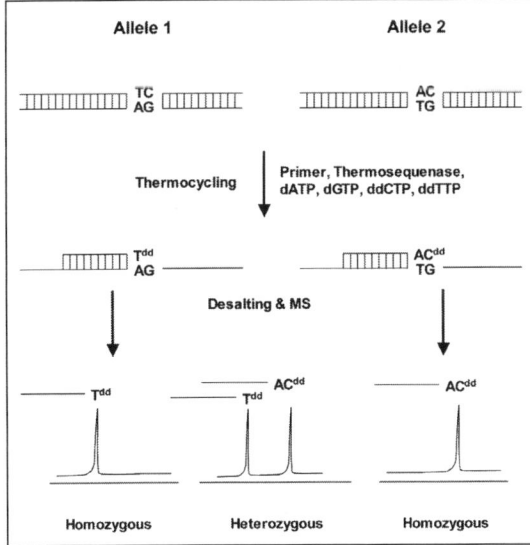

Figure 2. Genotyping for an A/T polymorphism. The primer is extended with ddT in one allele and dA + ddC in the other allele. The masses of the corresponding primer extension products are measured in MALDI-TOFMS. DNA duplexes are usually denatured under MALDI conditions. The long template does not interfere with signals obtained in the low mass region.

extension reaction.[46] The next critical step is the cocrystallization with matrix. It is at this step that chips are most commonly used, for sample concentration, accurate spot positioning and better crystallization.

Chip as a Sample Positioning and Concentration Device

The key feature for chips used for this purpose is the miniaturized sample spots in an array format with extremely hydrophobic surroundings. Typical sample spot size is $200 \times 200 \ \mu m^2$. There are several advantages for such design.

First, sample-matrix crystallization is better controlled in a miniaturized MALDI sample preparation. In the traditional dry-droplet preparation, sample and matrix are premixed in solution phase and a minimum of 0.2 μL or more of the mixture is dispensed with a conventional pipette tip. The droplet covers about 2 mm in diameter on un-treated metal or silicon surface. This area does not decease much during droplet evaporation. Since samples from genotyping reaction contain mainly aqueous solution and the commonly used matrix for nucleic acid, 3-hydroxypicolinic acid,[47] is only soluble in water, the drying process is relatively slow and crystal formation starts at the perimeter of the spot. The resulting sample crystal is usually in a donut shape. This makes it necessary to search for "hot" spot during data acquisition. Usually the XY stage of the mass spectrometer that carries the sample moves under manual control so that a "hot" spot is under the laser beam, which is usually 50-100 μm elliptical. In some latest models of mass spectrometers, image processing is used to direct the XY stage to move so that the laser hits where the crystals are. However, not all matrix crystals contain analyte molecules in a macro-preparation. Searching for "hot" spot is still necessary. This is incompatible with the high-throughput requirement for genotyping.

In the chip format, matrix is predispensed on the chip using high-speed piezo-electric pipette.[48] The solvent content, matrix concentration and dispense volume are optimized so that the crystallized matrix just covers the $200 \times 200 \ \mu m^2$ spot. Genotyping reactions are usually performed in 384-well microtiter plate before adding cation exchange resins for desalting. After centrifugation, the supernatant can be aspirated and dispensed on the chip preloaded with matrix using a pintool. In order to avoid destruction of matrix crystals upon contact dispensing, slot pins with openings greater than 200 μm are used so that when the pins touch the chip surface, the matrix spot at the center of the slot can be spared. The volume of liquid delivery is proportional to the speed of the pin upon contact and can be calibrated and precisely controlled. Usually, 15 to 20 nL of sample is dispensed. The aqueous sample solution partially redissolves the matrix crystal and the droplet quickly shrinks toward the $200 \times 200 \ \mu m^2$ center area during solvent evaporation. The matrix then recrystallized with the incorporation of analyte molecules. It has been found that such two-step dispensing and crystallization consistently yields better signal reproducibility in MALDI than those premixed and dispensed samples. Furthermore, crystal formation is more homogeneous in the $200 \times 200 \ \mu m^2$ area. Searching for "hot" spots is not necessary as signal can be obtained throughout the area.

Second, the hydrophobic surroundings serve as a sample-concentrating device. For sample concentration in the 0.1 to 1 μM range, traditional dry-droplet preparation allows for detection of DNA in the < 10 kDa range. However, when close to 0.1 μM, significant spot searching is necessary and poor signal-to-noise ratio may be observed. With the chip format, the entire sample is eventually concentrated to the small $200 \times 200 \ \mu m^2$ spot, although the droplet starts at a much bigger area. Sensitivity is significantly improved. Signal can be detected with better S/N ratio and without the need to search for "hot" spots. It is even possible to detect samples as low as 0.01 μM. Further improvement is also possible with smaller (e.g., $100 \times 100 \ \mu m^2$) sample spots.

Another advantage of using the chip is that spot locations in an array format can be accurately made on silicon chips or other kinds of conducting or semi conducting flat surfaces. The pintools can also be mounted accurately in an array format with $4500 \pm 20 \ \mu m$ spacing so that sample transfer from 384-well microtiter plates to chips can be performed in parallel. With a 4×6 pin array, the total transfer time is about 12 minutes including washing and drying of pins after each aspirating and dispensing.[31] Larger pin arrays can be made to further increase the throughput of

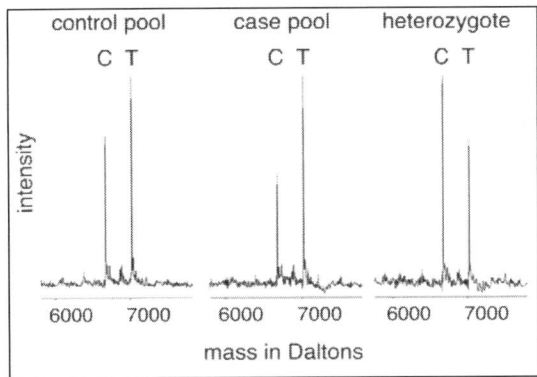

Figure 3. Allele frequency determination using pooled DNA samples. A control pool and a case pool are analyzed separately. The percentage of peak area for each allele specific peak represents its allele frequency. This allele frequency is further corrected by analyzing individual heterozygous samples for sequence specific ionization efficiency and detector response to obtain true allele frequency. In this example,[50] frequency estimates of the C allele are 0.424 and 0.355 in the control and case pools, respectively. Given the C allele frequency is overestimated in the heterozygote as 0.570, allele frequencies in the pools can be adjusted to 0.357 and 0.293, respectively. Copyright 2002 National Academy of Sciences, USA. Reproduced with permission from ref. 50.

sample transfer. After the chips are loaded with samples and introduced into the mass spectrometer, the entire array positions can be accurately generated with an automatic initial registration of several index positions. By using a solid-state laser with 200 Hz repetition rate and a high speed XY stage, MALDI-TOF genotyping of 384 samples on a chip can be achieved at a rate approaching 1 second/sample with real-time genotype calling. The key to the high-speed is based on the fact that there is no need to search for "hot" spots. This is not possible without miniaturizing sample preparation on a chip.

It should be noted that in order to determine allele frequency in pooled DNA samples[49,50] (e.g., comparing allele frequency in case and control populations as shown in Fig. 3), it is necessary to average signals from several different positions on the same $200 \times 200~\mu m^2$ spot in order to obtain better statistical data. The more homogeneous sample preparation on chip also ensures that each raster position producing a useful S/N ratio that can be used in allele frequency determination.

Chip as a Functionalized Capture Device

In an attempt to integrate genotyping reaction and MALDI sample preparation, a different type of chip format has been used.[51] The method is shown in Figure 4. It starts with PCR amplification with a thiolated primer and immobilization of the PCR product on aminated chip surface via N-succinimidyl (4-iodoacetyl)aminobenzoate (SIAB) chemistry.[52] The immobilized duplex is then denatured, leaving the covalently attached single-stranded DNA as template for genotyping reactions. The chip has shallow wells. Each well holds about 1 μL volume for primer extension reactions described previously. The chip is placed in a chamber made from a microscope slide and seal film. The chamber is positioned in an in situ PCR machine for primer extension reactions. Purification of the reaction products is accomplished simply by washing the chip wells. Matrix solution is then dispensed into the wells by a piezo-electric pipette. To ensure better crystallization, a second dispensing step with either matrix solution or water is used. A scanning electron microscope (SEM) photo of the matrix spot is shown in Figure 5. The chip can be analyzed directly by MALDI-TOFMS. Primer extension products with un-extended primer are usually detected. The covalently attached templates remain on the chip and can be re-used.

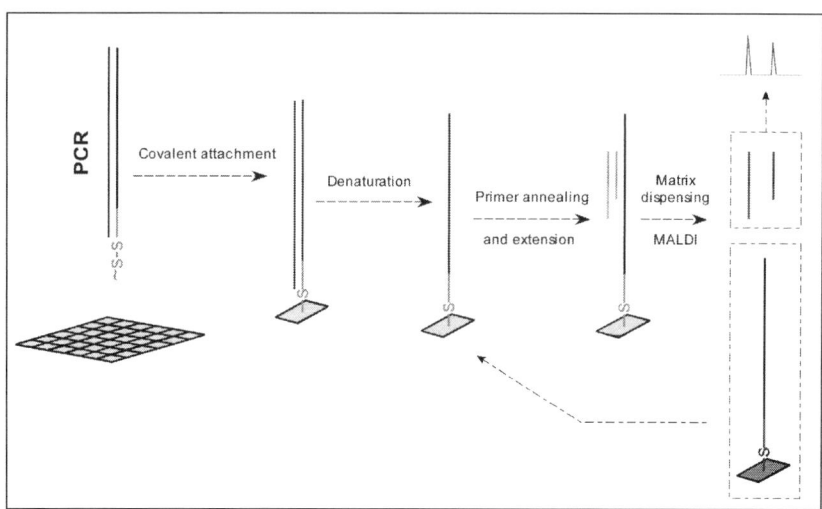

Figure 4. Primer extension reaction with covalently attached template on chip surface. PCR products are immobilized on chip surface with SIAB chemistry. Double-stranded DNAs are then denatured, leaving only one strand covalently attached on surface. Primer extension reactions are performed on chip followed by washing and subsequent MALDI-TOFMS on chip surface. Copyright 1999 National Academy of Sciences, USA. Reproduced with permission from ref. 51.

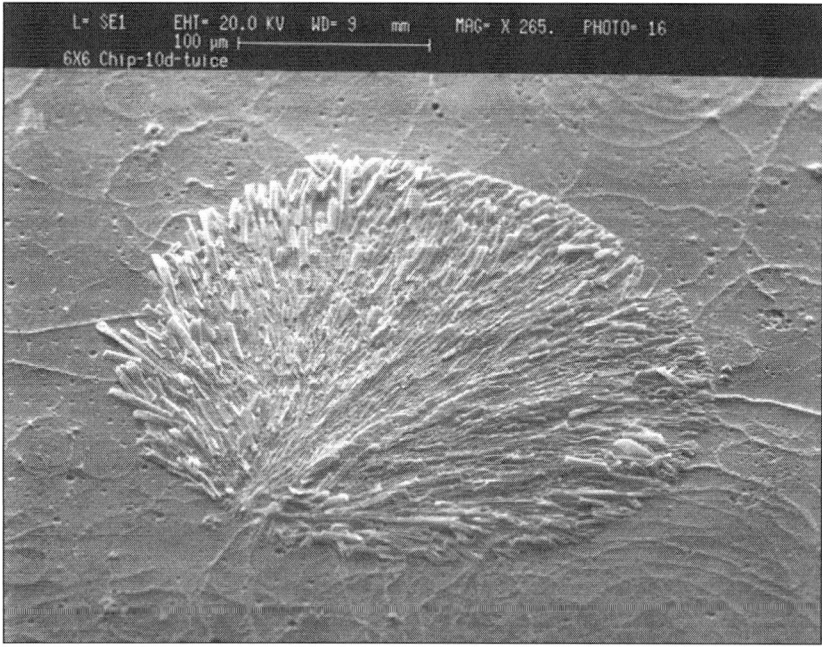

Figure 5. Scanning electron microscopy of one recrystallized matrix spot formed by dispensing 3 nl of matrix twice. Copyright 1999 National Academy of Sciences, USA. Reproduced with permission from ref. 51.

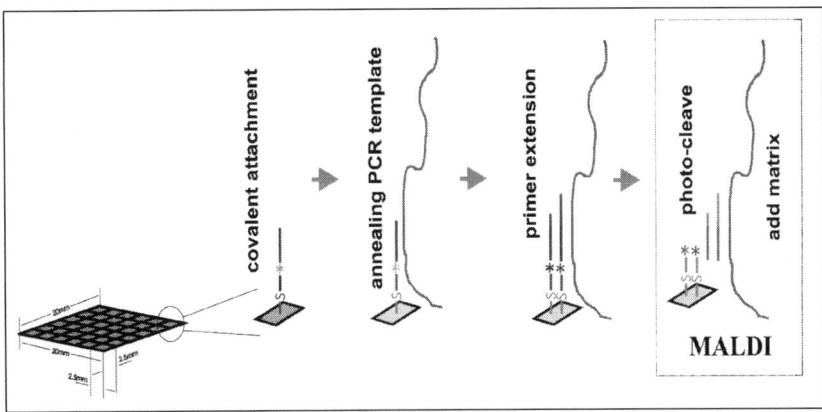

Figure 6. Primer extension reaction with covalently attached primer on chip surface. Genotyping primer is attached to the chip surface via a photo-cleavable or acid-liable linker and then used to capture the template. Primer extension can be performed on surface followed by denaturing and washing off the template. The primer extension products can be cleaved by exposing the chip to UV light or simply by adding acidic 3-HPA matrix.

An alternative approach can also be used, but has not been demonstrated.[53,54] This involves covalent attachment of the genotyping primer on chip surface via a photo-cleavable or acid liable linker, as shown in Figure 6. The PCR amplified template can be dispensed to hybridize with the primer. Primer extension reaction can be performed with the same setup on chip. Then the template can be denatured and washed away with buffers and salts, leaving only the extended primer on chip surface. The primer extension product can be cleaved from the linker after exposing the chip to UV lamp if photo-cleavable linker is used. Finally, matrix solution can be dispensed on chip for MALDI. When acid-liable linker is used, acidic matrix solution can be dispensed directly for cleavage and after crystallization, for MALDI-MS on chip.

Although functionalized chip formats could offer significant savings on cost by miniaturizing primer extension reaction on chip, major draw-backs include limited surface density of covalent attachment and lower percentage of primer extension compared to the same reaction performed in a test tube.[51,52] These result in lower detection sensitivity in MALDI-TOFMS. Others used biotinylated primer for genotyping reactions and streptavidin coated magnetic beads for purification.[55-57] However, genotyping cost using this format is significantly higher.

ESI Based Genotyping Methods

Electrospray ionization generates multiply charged ions. That makes it ideal for measuring large ions at low m/z value in mass spectrometers. When coupled with mass spectrometers with high mass accuracy (e.g., 50 ppm in Q-TOF or 2 ppm in FTICR), ESI can identify SNPs directly from PCR products.[58] However, that does not make it an ideal tool for genotyping SNPs. The main reason is that in order to determine allele frequencies (as low as 1% by definition of polymorphism) for pooled DNA samples, it is necessary to resolve the smallest mass difference in SNPs (9 Da in an A/T polymorphism) to baseline. For large ions (e.g., single-stranded DNA longer than 80 bases), there will be significant overlapping of isotope profiles of the two alleles. When the isotope distribution is not fully resolved, as in the case of Q-TOF, determination of allele frequencies is very difficult if not impossible. Even in a high resolution FTICR instrument, where isotopes can be fully resolved, isobars from the two alleles could overlap. Therefore, the technique may be useful for genotyping SNPs in individual samples, such as in clinical and forensic applications; it may not be applicable as a genomic tool. Moreover, throughput of ESI based MS is still lower than that of MALDI-TOF.

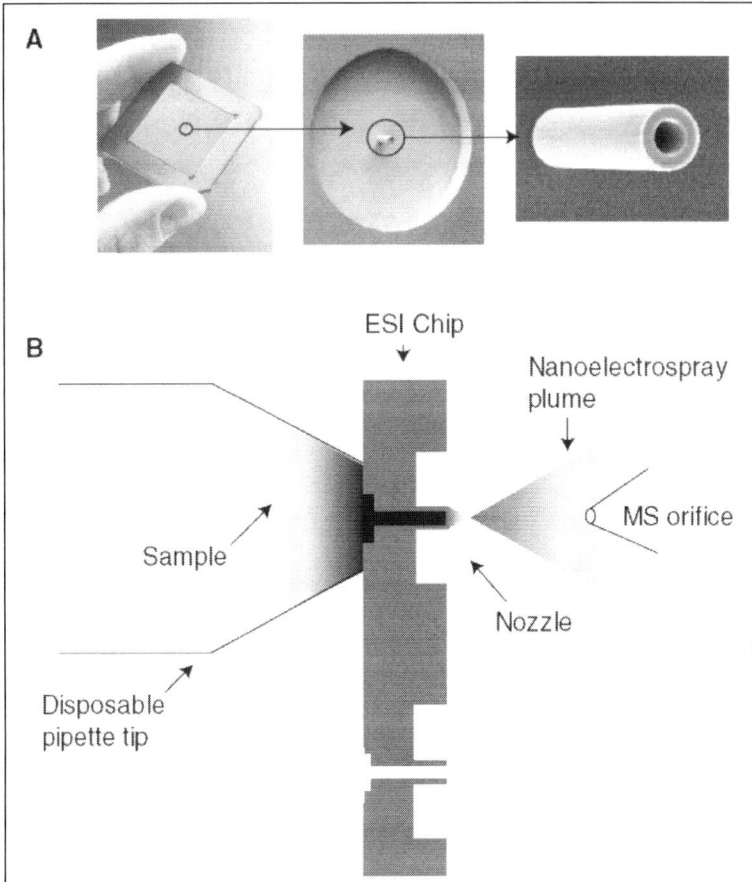

Figure 7. Nanoelectrospray chip. A) Photographs of the ESI chip showing the array of micro-fabricated nanoelectrospray nozzles with successive enlargements of an individual nozzle. This final scanning electron micrograph (SEM) shows a single nanoelectrospray emitter with dimensions of 10 μm ID and 20 μm OD. B) Schematic representation for a cross section of a portion of an ESI chip. A disposable pipette tip, containing sample, presses and seals against the inlet side of the chip with the nanoelectrospray plume spraying towards the mass spectrometer ion orifice entrance. Nanoelectrospray is initiated by applying a head pressure and voltage to the sample in the pipette tip. Copyright 2003 WILEY-VCH Verlag GmbH & Co., Reproduced with permission from ref. 63.

For genotyping of STRs, ESI coupled with FTICR has been demonstrated.[59] By direct measurement of double-stranded PCR products containing STRs with high mass accuracy (<50 ppm), base composition of simple STRs can be deduced with high confidence, given the sequences of PCR primers and the prerequisite condition that the two strands are complementary. The double-stranded PCR products can also be denatured, either by heating or by exonuclease digestion of one strand with 5'-phosphate, leaving both or one single-stranded amplicon for measurement. By using internal calibration, mass accuracy approaching 10 ppm can be achieved in FTICR, which makes it possible for genotyping compound repeats, where alleles may have the same number of repeats but with different base composition.[59] For more complex STRs, such as those repeats with microheterogeity, it is necessary to locate the heterogeneity within the tandem repeats. One of the methods is again primer extension reaction, as used for SNPs genotyping. Depending on the base composition of the heterogeneity, it might be possible to choose a dNTP and ddNTP combination so that primer extension terminates at the heterogeneous site.[38] The primer extension product can be measured with MALDI-TOF, ESI-QTOF or FTICR. When the appropriate nucleotide combination cannot be found, dissociation of the intact PCR product in FTICR can be used to generate sequence information in order to locate the heterogeneity.[60,61]

Nano-electrospray microchips have been developed for proteomics applications.[62] In one format, spray nozzles in a 10 × 10 array, each with an inner and outer diameter of 10 and 20 µm respectively, have been crated from a single silicon wafer using deep reactive ion etching.[63] A robotic liquid handling device sequentially delivers samples from a 96-well plate to the nano-ESI chip through a disposable conductive pipette tip. The pipette tip presses and seals against the inlet side of the nozzle. Nano-electrospray is initiated by applying a head pressure and voltage to the sample in the pipette tip. A schematic diagram is shown in Figure 7. The system provides fully automated nano-ESI infusion analysis without any carryover or cross contamination between samples. Low flow rate of nano-ESI reduces matrix suppression, improves overall transfer efficiency and yields higher sensitivity in mass spectrometry. Although it has not been demonstrated for analyzing PCR products, in principle, it can be used for genotyping of STRs.

Conclusion

Significant progresses have been made on chip-based SNP genotyping using MALDI-TOF mass spectrometry. However, much work is still needed to integrate front-end sample preparations on chip. STR genotyping with ESI mass spectrometry is still in its infancy and yet to be coupled with chip based format. Nevertheless, the unparalleled accuracy of mass spectrometry has already made it a gold standard for genotyping. Coupled with advanced chip based format, sensitivity and throughput of mass spectrometry based systems could be further improved for high throughput SNP and STR analyses.

References

1. Weber JL, May PE. Abundant class of human DNA polymorphisms which can be typed using the polymerase chain reaction. Am J Hum Genet 1989; 44(3):388-396.
2. Litt M, Luty JA. A hypervariable microsatellite revealed by in vitro amplification of a dinucleotide repeat within the cardiac muscle actin gene. Am J Hum Genet 1989; 44(3):397-401.
3. Edwards A, Civitello A, Hammond HA et al. DNA typing and genetic mapping with trimeric and tetrameric tandem repeats. Am J Hum Genet 1991; 49(4):746-756.
4. Tautz D. Hypervariability of simple sequences as a general source for polymorphic DNA markers. Nucleic Acids Res 1989; 17(16):6463-6471.
5. Lander ES, Schork NJ. Genetic dissection of complex traits. Science 1994; 265(5181):2037-2048.
6. Weber JL. Human DNA polymorphisms and methods of analysis. Curr Opin Biotechnol 1990; 1(2):166-171.
7. Hammond HA, Jin L, Zhong Y et al. Evaluation of 13 short tandem repeat loci for use in personal identification applications. Am J Hum Genet 1994; 55(1):175-189.
8. Fregeau CJ, Fourney RM. DNA typing with fluorescently tagged short tandem repeats: a sensitive and accurate approach to human identification. Biotechniques 1993; 15(1):100-119.
9. Zhu H, Clark SM, Benson SC et al. High-sensitivity capillary electrophoresis of double-stranded DNA fragments using monomeric and dimeric fluorescent intercalating dyes. Anal Chem 1994; 66(13):1941-1948.
10. Sprecher CJ, Puers C, Lins AM et al. General approach to analysis of polymorphic short tandem repeat loci. Biotechniques 1996; 20(2):266-276.
11. Lahiri DK, Zhang A, Nurnberger JI, Jr. High-resolution detection of PCR products from a microsatellite marker using a nonradioisotopic technique. Biochem Mol Med 1997; 60(1):70-75.
12. Cooper DN, Smith BA, Cooke HJ et al. An estimate of unique DNA sequence heterozygosity in the human genome. Hum Genet 1985; 69(3):201-205.
13. Sachidanandam R, Weissman D, Schmidt SC et al. A map of human genome sequence variation containing 1.42 million single nucleotide polymorphisms. Nature 2001; 409(6822):928-933.
14. Kwok PY. Methods for genotyping single nucleotide polymorphisms. Ann Rev Genomics Hum Genet 2001; 2:235-258.
15. Syvanen AC. Accessing genetic variation: Genotyping single nucleotide polymorphisms. Nature Rev Genet 2001; 2(12):930-942.

16. Kirk BW, Feinsod M, Favis R et al. Single nucleotide polymorphism seeking long term association with complex disease. Nucleic Acids Res 2002; 30(15):3295-3311.
17. Tanaka K, Waki H, Ido Y et al. Protein and polymer analysis up to m/z 100,000 by laser desorption time-of-flight mass spectrometry. Rapid Comm Mass Spectrom 1988; 2:151-153.
18. Karas M, Hillenkamp F. Laser desorption ionization of proteins with molecular masses exceeding 10,000 daltons. Anal Chem 1988; 60(20):2299-2301.
19. Fenn JB, Mann M, Meng CK et al. Electrospray ionization for mass spectrometry of large biomolecules. Science 1989; 246(4926):64-71.
20. Henry KD, Williams ER, Wang BH et al. Fourier-transform mass spectrometry of large molecules by electrospray ionization. Proc Natl Acad Sci USA 1989; 86(23):9075-9078.
21. Mclafferty FW. High-Resolution Tandem FT Mass-Spectrometry above 10-Kda. Acc Chem Res 1994; 27(11):379-386.
22. Comisarow MB, Marshall AG. The early development of Fourier transform ion cyclotron resonance (FT-ICR) spectroscopy. J Mass Spectrom 1996; 31(6):581-585.
23. He F, Hendrickson CL, Marshall AG. Baseline mass resolution of peptide isobars: a record for molecular mass resolution. Anal Chem 2001; 73(3):647-650.
24. Gut IG. Automation in genotyping of single nucleotide polymorphisms. Hum Mutat 2001; 17(6):475-492.
25. Lechner D, Lathrop GM, Gut IG. Large-scale genotyping by mass spectrometry: experience, advances and obstacles. Curr Opin Chem Biol 2002; 6(1):31-38.
26. Sauer S, Gut IG. Genotyping single-nucleotide polymorphisms by matrix-assisted laser-desorption/ionization time-of-flight mass spectrometry. J Chromatogr B Analyt Technol Biomed Life Sci 2002; 782(1-2):73-87.
27. Tost J, Gut VG. Genotyping single nucleotide polymorphisms by mass spectrometry. Mass Spectrom Rev 2002; 21(6):388-418.
28. Chen X, Sullivan PF. Single nucleotide polymorphism genotyping: biochemistry, protocol, cost and throughput. Pharmacogenomics J 2003; 3(2):77-96.
29. Kim S, Ruparel HD, Gilliam TC et al. Digital genotyping using molecular affinity and mass spectrometry. Nature Rev Genet 2003; 4(12):1001-1008.
30. Marvin LF, Roberts MA, Fay LB. Matrix-assisted laser desorption/ionization time-of-flight mass spectrometry in clinical chemistry. Clin Chim Acta 2003; 337(1-2):11-21.
31. Tang K, Opalsky D, Abel K et al. Single nucleotide polymorphism analyses by MALDI-TOF MS. Int J Mass Spectrom 2003; 226(1):37-54.
32. Gut IG. DNA analysis by MALDI-TOF mass spectrometry. Hum Mutat 2004; 23(5):437-441.
33. Jurinke C, Oeth P, van den Boom D. MALDI-TOF mass spectrometry-A versatile tool for high-performance DNA analysis. Mol Biotechnol 2004; 26(2):147-163.
34. Tost J, Gut IG. Genotyping single nucleotide polymorphisms by MALDI mass spectrometry in clinical applications. Clin Biochem 2005; 38(4):335-350.
35. Null AP, Muddiman DC. Perspectives on the use of electrospray ionization Fourier transform ion cyclotron resonance mass spectrometry for short tandem repeat genotyping in the postgenome era. J Mass Spectrom 2001; 36(6):589-606.
36. Hofstadler SA, Sannes-Lowery KA, Hannis JC. Analysis of nucleic acids by FTICR MS. Mass Spectrom Rev 2005; 24(2):265-285.
37. Ross PL, Belgrader P. Analysis of short tandem repeat polymorphisms in human DNA by matrix-assisted laser desorption ionization mass spectrometry. Anal Chem 1997; 69(19):3966-3972.
38. Braun A, Little DP, Reuter D et al. Improved analysis of microsatellites using mass spectrometry. Genomics 1997; 46(1):18-23.
39. Butler JM, Li J, Shaler TA et al. Reliable genotyping of short tandem repeat loci without an allelic ladder using time-of-flight mass spectrometry. Int J Legal Med 1998; 112(1):45-49.
40. Ross PL, Davis PA, Belgrader P. Analysis of DNA fragments from conventional and microfabricated PCR devices using delayed extraction MALDI-TOF mass spectrometry. Anal Chem 1998; 70(10):2067-2073.
41. Taranenko NI, Potter NT, Allman SL et al. Detection of trinucleotide expansion in neurodegenerative disease by matrix-assisted laser desorption/ionization time-of-flight mass spectrometry. Genet Anal 1999; 15(1):25-31.
42. Krebs S, Seichter D, Forster M. Genotyping of dinucleotide tandem repeats by MALDI mass spectrometry of ribozyme-cleaved RNA transcripts. Nature Biotechnol 2001; 19(9):877-880.
43. Paris M, Jones MGK. Microsatellite genotyping by primer extension and MALDI-ToF mass spectrometry. Plant Mol Biol Rep 2002; 20(3):259-263.
44. Braun A, Little DP, Koster H. Detecting CFTR gene mutations by using primer oligo base extension and mass spectrometry. Clin Chem 1997; 43(7):1151-1158.

45. Shaler TA, Wickham JN, Sannes KA et al. Effect of impurities on the matrix-assisted laser desorption mass spectra of single-stranded oligodeoxynucleotides. Anal Chem 1996; 68(3):576-579.

46. Nordhoff E, Ingendoh A, Cramer R et al. Matrix-assisted laser desorption/ionization mass spectrometry of nucleic acids with wavelengths in the ultraviolet and infrared. Rapid Comm Mass Spectrom 1992; 6(12):771-776.

47. Wu KJ, Steding A, Becker CH. Matrix-assisted laser desorption time-of-flight mass spectrometry of oligonucleotides using 3-hydroxypicolinic acid as an ultraviolet-sensitive matrix. Rapid Comm Mass Spectrom 1993; 7(2):142-146.

48. Little DP, Cornish TJ, Odonnell MJ et al. MALDI on a chip: Analysis of arrays of low femtomole to subfemtomole quantities of synthetic oligonucleotides and DNA diagnostic products dispensed by a piezoelectric pipet. Anal Chem 1997; 69(22):4540-4546.

49. Ross P, Hall L, Haff LA. Quantitative approach to single-nucleotide polymorphism analysis using MALDI-TOF mass spectrometry. Biotechniques 2000; 29(3):620-629.

50. Mohlke KL, Erdos MR, Scott LJ et al. High-throughput screening for evidence of association by using mass spectrometry genotyping on DNA pools. Proc Natl Acad Sci USA 2002; 99(26):16928-16933.

51. Tang K, Fu DJ, Julien D et al. Chip-based genotyping by mass spectrometry. Proc Natl Acad Sci USA 1999; 96(18):10016-10020.

52. ODonnell MJ, Tang K, Koster H et al. High density, covalent attachment of DNA to silicon wafers for analysis by MALDI-TOF mass spectrometry. Anal Chem 1997; 69(13):2438-2443.

53. Kang C, Kwon Y-S, Kim YT et al. Mass spectrometric methods for sequencing nucleic acids. US Patent 6,268,131, 2001.

54. Koster H, Little DP, Braun A et al. DNA diagnosis based on mass spectrometry. US Patent Application 20020042112 2002.

55. Wenzel T, Elssner T, Fahr K et al. Genosnip: SNP genotyping by MALDI-TOF MS using photocleavable oligonucleotides. Nucleosides Nucleotides Nucleic Acids 2003; 22(5-8):1579-1581.

56. Sauer S, Lehrach H, Reinhardt R. MALDI mass spectrometry analysis of single nucleotide polymorphisms by photocleavage and charge-tagging. Nucleic Acids Res 2003; 31(11).

57. Vallone PM, Fahr K, Kostrzewa M. Genotyping SNPs using a UV-photocleavable oligonucleotide in MALDI-TOF MS. Methods Mol Biol 2005; 297:169-178.

58. Oberacher H, Niederstatter H, Casetta B et al. Detection of DNA sequence variations in homo- and heterozygous samples via molecular mass measurement by electrospray ionization time-of-flight mass spectrometry. Anal Chem 2005; 77(15):4999-5008.

59. Null AP, Hannis JC, Muddiman DC. Genotyping of simple and compound short tandem repeat loci using electrospray ionization Fourier transform ion cyclotron resonance mass spectrometry. Anal Chem 2001; 73(18):4514-4521.

60. Little DP, Speir JP, Senko MW et al. Infrared Multiphoton Dissociation of Large Multiply-Charged Ions for Biomolecule Sequencing. Anal Chem 1994; 66(18):2809-2815.

61. Little DP, Aaserud DJ, Valaskovic GA et al. Sequence information from 42-108-mer DNAs (complete for a 50-mer) by tandem mass spectrometry. J Am Chem Soc 1996; 118(39):9352-9359.

62. Zhang S, Van Pelt CK. Chip-based nanoelectrospray mass spectrometry for protein characterization. Expert Rev Proteomics 2004; 1(4):449-468.

63. Zhang S, Van Pelt CK, Henion JD. Automated chip-based nanoelectrospray-mass spectrometry for rapid identification of proteins separated by two-dimensional gel electrophoresis. Electrophoresis 2003; 24(21):3620-3632.

CHAPTER 10

Analyzing DNA-Protein Interactions on a Chip

Limin Lin and James P. Brody*

Abstract

Gene expression is regulated by multi-protein complexes binding to short noncoding regions of genomic DNA, called *cis*-regulatory elements. A long-term goal of genomics is to identify and annotate all essential elements of the genome. Most coding elements of the genome have been identified and annotated; however, most regulatory elements have not. This chapter describes a method for the high throughput identification of DNA-protein interactions to identify *cis*-regulatory elements.

Introduction

DNA Protein Interactions Control Gene Expression

An important aspect of the analysis of genomic DNA is not only identifying its structure or sequence, but also its function. If its function is to encode sequences of amino acids, decoding it is trivial. The human genome encodes at least two important functions. The first, DNA sequence that encodes protein sequence, has now been almost completely identified. The second, DNA sequence that binds specific proteins, is almost completely unknown. In eukaryotes, the DNA sequence that binds specific proteins is known as *cis*-regulatory elements. Multi-protein complexes bind to these short, specific DNA sequences regulating the transcription process.

The human genome contains vast amounts of *cis*-regulatory elements, most of which have not yet been identified. These regulatory sequences are responsible for directing spatial and temporal patterns of gene expression, which affect metabolic requirements and developmental programs.[1] *Cis*-regulatory elements are generally less than 20 base pairs and usually located in the promoter region of the regulated gene. The regulatory elements can be located upstream or downstream from the transcription start site.

The simplest binding model involves a molecule (a particular region in a single protein tertiary structure, P) that can bind to a specific binding site (a specific sequence of DNA, D) and form a resulting complex, C, which can enhance or repress expression.

$$P + D \rightarrow C$$

Their interactions with the targets range from the highly specific (P binds a single site) to the nonspecific (P binds most sites).[2] Gene expression may be regulated through a combination of multi-protein complexes binding to several distinct elements of the promoter and cooperative interactions, as shown in Figure 1. A comprehensive identification of the location and relative strength

*Corresponding Author: James P. Brody—Department of Biomedical Engineering, University of California, Irvine, California U.S.A. Email: jpbrody@uci.edu

Integrated Biochips for DNA Analysis, edited by Robin Hui Liu and Abraham P. Lee.
©2007 Landes Bioscience and Springer Science+Business Media.

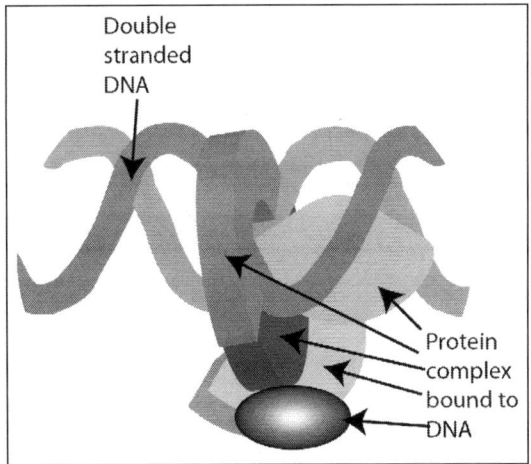

Figure 1. This schematic diagram depicts a multi protein complex binding to a short, specific sequence of genomic DNA known as a *cis*-regulatory element. This binding enhances or represses gene expression. Identifying these *cis*-regulatory elements is a challenge, since the identity of the proteins in the complex are unknown.

of all these *cis*-regulatory elements on every gene is key to understanding the composition and function of the regulation networks that carry out the essential processes of living organisms.[3]

As a result, the development of general and efficient assays of DNA-protein interactions has been extensively studied.[4] However, the ability to identify and predict functions of the *cis*-regulatory elements is limited. Many different approaches to this challenging problem have been tried.

Approaches to Identify *cis*-Regulatory Elements

In general, the approaches developed to identify and predict functions of *cis*-regulatory elements fall into two classes: experimental and computational. Experimental approaches focus on the identification of regulatory elements for individual genes. Computational approaches focus on scanning large data sets to identify *cis*-regulatory elements.

Experimental Strategies for Identification of cis-*Regulatory Elements*

The search for *cis*-regulatory sequences generally involves various trial-and-error strategies. Traditional experimental approaches to identify regulatory elements for individual genes include: (1) Generation of deletion constructs to determine the minimal sequences necessary for transcription in cell-culture-based systems; (2) Gel shift assays, in which DNA molecules binding to proteins migrate more slowly in the gel relative to the samples with no protein[5]; to determine the sequences that bind various regulatory proteins (3) DNA footprinting assays to identify a region of DNA protected from digestion by DNAse I by a bound protein (usually a transcription factor).[6] These methods generally require labeling with a fluorescent or radioactive tag. The procedures are complicated and time-consuming.

Surface Plasmon Resonance-Based Assay to Identify cis-*Regulatory Elements*

With a purely computational approach, uncertainty remains as to whether a predicted *cis*-regulatory element actually possesses the expected function.[7,8] With a purely experimental method, it is difficult to predict *cis*-regulatory elements on a large scale. The union of experimental and computational strategies is a new trend in deciphering *cis*-regulatory codes.[9] The most successful of these approaches to date appear to be those that rely on gene expression profiles from DNA microarrays.[10,11]

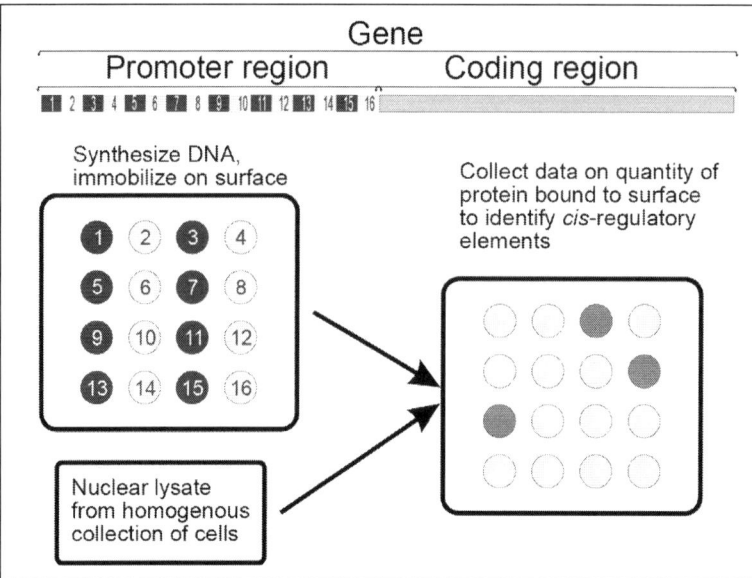

Figure 2. A schematic diagram showing a strategy for identifying *cis*-regulatory elements. The promoter region of a gene is split into small overlapping subsections, each about 20-40 bp long. These DNAs are synthesized and immobilized onto physically distinct areas of a surface. Nuclear lysate from a homogeneous collection of cells is incubated on the surface. Simultaneously, the amount of protein bound to the surface is monitored. Surface areas exhibiting high levels of bound protein will indicate regions of the promoter where protein complexes bind. These are *cis*-regulatory elements.

However, these approaches are all indirect methods. A direct approach, which identifies DNA elements that bind to nuclear protein, is better, because this approach is more likely to provide an accurate measurement. Many different assays exist for measuring DNA/protein binding, but the identification of *cis*-regulatory elements has specific constraints:

1. The assay needs to work with raw nuclear extract rather than an isolated and purified protein.
2. The assay should allow parallelization or use other approaches to enable high throughput measurements.

A surface plasmon resonance based assay satisfies these requirements. In the surface plasmon resonance assay, the first biomolecule, DNA, is immobilized onto a surface and a change in signal occurs if a second molecule or molecular complex binds to the first as shown in Figure 2. A surface plasmon resonance-based assay offers several advantages over other biomolecular binding assays: the assay can monitor binding rapidly, in real time and without any labels; and information about specificity, affinity and kinetics can be extracted from a surface plasmon resonance analysis.

Surface Plasmon Resonance-Based Assays

The physical principle known as surface plasmon resonance can be used to detect interactions of biomolecules. For instance, surface plasmon resonance has been successfully used to detect interactions between biomolecules with applications in various fields such as drug discovery,[12] cell signaling,[13] immunoassays[14,15] and virology.[16]

Surface plasmon resonance signals are proportional to the refractive index close to the sensor surface and are related to the amount of bound macromolecules.[17] The main advantage of surface plasmon resonance over the conventional assays of molecular recognition, such as the enzyme-linked immunosorbent assay (ELISA), is its ability to provide continuous real-time monitoring and its ability to work without chemical labels. The specificity and kinetics of the interaction of the biomolecules can be extracted from a surface plasmon resonance analysis.

Several commercial surface plasmon resonance instruments are currently available.[18] The first and best known is the BIAcore biosensor made by BIAcore AB (Uppsala, Sweden).[19-21] However, its size and cost (more than $100,000) limit its use. In the late 1990s Texas Instruments developed the Spreeta, an integrated electro-optical sensor package.[18,22,23] Spreeta utilizes surface plasmon resonance to measure the refractive index (RI) of liquid to monitor precisely the change in its composition. It can be adapted to detect DNA-protein interactions and obtain kinetic rate constants. This sensor is cost-effective (a single Spreeta sensor package costs only $20) and portable.

Using Surface Plasmon Resonance to Measure DNA-Protein Interaction

The Physical Principle of Surface Plasmon Resonance

Surface plasmon resonance is based on a unique optical phenomenon.[24] The underlying physical principle is that at an interface between two transparent media (e.g., glass and water) with different refractive indices, the light coming from the side of higher refractive index is partly reflected and partly refracted.

Above a certain critical angle of incidence, total internal reflection is obtained without any refraction across the interface. In this situation, the electromagnetic field component penetrates a short (hundreds of nanometers) distance into a medium of a lower refractive index, generating an exponentially attenuated evanescent wave. If the interface between the media is coated with a thin layer of metal (e.g., gold), then waves, called surface plasmons, form from the oscillation of mobile electrons at the surface of the thin metal layer. The mechanism of surface plasmon resonance is related to the resonance energy transfer between the evanescent wave and the surface plasmon. If the light is monochromatic and p-polarized (meaning the electric field of the light wave lies in the same plane as the incident wave and the surface normal) and the wave vector of the incident light matches the wavelength of the surface plasmons, then surface plasmon resonance occurs.

Some of the energy from the incident light is transferred to the electron density waves in the metallic film and therefore the intensity of the reflected light is reduced. The specific incident angle at which this occurs is called the surface plasmon resonance angle. This surface plasmon resonance angle can be measured by observing the intensity of the reflected light at a range of different angles and then identifying the angle at which the intensity is minimized.

The surface plasmon resonance angle is affected by the amount and type of materials adsorbed onto the thin metal film. A satisfactory linear relationship exists between this angle and the mass concentration of biochemically relevant molecules such as proteins, sugars and DNA adsorbed to the surface. The angle can be monitored in real time. Therefore, it is possible to determine the analyte and ligand association and dissociation rate and simultaneously detect interactions between unmodified proteins and directly measure kinetic parameters of the interaction. The surface plasmon resonance angle is also a function of the refractive index adjacent to the sensor surface and is often reported in units of refractive index.

Although surface plasmon resonance-based assays are typically performed one assay at a time, work is progressing on making these measurements amenable to parallelization. One approach is to immobilize different elements in physically distinct locations on a surface and then to acquire a surface plasmon resonance image of the surface. This approach is feasible; several different surface plasmon resonance imagers have been developed.[25-27] The primary limitation to this approach is sensitivity.

Surface plasmon resonance imagers usually work fundamentally different from the Kretschmann geometry shown in Figure 3. Imagers generally measure changes in signal through changes in optical intensity. Most other surface plasmon resonance instruments use the Kretschmann geometry, which measures a change in angle. It is easier to measure a very small change in angle as compared to a small change in intensity. Hence, the Kretschmann geometry is significantly more sensitive

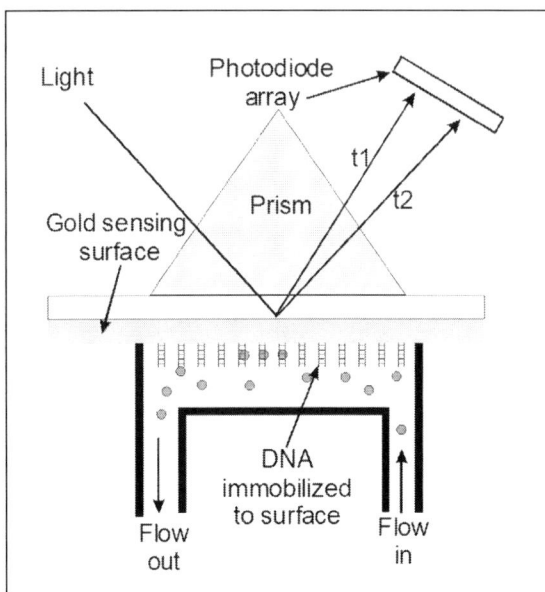

Figure 3. A schematic describing surface plasmon resonance. Light is aimed through a prism and reflects off the surface to a photodiode array. The line projected at time = t1 corresponds to the situation before binding of protein to the DNA on the surface and the line projected at time = t2 corresponds to the position of resonance after binding. The binding causes a shift in the angle of minimum reflection, where the resonance condition is satisfied.

than most current surface plasmon resonance imagers are. A challenge, then, is to design surface plasmon resonance imagers with sensitivity close to the Kretschmann geometry.

Tests Using a Spreeta Biosensor

To test this concept in a low throughput experiment, we used the Spreeta biosensor made by Texas Instruments to do real-time measurement of the interaction between protein and DNA. Spreeta was designed for applications in diagnostics, food/beverage quality and safety.[19] As compared with other instruments, Spreeta's main advantages are low cost and compact format. Figure 4 is a sketch of the Spreeta biosensor.

A Spreeta biosensor occupies a few square centimeters. It consists of a light-emitting diode (LED), a sensing surface and a light detector, which are integrated into a compact electro-optical package. Electrical connections are made to the sensor *via* pins at the bottom of the device.

When the liquid contacts the gold surface plasmon layer and the appropriate signals are applied to the pins, the sensor generates an output that corresponds to the refractive index of the liquid or the mass adsorbed on the surface. The output of the Spreeta sensor is a series of analog voltages, one per clock pulse, from which the refractive index of the liquid is derived when the voltages are digitized and processed.

Tests using a Spreeta biosensor (the miniaturized surface plasmon resonance system) were performed to evaluate the feasibility of the concept presented in Figure 1. A Spreeta evaluation module (EVM), which consists of a Spreeta biosensor, a flow cell with temperature compensation and an electronic PC interface control along with comprehensive software, was used. The Spreeta biosensor is made by Texas Instruments and the other three parts are made by Nomadics.

Experimental Set up

A schematic diagram of the Spreeta is shown in Figure 4. All of the components are immobilized in an optically transparent material. Near infrared light generated from a light-emitting diode (LED) passes through a polarizer, reflects off the back of the gold sensing surface and is then directed onto a linear array of silicon photodiodes. Each detection pixel corresponds to a narrow range of incident angles. The signals arising from the reflected light are monitored to determine the minimum signal intensity versus scattering angle, which occurs at the surface plasmon resonance angle.

In the sample handling and flow system of Spreeta, solutions are imported to a flow cell that is attached to the front of the sensor and then flow across the gold sensing surface. A rectangular channel cut into a rubber gasket defines the volume of the cell. The volume of the flow cell we used is 8 μL. The flow cell has an internal thermistor that is used for temperature compensation. Figure 5 shows the mechanism of how the flow cell works in the Spreeta evaluation module (EVM).

Our laboratory test bench consists of a computer, a VWR peristaltic pump and the EVM, which includes a Spreeta biosensor, a Spreeta interface control box that communicates between the host computer and sensor and an integrated flow cell mounted onto the Spreeta. The interface control module has a 12-bit analog to digital converter. It digitizes the analog signal from each pixel. The whole system has a resolution of about 8×10^{-7} refractive index units (RIU). According to the manufacturer, 11×10^{-3} RIU is equal to a change in the surface plasmon resonance angle of $1°$. Generally, 1 pg/mm^2 of adsorbed protein results in a change of 1×10^{-6} RIU (using Spreeta's measurement units) or a change of 1 RU (using the Biacore's measuring units).

Refractive index is a unitless number, but by convention and for clarity, it is given the dimensionless unit RIU. The resolution and noise level of this system is comparable to other commercially available surface plasmon resonance systems. The analysis software identifies the angle with the least intense reflected light. This angle is converted into RIU, which is recorded every ten seconds.

As compared with traditional time-consuming and costly methods, Spreeta has the advantages of simplicity, low cost, high efficiency and real-time measurement. There are several reports on the use of Spreeta to obtain information on biological interactions.[17,28-31]

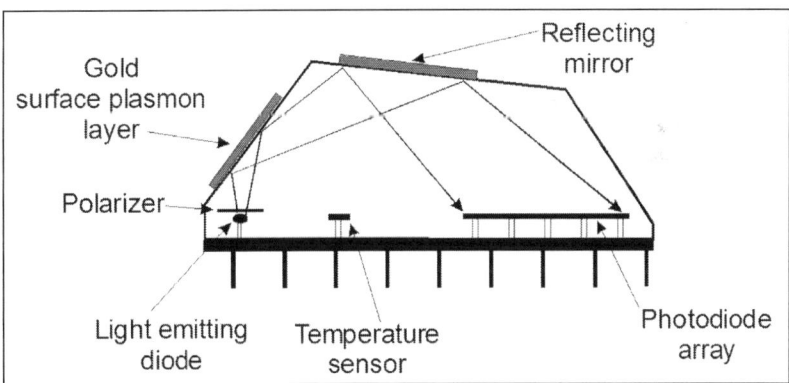

Figure 4. Sketch of Spreeta biosensor. The Spreeta sensor incorporates a LED, polarizer, thermistor and two silicon photodiode arrays, all mounted on a printed circuit board. Components are encapsulated with a clear, optical epoxy. A surface plasmon layer (the sensing surface) and reflecting mirror surface are coated on the epoxy. The photodiode array measures intensity of light reflected at different angles. The integrated temperature sensor is necessary since the refractive index varies significantly with temperature, for water it varies by about 1 part in 10,000 per degree C.

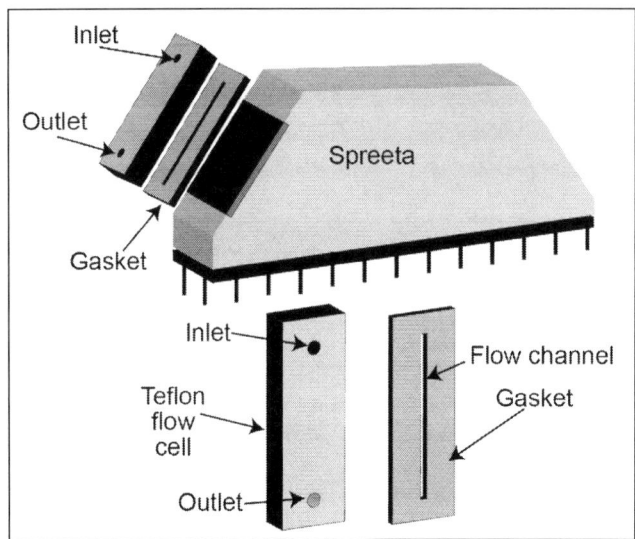

Figure 5. This drawing is a sketch of the fluidics in the Spreeta Evaluation Modules (EVM) made by Nomadics. A Teflon flow cell with two holes drilled through it is clamped onto a gasket to form a flow channel across the sensing surface of the Spreeta.

Measuring DNA-Protein Interactions Using the Surface Plasmon Resonance Sensor

The utilization of surface plasmon resonance measurements of DNA/protein binding has been previously published.[32,33] In principle, the Spreeta should be able to detect transcription factor-DNA binding and other related information. We used the Spreeta to detect transcription factor binding and combine these results with conventional biochemical assays that measure actual transcription rates. In addition, the measurements were performed not only of the interactions between purified protein and immobilized DNA, but also of the detection of the binding between DNA and a multi-protein complex within the raw nuclear lysate.

Surface Preparation

The method involves immobilizing a 16-mer biotinylated oligonucleotide on the sensor chip of Spreeta. The general immobilization scheme consists of depositing layers of biotin-BSA, streptavidin, DNA and protein, as shown in Figure 6.

The streptavidin-biotin coupling is specific due to its extremely high binding affinity, K_a, which is about 10^{13} M.[34] Each streptavidin has four equivalent binding sites for biotins. Streptavidin binds biotin rapidly and the bound complex tolerates a range of temperature and pH conditions. Because of the ease with which a wide variety of molecules can be chemically modified with biotin, the streptavidin-biotin coupling has been extensively used in many different applications in biotechnology.[35]

The utilization of surface plasmon resonance enables the real-time monitoring of changes in the refractive index of a thin film close to the sensing surface. The evanescent field created at the surface decays exponentially from the surface and falls to one third of its maximum intensity at approximately 200 nm from the surface. In our experiments, the biotinylated BSA layer is about 14.1 nm; streptavidin is about 5.8 nm; and the biotinylated DNA is about 8.0 nm. The total thickness of all the layers for surface preparation is in the range of sensing.

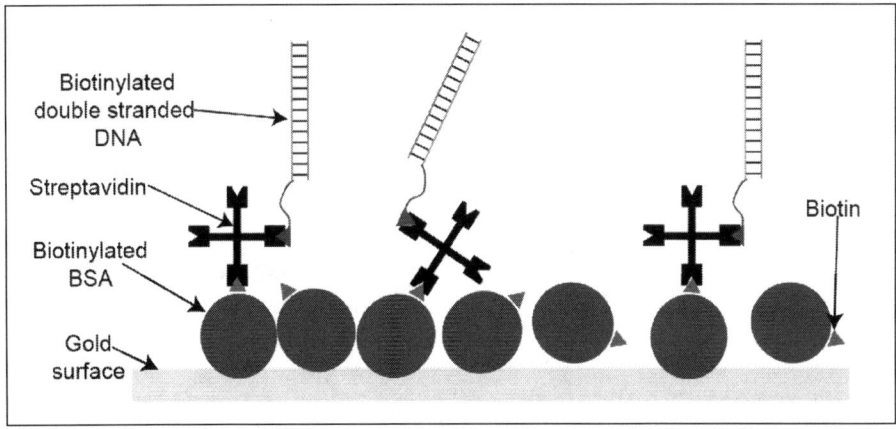

Figure 6. A cartoon schematic of the protein binding layers for the surface preparation chemistry. Biotinylated bovine serum albumin (BSA) binds nonspecifically to the gold surface. Streptavidin, which has four biotin binding sites, acts as a linker between the biotinylated BSA and the biotinylated DNA. This prepares the sensor to detect transcription factor binding.

Measurements of Double-Stranded DNA Surface Adsorption

We measured the change in surface plasmon resonance signal when streptavidin and double-stranded DNA were adsorbed to the surface. Based on the usual approximation (a change in refractive index of 1×10^{-6} is the equivalent of 1 pg/mm^2 of adsorbed mass), we calculated the surface density of streptavidin to be $2.6 \ (\pm 0.3) \times 10^{11}$/cm^2, whereas the double-stranded DNA had a surface density of $1.0 \ (\pm 0.1) \times 10^{11}$/cm^2. The ratio of double-stranded DNA to streptavidin was measured at $0.38 \ (\pm 0.06)$.

Sensitivity of the Surface Plasmon Resonance Based Sensor

Surface plasmon resonance based sensors rely upon an evanescent electromagnetic field that decays exponentially from the surface. Hence, two factors affect the ultimate limit of detection: the surface attachment chemistry and the optical data collection system. The signal could be enhanced by minimizing the attachment scheme. The surface attachment chemistry should present the capture site as close to the surface as possible. We have used an elaborate, but robust, attachment scheme: gold surface-BSA/biotin-streptavidin-biotin/DNA. This is not optimal and our signal could be increased by using a simpler scheme: attaching thiol/DNA to the gold surface. This, however, presents more problems; the thiol/DNA can form disulfide bonds (dimers) in solution.

The signal could be further enhanced by using a three dimensional attachment scheme. For instance, various investigators have shown the feasibility of coating the gold surface with a thin hydrogel, then attaching biomolecules to attachment sites. This provides enhanced signal as compared to a monolayer.

Surface plasmon resonance array imagers measure changes in signal through changes in optical intensity. Most other surface plasmon resonance instruments use the Kretschmann geometry, which measures a change in angle. A surface plasmon resonance array imager using the Kretschmann geometry would have enhanced sensitivity compared to other imagers. This would also be much slower than other surface plasmon resonance array imagers. It should have sensitivity equivalent to that achieved in these experiments.

Projected Sensitivity Based on Preliminary Experiments

To determine the sensitivity of the assay, we measured binding between a known protein, MutS and an immobilized piece of DNA. This was using the Spreeta, which has an active area of

30 mm². We measured a difference of about 0.00005 refractive index units (RIU) that could be attributable to this binding. Using the standard conversion, a change of 1×10^{-6} RIU $\approx$ 1pg/mm², this corresponds to about 9×10^9 copies of MutS, which is a 90 kDa protein. This was well within our detection limit; we could detect 9×10^8 copies, but not 9×10^7. For comparison, one liter of yeast culture can produce approximately 10^{10} cells. This could be used for 3-5 experiments and hence each experiment will have nuclear extract from about 2×10^9 cells. If we can detect 9×10^8 copies of a DNA-bound protein complex, we will be able to detect protein complexes that are present at the single copy per cell level.

Surface Plasmon Resonance Is Sensitive Enough to Detect Physiologically Relevant DNA-Protein Interactions

Surface plasmon resonance has sufficient sensitivity to accomplish our main goal. We had previously shown that the promoter of the SQSTM1 gene contained a single base pair (out of 1800) that was responsible for 80% of its transcriptional activity in a cancer cell line.[36] First, we quantified the physiological effect of this single base pair. We cloned the promoter and constructed a mutated version that differed by that single key base pair. We inserted these into luciferase vectors, then (in separate experiments) transfected them into NIH/3T3 cells. We measured the luciferase activity, which is a measure of the transcriptional activity of the vectors. We found a significant decrease in the luciferase activity when the key single base pair is mutated (see Fig. 7). This established that this single base pair is responsible for a physiological change.

Next, we measured the amount of nuclear protein that binds to this piece of DNA. We made two short (16 bp) pieces of DNA from the SQSTM1 promoter. These were centered on

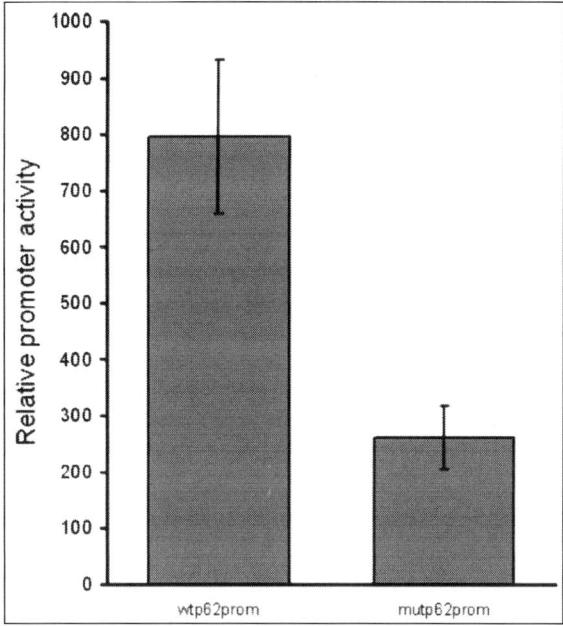

Figure 7. A single base mutation in the SQSTM1 promoter leads to significantly lower promoter activity in NIH/3T3 cells. Two luciferase reporter constructs were made: the first contained the wild-type 1800-bp promoter region from the SQSTM1 gene, the second differed by a single base pair 512 bases before the transcription start site. Each construct was inserted into a luciferase reporter vector and transfected into NIH/3T3 cells and the luciferase activity was measured.

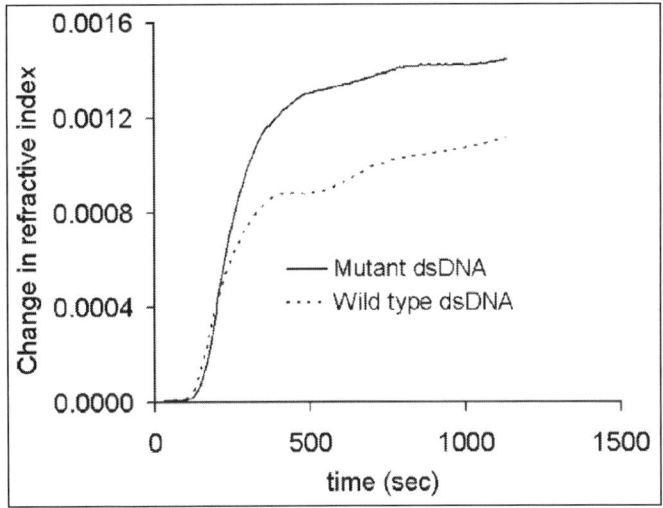

Figure 8. Surface plasmon resonance recordings can detect binding between DNA and proteins present in nuclear lysate. The surface of the sensor chip was coated with a 16-bp sequence of DNA from a region of the SQSTM1 promoter. As a control, a similar sequence containing a single-base-pair mutation was also immobilized in a different experiment. In each case, the surface was exposed to 0.33 mg/mL of nuclear lysate from NIH/3T3 cells. This lysate contains a complex mixture of proteins present within the nucleus. The relative change in refractive index is shown. The mutant and wild type were normalized to the same levels before the introduction or removal of protein. Repeated measurements showed similar behavior.

the key base pair and differed only in the key base pair. Using surface plasmon resonance, we measured the binding of protein complexes present in raw nuclear lysate to these two pieces of DNA immobilized on the surface of the sensor. We found a significant increase in binding of the modified piece of DNA (see Fig. 8).

Conclusion

Based upon these experiments we conclude that this surface plasmon resonance-based system was able to identify physiologically relevant *cis*-regulatory elements. Hence, the sensitivity is sufficient to accomplish the main goal of this project. Based on its design, we expect the high throughput version to have similar sensitivity.

References

1. Lemon B, Tjian R. Orchestrated response: a symphony of transcription factors for gene control. Genes Dev 2000; 14(20):2551-2569.
2. Wilson WD. Tech Sight. Analyzing biomolecular interactions. Science 2002; 295(5562):2103-2105.
3. Davidson EH. Genomic regulatory systems: development and evolution. San Diego: Academic Press 2001.
4. Travers AA, Buckle M. DNA, protein interactions: a practical approach. New York: Oxford University Press 2000.
5. Ausubel FM. Current protocols in molecular biology. Media, Pa.: Greene Publishing Associates 1987.
6. Galas DJ. The invention of footprinting. Trends Biochem Sci 2001; 26(11):690-693.
7. Michelson AM. Deciphering genetic regulatory codes: a challenge for functional genomics. Proc Natl Acad Sci USA 2002; 99(2):546-548.
8. Tompa M, Li N, Bailey TL et al. Assessing computational tools for the discovery of transcription factor binding sites. Nat Biotechnol 2005; 23(1):137-144.
9. Hasty J, McMillen D, Isaacs F et al. Computational studies of gene regulatory networks: in numero molecular biology. Nat Rev Genet 2001; 2(4):268-279.

10. Pilpel Y, Sudarsanam P, Church GM. Identifying regulatory networks by combinatorial analysis of promoter elements. Nat Genet 2001; 29(2):153-159.
11. Qian J, Lin J, Luscombe NM et al. Prediction of regulatory networks: genome-wide identification of transcription factor targets from gene expression data. Bioinformatics 2003; 19(15):1917-1926.
12. Cooper MA. Optical biosensors in drug discovery. Nat Rev Drug Discov 2002; 1(7):515-528.
13. Mozsolits H, Thomas WG, Aguilar MI. Surface plasmon resonance spectroscopy in the study of membrane-mediated cell signalling. J Pept Sci 2003; 9(2):77-89.
14. Mullett WM, Lai EP, Yeung JM. Surface plasmon resonance-based immunoassays. Methods 2000; 22(1):77-91.
15. Gonzales NR, Schuck P, Schlom J et al. Surface plasmon resonance-based competition assay to assess the sera reactivity of variants of humanized antibodies. J Immunol Methods 2002; 268(2):197-210.
16. Rich RL, Myszka DG. Spying on HIV with SPR. Trends Microbiol 2003; 11(3):124-133.
17. Kukanskis K, Elkind J, Melendez J et al. Detection of DNA hybridization using the TISPR-1 surface plasmon resonance biosensor. Anal Biochem 1999; 274(1):7-17.
18. Baird CL, Myszka DG. Current and emerging commercial optical biosensors. J Mol Recognit 2001; 14(5):261-268.
19. Jonsson U, Fagerstam L, Ivarsson B et al. Real-time biospecific interaction analysis using surface plasmon resonance and a sensor chip technology. Biotechniques 1991; 11(5):620-627.
20. Sjolander S, Urbaniczky C. Integrated fluid handling system for biomolecular interaction analysis. Anal Chem 1991; 63(20):2338-2345.
21. Malmqvist M. Biospecific interaction analysis using biosensor technology. Nature 1993; 361(6408):186-187.
22. Melendez J, Carr R, Bartholomew DU et al. A commercial solution for surface plasmon sensing. Sens Actuators B Chem 1996; 35(1-3):212-216.
23. Elkind JL, Stimpson DI, Strong AA et al. Integrated analytical sensors: the use of the TISPR-1 as a biosensor. Sens Actuators B Chem 1999; 54(1-2):182-190.
24. Liedberg B, Nylander C, Lundstrom I. Surface-Plasmon Resonance for Gas-Detection and Biosensing. Sens Actuators 1983; 4(2):299-304.
25. Rothenhausler B, Knoll W. Surface-Plasmon Microscopy. Nature 1988; 332(6165):615-617.
26. Fu E, Foley J, Yager P. Wavelength-tunable surface plasmon resonance microscope. Rev Sci Instrum 2003; 74(6):3182-3184.
27. Wegner GJ, Lee HJ, Marriott G et al. Fabrication of histidine-tagged fusion protein arrays for surface plasmon resonance imaging studies of protein-protein and protein-DNA interactions. Anal Chem 2003; 75(18):4740-4746.
28. Naimushin AN, Soelberg SD, Nguyen DK et al. Detection of Staphylococcus aureus enterotoxin B at femtomolar levels with a miniature integrated two-channel surface plasmon resonance (SPR) sensor. Biosens Bioelectron 2002; 17(6-7):573-584.
29. Whelan RJ, Wohland T, Neumann L et al. Analysis of biomolecular interactions using a miniaturized surface plasmon resonance sensor. Anal Chem 2002; 74(17):4570-4576.
30. Matejka P, Hruby P, Volka K. Surface plasmon resonance and Raman scattering effects studied for layers deposited on Spreeta sensors. Anal Bioanal Chem 2003; 375(8):1240-1245.
31. Yi SJ, Yuk JS, Jung SH et al. Investigation of selective protein immobilization on charged protein array by wavelength interrogation-based SPR sensor. Mol Cells 2003; 15(3):333-340.
32. Kyo M, Yamamoto T, Motohashi H et al. Evaluation of MafG interaction with Maf recognition element arrays by surface plasmon resonance imaging technique. Genes Cells 2004; 9(2):153-164.
33. Shumaker-Parry JS, Zareie MH, Aebersold R et al. Microspotting streptavidin and double-stranded DNA arrays on gold for high-throughput studies of protein-DNA interactions by surface plasmon resonance microscopy. Anal Chem 2004; 76(4):918-929.
34. Green NM. Avidin. Adv Protein Chem 1975; 29:85-133.
35. Jung LS, Nelson KE, Campbell CT et al. Surface plasmon resonance measurement of binding and dissociation of wild-type and mutant streptavidin on mixed biotin-containing alkylthiolate monolayers. Sens Actuators B Chem 1999; 54(1-2):137-144.
36. Thompson HG, Harris JW, Wold BJ et al. p62 overexpression in breast tumors and regulation by prostate-derived Ets factor in breast cancer cells. Oncogene 2003; 22(15):2322-2333.

CHAPTER 11

Single Molecule DNA Detection

Tza-Huei Wang,* Christopher M. Puleo and Hsin-Chih Yeh

Abstract

The development of single-molecule detection (SMD) technologies that allow measurements of intra and intermolecular interactions at the most fundamental level has greatly advanced basic research in biophysics and biochemistry over the past two decades. Incorporated with the use of a variety of new molecular probes, probe strategies and functional nanomaterials, SMD has recently been demonstrated as a powerful tool in medical applications. The use of SMD in quantitative study of both specific DNA sequences and mutations has resulted in extraordinary performance characteristics (sensitivity, specificity and accuracy) unmatched by conventional ensemble detection methods. The rapid advances in microfluidics and MEMS technologies have further made possible on-chip SMD analysis in a high-throughput and cost-effective manner. An integrated molecular analysis platform based on a combination of SMD, nanotechnology and microfluidics promises to tackle the current technological challenges in genomic analysis such as amplification-free detection of genomic DNA and accurate quantification of minute gene expression changes.

Introduction

Quantitative and qualitative analysis of DNA/RNA sequences provides information essential to a variety of fields such as disease diagnostics, medical therapeutics, gene expression profiling, environmental analysis, drug discovery, as well as biowarfare defense. Routine laboratory methods, such as Southern Blot and Northern Blot require a large amount of samples for standard analysis and are not amenable to applications where only a limited amount of samples are available for analysis; for example, early disease diagnosis, single cell gene expression analysis and biowarfare agent detection. In order to detect low-abundance nucleic acid targets, conventional methods predominantly rely on the use of sequence amplification techniques such as PCR or RT-PCR to amplify the quantity of DNA or RNA to a detectable level. However, the amplification process is expensive and time-consuming and tends to be complicated by false-positive and -negative results due to contamination. Furthermore, quantitative PCR is limited by the lack of standardized protocols and sensitivity to reaction conditions and analytical methods. For these reasons, single molecule detection (SMD) is receiving increased attention in efforts to enhance the sensitivity of nucleic acid detection, as well as, validate currently available techniques by means of its PCR-like sensitivity and potential to achieve high accuracy quantification of low abundant targets.

In the last two decades a variety of optical SMD techniques, which measure single molecules either in solution or on a solid surface, have been developed.[1-4] These techniques have made possible the exploration of the underlying processes of molecular bindings and reactions at the most fundamental level.[5-7] SMD in solutions is mainly based on the use of confocal fluorescence spectroscopy, which detects the photons emitted from a fluorescence molecule when it passes through a minuscule measurement volume (~ femtoliter) defined by a high-numerical-aperture

*Corresponding Author: Tza-Huei Wang—3400 N. Charles St. Latrobe Hall Rm. 108, Baltimore, MD 21218, U.S.A. Email: thwang@jhu.edu

Integrated Biochips for DNA Analysis, edited by Robin Hui Liu and Abraham P. Lee.
©2007 Landes Bioscience and Springer Science+Business Media.

objective (typically N.A. > 1.2) and a confocal pinhole (~ 50-100 μm; Fig. 1). Conducting biological measurements within such a small volume greatly reduces the background noise coming from spurious fluorescence of impurities and Raman scattering of solvent molecules, so that the fluorescence signal from even a single fluorescently labeled target can be effectively distinguished from the background. When measuring samples containing low-concentration targets (< 1 nM), the detected fluorescence signal becomes digital because the molecular occupancy (the average number of molecules residing within the detection volume at any time) is smaller than unity. A fluorescence burst is detected only when a molecule passes through the detection volume. Since the advent of SMD, this technology has greatly advanced fundamental study in molecular bindings and interactions. Recently, by incorporating new probes or probe strategies, such as fluorescence resonance energy transfer (FRET)-based probes[8-10] and dual-color fluorescence coincidence analysis,[11-15] SMD has been utilized in genomic analysis including detection of DNA of specific sequences and genetic variations, such as single-nucleotide polymorphisms (SNPs).[8,9]

When applying SMD for quantitative analysis, the amount of molecules in a sample is measured according to the number of single molecules being detected. Since the signal-to-noise ratio (SNR) of the single-molecule fluorescence bursts remain unchanged despite decreasing the concentration of the molecules, SMD enables an ideal platform for quantitative detection of low-concentration molecules. The small detection volume associated with SMD also confers the unique advantage of requiring small sample volumes for genomic analysis. Indeed, the femtoliter detection element allows the use of one nanoliter sample volumes, while retaining high statistical confidence in measurements. Still, this small detection volume results in low molecular detection efficiency during

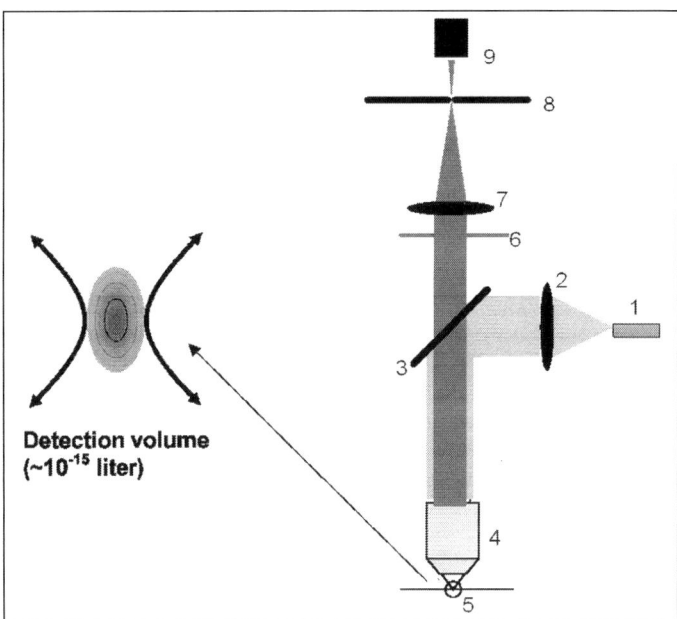

Figure 1. A schematic view of the detection volume associated with a confocal SMD spectroscope. The detection volume (gray shaded area; left) is defined by the focused excitation laser beam (black line; left) and the collection efficiency of the microscope. The confocal pinhole in this setup (typically 50-100 μm) serves to reject out-of-focus emission from the sample, further defining the detection volume in the vertical axis. Components of a typical SMD spectroscope (right) are: (1) laser source, (2) Collimating lens for excitation source, (3) dichroic mirror, (4) objective, (5) sample, (6) emission bandpass filter, (7) focusing lens, (8) pinhole and (9) single photon detector.

measurement and brings about the need for a mechanism to guide molecules within a sample to this volume element. Manipulation of biomolecules using electrokinetic[9,16-19] or hydrodynamic forces[20,21] affords the opportunity to increase mass transport efficiency in micro- and nanofluidic technologies. Therefore, the high detection sensitivity of SMD integrated with the high mass transport efficiency enabled by the use of microfluidic devices makes possible the accurate and quantitative analysis of rare and low abundant target molecules.

Single-Molecule DNA Detection Based on Use of Molecular FRET Probes

FRET is a process in which energy from an excited donor transfers to an acceptor molecule through long range dipole interactions. This process is extremely distance dependent and typically occurs at a length scales less than 10 nm. The energy transfer efficiency *(E)* inversely depends on the sixth-power of the separation distance *(r)* of a donor and an acceptor:

$$E = \frac{R_0^6}{R_0^6 + r^6}$$

where the Förster distance (R_0) is the donor-acceptor distance at which the energy transfer efficiency is 50%. FRET has been used to design a number of molecular probes, such as Taqman[22] and molecular beacons[23] that rely on changes in intermolecular distances between fluorophores to detect specific nucleic acid sequences.[10,22-25] Upon interaction with target sequences these probes undergo conformational changes; thereby, allowing detection of the target molecules through changes in the energy transfer efficiency between the excited donor and the acceptor fluorophore. The use of FRET-based probes can eliminate complicated steps in detection protocols, such as the removal or separation of unbound probes, which is usually carried out through immobilization and washing. These steps become unnecessary and thus DNA detection is conducted in a homogenous format; therefore, one can expect more efficient probe-target binding kinetics and improved detection speed and throughput.

Recently, SMD methods for detection of low-concentration DNA sequences have been developed by incorporating the use molecular beacons.[16,26,27] Molecular beacons are hairpin oligonucleotide probes, each of which is labeled with a fluorophore on one end and a quencher on the other end. The hairpin structure results in quenching of the fluorophore due to proximity of the quencher, rendering molecular beacons nonfluorescent. When hybridized with complementary sequences, molecular beacons restore fluorescence due to conformational changes that separate the quencher and the fluorophore. By using SMD spectroscopy for fluorescence measurements, DNA hybridization can be detected at the single-molecule level through detection of digital fluorescence bursts occurring when individual molecular beacon-target hybrids pass through the detection volume of a SMD spectroscope.

A dual-color SMD method has been developed for comparative quantification of specific nucleic acids using molecular beacons.[26] In this method, two different color molecular beacons were used to perform a comparative hybridization assay for simultaneous quantification of both target and control strands. Techniques for gene expression analysis through comparative hybridization often rely on dual-color fluorescence techniques for quantification. However, traditionally dual-color fluorescence analysis has been plagued by complicated data analysis. For instance, dual-color microarray technologies require normalization of fluorescence intensity and background subtraction to compensate for variance in fluorescence due to dissimilarities in photophysical or photochemical properties of different fluorophores.[28] These complexities reduce the quantification accuracy of analyzing low-abundant targets due to the inherent limitations of low SNR and varying signal levels. In contrast, dual-color fluorescence coincidence-based SMD methods obviate such technical complications through quantification schemes that rely on counting of high SNR single-molecule fluorescent bursts in different wavelengths. As this method retains high SNR even with low target concentration, comparative quantification of low-abundant (picomolar or lower) nucleic acid targets is possible, allowing discrimination of as small as 2-fold differences in target quantity at these concentration levels.[26]

Single-Quantum-Dot-Based FRET Nanosensor for DNA Detection

Recently, it has been shown that QDs can undergo FRET phenomena and can be used to investigate interactions between biomolecules.[29-31] The FRET process requires a nonzero integral of the spectral overlap between donor emission and acceptor excitation. Proper selection of the donor-acceptor pair is required to ensure high energy transfer efficiency which can be determined according to quenched donor fluorescence emission and enhanced acceptor fluorescence. Although good spectral overlap is important to high energy transfer efficiency, the donor and acceptor emission must also be well resolved in order to extract accurate experimental information. Moreover, the wavelength of excitation needs to be selected close to the minimal absorption spectrum of the acceptor to minimize the complications from direct acceptor excitation. Satisfying all these conditions simultaneously is inherently difficult in conventional molecular FRET systems because of the narrow absorption spectra, wide emission wavelengths and small Stokes shift (separation between peaks of absorption and emission spectra) of the organic fluorophores.[32] Using QDs as energy donors for FRET study shows great promise to overcome the complications encountered by the molecular FRET system because of their unique photophysical properties, such as size-tunable spectral properties, broad excitation spectra, narrow emission spectra and high quantum yields.[33-35]

A novel nanosensor design based on the use of QDs as FRET donors has recently been demonstrated for application in DNA and point mutation detection.[8] In this design (Fig. 2), each DNA nanosensor is comprised of two target-specific oligonucleotide probes. One is a reporter probe

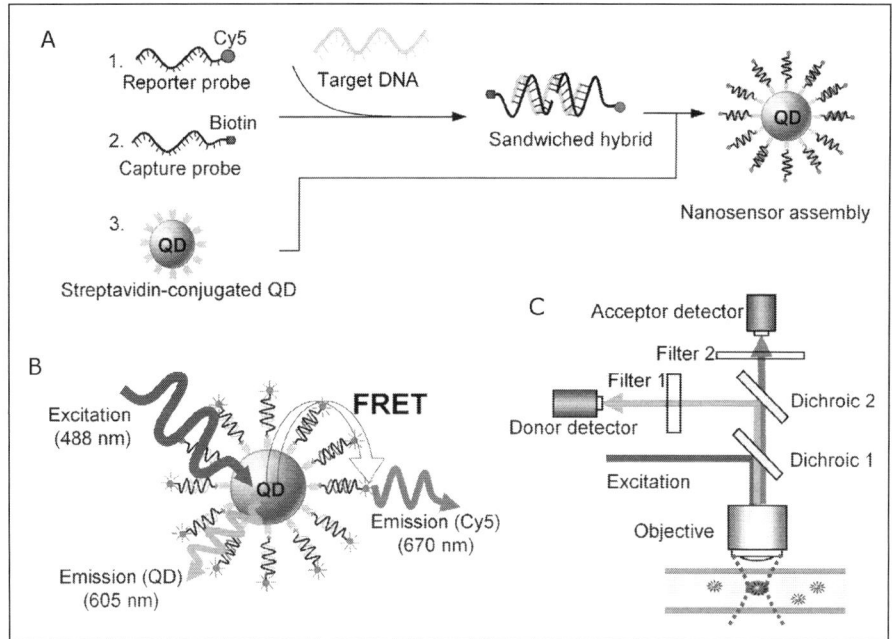

Figure 2. Schematic of single quantum dot-based DNA nanosensors A) Conceptual representation of hybridization-based capture of DNA targets via formation of nanosensor assembly. B) Visualization of FRET-based emission from nanosensor assembly after successful capture of DNA target sequence. C) Typical instrumentation for SMD setup used for detection of hybridized DNA/QD nanoassembly. Copyright 2005 Nature Publishing Group.[8]

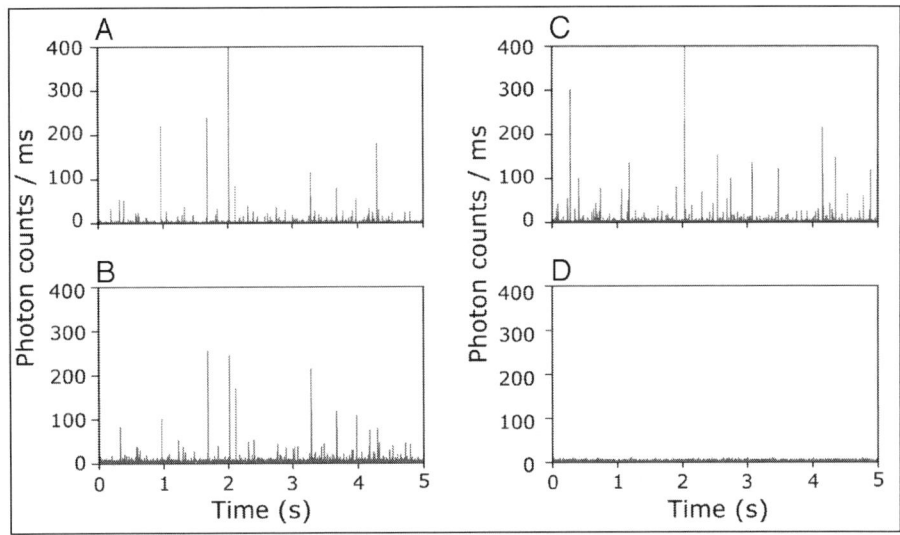

Figure 3. SMD of QD FRET signal (quantum dot-based DNA nanosensors). A-B) Detection of fluorescent bursts from both donor (A) and acceptor (B) in the presence of DNA target sequences. c-d) Isolated fluorescent bursts from donor alone (C) in the absence of target DNA sequences (acceptor detector (D) shows no fluorescent bursts). Copyright 2005 Nature Publishing Group.[8]

labeled with an organic fluorophore (Cy5, emission peak at 670 nm) and the other is a biotinylated capture probe. A streptavidin-conjugated QD (emission peak at 605 nm) serves as both a target concentrator and a FRET donor. When a target DNA strand is present in solution, it is sandwiched by the two probes and the resulting hybrid is captured by a QD through biotin/streptavidin binding. The QD functions as a concentrator that captures multiple probe-target hybrids within a nanoscale domain, resulting in strong fluorescence emission from the Cy5 via resonance energy transfer upon illumination of the QD. Excitation of QDs and detection of acceptor emission indicates the presence of hybridization targets (Fig. 3). Unbound nanosensors were found to produce near-zero background fluorescence, but generated a very distinct FRET signal upon binding to even a small amount of target DNA (~50 copies or less). By incorporating the QD nanosensor scheme with oligonucleotide ligation, the nanosensors have been successfully applied to detection of a point mutation in KRAS oncogene typical in clinical samples of some ovarian tumors.

Single-Molecule Fluorescence Burst Coincidence Detection

The minute detection volume of a SMD spectroscope also allows analysis of coincident single-molecule fluorescence signals that occur when two or more fluorescently labeled molecules simultaneously pass through the detection volume. This technique is referred to as single-molecule fluoresce burst coincidence detection, which has been applied to detect low-abundance bimolecular targets including DNA[12,15,36] and proteins.[36,37] As shown in Figure 4A, the coincidence detection method for DNA detection uses two probes, each labeled with a different fluorescent tag, to bind a specific DNA target. Emissions of the two different fluorescent tags can be detected by dual-color confocal spectroscopy. When the targets of interest are absent in the solution, the two fluorescent probes are not associated with each other and move independently. Thus, the fluorescence bursts, measured in the two different-wavelength detection channels, are stochastic and not correlated (Fig. 4B). On the other hand, when used to analyze sample containing the specific targets, the two probes bind to a target and a doubly-bound hybrid forms. Coincident fluorescence bursts occur

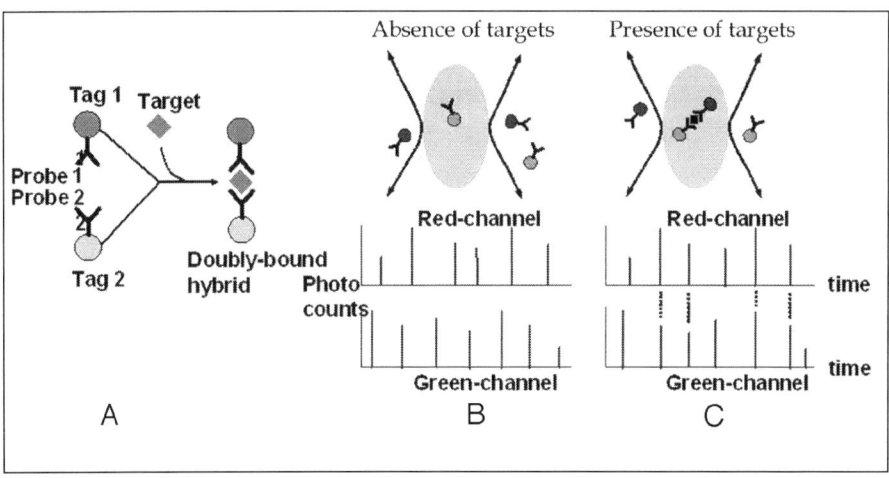

Figure 4. Coincidence detection of DNA targets using two probes/tags. A) Schematic of individual probes binding to DNA target molecule to form a doubly-bound hybrid assembly. B) In absence of target molecules, fluorescent tags make uncorrelated entrances into SMD detection volume and show no coincident fluorescent bursts. C) In the presence of target DNA, fluorescent tags make a correlated entrance into SMD detection volume, resulting in coincident emission from two channels (dashed lines show coincident fluorescent peaks). Copyright 2005 Elsevier.[15]

and can be detected in the two detection channels as the individual doubly-bound hybrids flow through the detection volume (Fig. 4C).

Assays based on single-molecule fluorescence coincidence detection require the use of a low probe concentration in order to minimize the probability of statistic coincident signals from two unbound probes simultaneously passing through the detection volume. The probability of having more than one unbound probe simultaneously present in the detection volume due to stochastic events can be estimated using Poisson statistics:

$$P_x = \frac{\exp(-\lambda) \cdot \lambda^x}{x!}$$

where P_x stands for the probability of finding exactly x molecules in the focal volume at a given time and λ is the average number of molecules in the detection volume. As an example, λ is only $\sim 10^{-4}$ at a probe concentration of 1 pM. Calculation shows the probability of finding one (P_1) and two (P_2) molecules in the detection volume is 10^{-4} and 5.0×10^{-9}, respectively. P_2/P_1 is thus 5.0×10^{-5}, implying a low probability of coincident signals detected from stochastic events compared to binding events; thereby a low level of background noise can be achieved.

Detection specificity in single-molecule coincidence detection can be further improved by using peptide nucleic acid (PNA) probes. PNAs are base sequences attached to a N-(2-aminoethyl)-glycine backbone that is linked by peptide bonds. This DNA analog contains an uncharged backbone, which confers stronger binding between complementary PNA/DNA sequences, compared to DNA/DNA counterparts. Castro and coworkers demonstrated SMD of specific DNA sequences in unamplified genomic DNA samples, utilizing this stronger binding property of PNA probes.[12,38,39] Li and coworkers investigated alternative methods of probing the limits of detection sensitivity in coincidence detection by integrating a dual-laser excitation confocal setup and detecting within samples having a large excess of singly-covalently-labeled DNA molecules.[14] Using this dual-laser excitation with its subfemtoliter-sized detection volume proved to further reduce both background noise and spectral cross-talk.

One challenging aspect of dual-color coincidence detection is to minimize cross-talk induced false coincidences that result from using fluorescent tags with overlapping emission. The small (~20-30 nm) Stokes' shift of organic fluorophores creates further complications since two fluorophores that can be excited by the same laser will have close peak emission wavelengths. The complication of cross-talk can be overcome by using dual-excitation confocal spectroscopy as it allows use of two fluorophores with distinct excitation and emission spectra. Still, good coincidence detection results from dual-excitation setups necessitate good overlap between the illumination volumes of both lasers, which requires complicated optical alignment and highly chromatic aberration-corrected objectives.[40]

However, dual-color coincidence detection of DNA targets has been demonstrated using single-laser excitation and QDs as fluorescent tags. As aforementioned, QDs have great photophysical properties such as size-tunable spectra, broad absorption and narrow emission spectra and

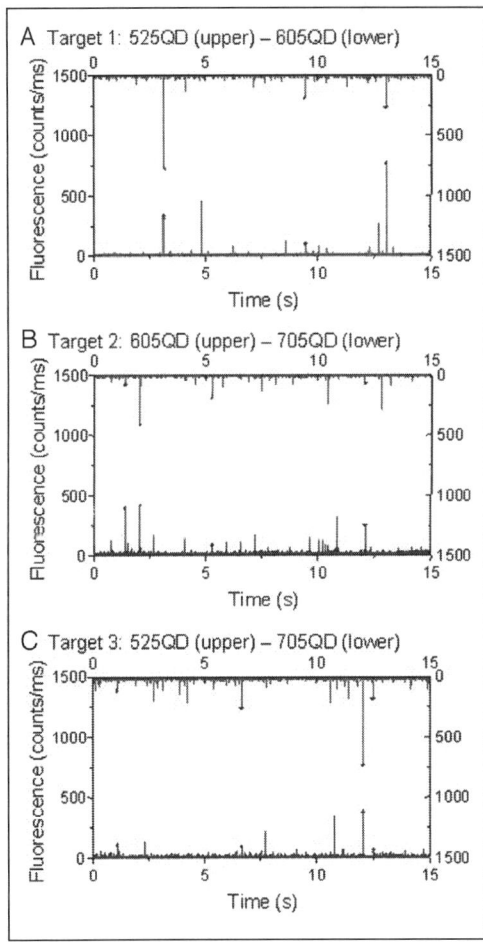

Figure 5. Identification of three DNA targets using single laser excitation and combinatorial dual-color coincidence detection. A) Coincident peaks using emission wavelength 525 and 605 nm QDs, surface functionalized for hybridization to first DNA sequence. B) Coincident peaks using emission wavelength 605 and 705 QDs, surface functionalized for hybridization to second DNA sequence. C) Coincident peaks using emission wavelength 525 and 705 nm QDs, surface functionalized for hybridization to third DNA sequence. Copyright 2005 Elsevier.[15]

large Stokes' shifts. As a result, QDs are prime candidates for fluorescent tags that can be efficiently used in multiplexed detection and excited by single laser sources. Figure 5 shows an application of this multiplexed detection using three QDs (peak emission wavelengths at 525, 605 and 705 nm). This principle was demonstrated using a combinatorial dual-color coincidence detection scheme and a single-laser excitation confocal spectroscope to identify three DNA targets.[15]

Multiplexed DNA Detection Based on Multicolor Colocalization Analysis of Quantum Dot Nanoprobes

QDs are well suited as optical labels in multiplexed detection due to their narrow emission spectra and size tunable emission. Their high brightness and photo stability have also allowed measurements of single DNA molecules with high SNR using a standard CCD-coupled, wield-field microscope instead of a sophisticated confocal optical setup.[41,42] In a recent work that demonstrated the detection of single-molecule hybridization, two different-color oligonucleotide-conjugated QD nanoprobes were designed to bind in juxtaposition to the same target DNA and to form a

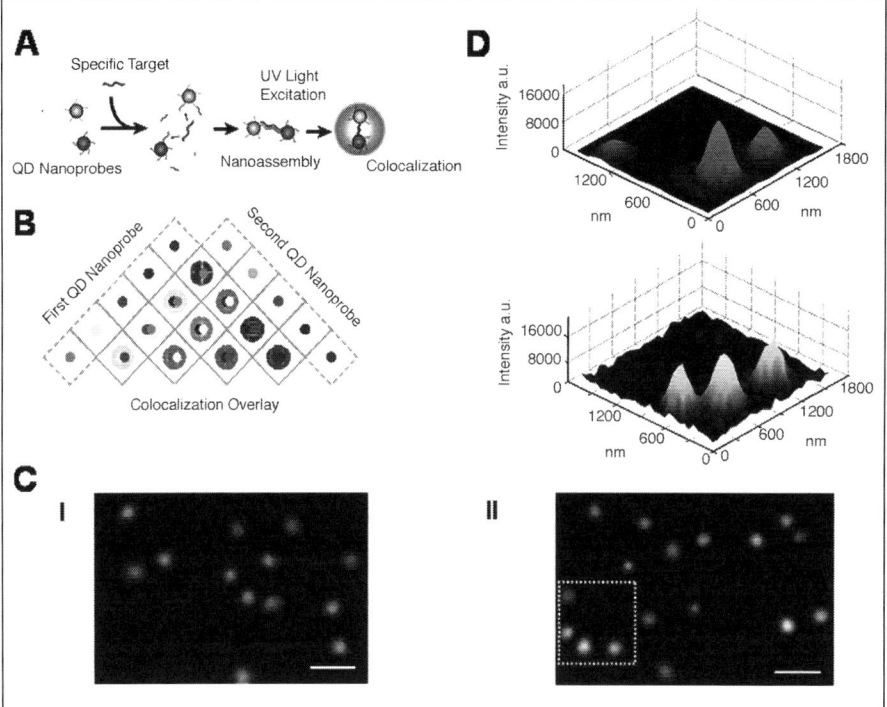

Figure 6. Multiplexed Hybridization Detection with QD Colocalization A) Two target specific oligonucleotide-functionalized QD nanoprobes bind target, forming QD/DNA nanoassembly. The nanoassembly is detected as a blended color (orange) due to colocalization of multiple emission wavelength QD probes. B) Color combination scheme for multiplexed colocalization detection. C) Image I shows discrete green and red fluorescent spots from unhybridized QD nanoprobes in a negative control experiment (genomic bacterial DNA). However, Image II demonstrates colocalization events (orange pseudo-color) upon introduction of specific oligo-nucleotide targets, indicating the formation of QD/DNA nanoassemblies. D) Three-dimensional contour plots from the range of interest (ROI) in (C); the dashed lines indicate colocalization events of the green and red fluorescent signals. Copyright 2005 American Chemical Society.[42] A color version of this figure is available at www.eurekah.com.

sandwiched nanoassembly.[42] The nanoassembly allowed colorimetric measurements upon imaging with a fluorescence microscope, due to the colocalization of QDs after DNA binding (Fig. 6A).

Specifically, green (525 QD) and red (605 QD) QD nanoprobes were used and combined for a pseudo orange color fluorescent spot, which indicated QD-DNA nanoassemblies and the presence of DNA targets (Fig. 6C). The presence of target DNA was also verified by analyzing 3-D contour plots of fluorescent signal from the two different channels (Fig. 6D). A negative control using genomic DNA from *Escherichia Coli* demonstrated the high specificity and low background noise using this detection method, as colocalization events were rare (Fig. 6C). The utility of this method for multiplexed analysis is evident, as ½ n(n-1) targets can be detected using n QDs for conjugation of (n-1) oligonucleotide probes to prepare n(n-1) QD nanoprobes (Fig. 6B). To prove this concept, three *B. anthracis*-related gene sequences were detected with high sensitivity and specificity.[42]

Manipulation and Focusing of Single Molecules

The miniscule detection volume utilized in confocal SMD spectroscopy brings with it duplicitous results. First, the small laser illuminated region renders confocal fluorescent spectroscopy measurements highly sensitive and well suited for biological detection due to a maximized SNR.[43] However, traditionally this advantage comes at a cost, owing to the small fraction of sample that can be interrogated in reasonable measurement times. Therefore, in order to realistically apply SMD to detect rare or limited samples, mechanisms of transporting target molecules within samples to the small SMD detection region must be developed. A number of molecular confinement techniques have been developed and applied to flow cytometry and capillary electrophoresis using hydrodynamic focusing,[20,21] electrokinetic focusing[16,19,44] and physical confinement via nanometer-sized fluidic channels.[1,45-47] However, application of confinement methods to SMD detection holds several important challenges. For example, physical confinement in nanometer structures often leads to high background levels caused by increased surface reflection. The use of submicron channels for sample confinement is also limited due to clogging and increase throughput. In addition, hydrodynamic and electrokinetic focusing schemes often suffer from off-centered focusing with respect to the SMD detection volume.

Recently, a microfluidic device capable of manipulating single DNA molecules (Fig. 7) was developed and used to facilitate the quantitative detection of low-abundance DNA using SMD.[16] An electrode-embedded microfluidic channel enabled electrokinetic focusing for the transport of molecules through a laser-focused detection element (Fig. 7A,B). Fluorescent particles were guided toward a region of minimum energy within the electrode section of the microchip, where the focused laser beam of the confocal fluorescence spectroscope was positioned for measurement of single molecule fluorescent bursts. As shown in Figure 7C, ac and dc electric fields were applied to the electrodes so that the induced electrophoretic (EP) and dielectrophoretic (DEP) forces effectively brought DNA molecules (T2 DNA) to the centre of the microchannel (a real-time video showing the molecular focusing process can be accessed through the American Chemical Society website).[16] This integrated micro-electro-mechanical system (MEMS) and SMD detection scheme was implemented using a variety of bioparticles, ranging in size from micro (T2 DNA and 1μm latex spheres) to nanoscales (25-mer molecular beacons; Fig. 7D). As designed, the electro-molecular focusing device allowed measurements of DNA at a concentration of 0.7 nM with 99-percent accuracy in less than one second, proving the rapid, accurate and quantitative detection of this integrated technique.[16]

Conclusion and Prospectives

Recent advances in nanotechnology has led to the development of a variety of functional nanomaterials, such as metallic nanoparticles and semiconductor quantum dots (QDs), that can serve as molecular tags with superior performance unmatched by conventional molecular tags. For example, QDs have been demonstrated 20 times as bright and 100 times as stable against photobleaching, compared with Rhodamine 6G. In addition, nanoparticles have been shown to transduce

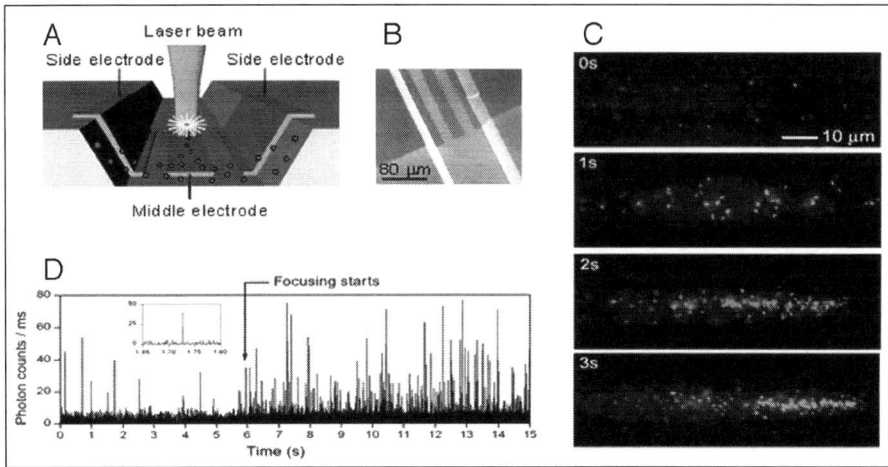

Figure 7. On-chip single-molecule detection with electro-molecular focusing. A) Schematic diagram of molecular focusing within the electro-embedded microchannel. Individual molecules are sequentially excited in the laser-focused detection volume after applying electric potential. Individual targets emit fluorescence and are counted as they flow through the SMD detection region that is aligned to an energy minimum on the middle electrode. B) Image of actual electrode-embedded microchip. C) Time series epifluorescent images of electrokinetic focusing of T2 DNA within the electrode-embedded microchip. D) Detection of molecular beacon-target hybrids within the electrode-embedded microchip. The measured solution contained 70 pM MB and a 10-fold excess of targets. Note that the onset of electro-molecular focusing at 6 seconds drastically increased fluorescent burst count rate. Copyright 2005 American Chemical Society.[16]

signals large enough in amplitude and discrete enough in identity to be positively identified at the single-molecule level. The extraordinary performance characteristics of nanoparticles used as molecular tags or transducers in biological detection enable the use of conventional, arc-lamp wide-field fluorescence microscopy setups. The combination of nanotechnology and single-molecule detection is expected to enable the development of simple, cost-effective and high-performance biological detection platforms for routine tests in clinically relevant laboratory settings.

Indeed, rapid, quantitative and high-throughput SMD analysis has been achieved by leveraging advances in microfabrication and microfluidic technologies. In the near future, critical challenges in genomic analysis, such as amplification-free detection of rare DNA targets and genetic defects, accurate quantification of low-abundance targets and differentiation of minute differences in molecular quantity, are poised to be resolved through innovative bioanalytical platforms based on integrated SMD, microfluidics and nano-technologies.

References

1. Nie SM, Zare RN. Optical detection of single molecules. Annu Rev Biophys Biomol Struct 1997; 26:567-596.
2. Kulzer F, Orrit M. Single-molecule optics. Annu Rev Phys Chem 2004; 55:585-611.
3. Moerner WE, Fromm DP. Methods of single-molecule fluorescence spectroscopy and microscopy. Rev Sci Instrum 2003; 74:3597-3619.
4. Weiss S. Fluorescence spectroscopy of single biomolecules. Science 1999; 283:1676-1683.
5. Lu HP, Xun L, Xie XS. Single-molecule enzymatic dynamics. Science 1998; 282:1877-1882.
6. Levene MJ, Korlach J, Turner SW et al. Zero-mode waveguides for single-molecule analysis at high concentrations. Science 2003; 299:682-686.
7. Edman L, Foldes-Papp Z, Wennmalm S et al. The fluctuating enzyme: a single molecule approach. Chem Phys 1999; 247:11-22.

8. Zhang CY, Yeh HC, Kuroki MT et al. Single-quantum-dot-based DNA nanosensor. Nat Mater 2005; 4:826-831.
9. Wabuyete MB, Farquar H, Stryjewski W et al. Approaching real-time molecular diagnostics: single-pair fluorescence resonance energy transfer (spFRET) detection for the analysis of low abundant point mutations in K-ras oncogenes. J Am Chem Soc 2003; 125:6937-6945.
10. Knemeyer JP, Marmé N, Sauer M. Probes for detection of specific DNA sequences at the single-molecule level. Anal Chem 2000; 72:3717-3724.
11. Nolan RL, Cai H, Nolan JP et al. A simple quenching method for fluorescence background reduction and its application to the direct, quantitative detection of specific mRNA. Anal Chem 2003; 75:6236-6243.
12. Castro A, Williams JGK. Single-molecule detection of specific nucleic acid sequences in unamplified genomic DNA. Anal Chem 1997; 69:3915-3920.
13. Korn K, Gardellin B, Liao M et al. Gene expression analysis using single molecule detection. Nucleic Acids Res 2003; 31:e89.
14. Li HT, Ying LM, Green JJ et al. Ultrasensitive coincidence detection of single DNA molecules. Anal Chem 2003; 75:1664-1670.
15. Yeh HC, Ho YP, Yang TH. Quantum-dot mediated biosensing assays for specific nucleic acid detection. Nanomedicine 2005; 1:115-121.
16. Wang TH, Peng YH, Zhang PK et al. Single molecule tracing on a fluidic microchip for quantitative detection of low-abundance nucleic acids. J Am Chem Soc 2005; 127:5354-5359.
17. Schrum DP, Culberston CT, Jacobson SC et al. Microchip flow cytometry using electrokinetic focusing. Anal Chem 1999; 71:4173-4177.
18. Jacobson SC, Ramsey JM. Electrokinetic focusing in microfabricated channel structures. Anal Chem 1997; 69:3212-3217.
19. Haab BB, Mathies RA. Single-molecule detection of DNA separations in microfabricated capillary electrophoresis chips employing focused molecular streams. Anal Chem 1999; 71:5137-5145.
20. Nguyen DC, Keller RA, Jett JH et al. Detection of single molecules of phycoerythrin in hydrodynamically focused flows by laser-induced fluorescence. Anal Chem 1987; 59:2158-2161.
21. Castro A, Farifield FR, Shera EB. Fluorescence detection and size measurement of single DNA molecules. Anal Chem 1993; 65:849-852.
22. Holland PM, Abramson RD, Watson R et al. Detection of specific polymerase chain-reaction product by utilizing the 5'-3' exonuclease activity of thermos-aquaticus DNA-polymerase. Proc Natl Acad Sci USA 1991; 88:7276-7280.
23. Tyagi S, Kramer FR. Molecular beacons: probes that fluoresce upon hybridization. Nat Biotech 1996; 14:303-308.
24. Cardullo RA, Agrawal C, Flores C et al. Detection of nucleic-acid hybridization by nonradiative fluorescence resonance energy-transfer. Proc Natl Acad Sci USA 1988; 85:8790-8794.
25. Durbertret B, Calame M, Libchaber AJ. Single-mismatch detection using gold-quenched fluorescent oligonucleotides. Nat Biotech 2001; 19:365-370.
26. Zhang CY, Chao SY, Wang TH. Comparative quantification of nucleic acids using single-molecule detection and molecular beacons. Analyst 2005; 130:483-488.
27. Foldes-Papp Z, Kinjo M, Tamura M et al. A new ultrasensitive way to circumvent PCT-based allele distinction: direct probing of unamplified genomic DNA by solution-phase hybridization using two-color fluorescence cross-correlation spectroscopy. Mol Pathol 2005; 78:177-189.
28. Duggan DJ, Bittner M, Chen YD et al. Expression profiling using cDNA microarrays. Nat Gen 1999; 21:10-14.
29. Medintz IL, Konnert JH, Clapp AR et al. A fluorescence resonance energy transfer-derived structure of a quantum dot-protein bioconjugate nanoassembly. PNAS 2004; 101:9612-9617.
30. Medintz IL, Clapp AR, Mattoussi H et al. Self-assembled nanoscale biosensors based on quantum dot FRET. Nat Mater 2003; 2:630-638.
31. Willard DM, Carillo LL, Jung J et al. CdSe-ZnS quantum does as a resonance energy transfer donors in a model protein-protein binding assay. Nano Lett 2001; 1:469-474.
32. Lakowicz JR. Principles of fluorescence spectroscopy, 2nd ed. New York: Kluwer Academic/Plenum, 1999.
33. Chan WCW, Nie SM. Quantum dot bioconjugates for ultrasensitive non-isotropic detection. Science 1998; 281:2016-2018.
34. Han MY, Gao XH, Su JZ et al. Quantum-dot-tagged microbeads for multiplexed optical coding of biomolecules. Nat Biotech 2001; 19:631-635.
35. Parak WJ, Gerion D, Zanchet AS et al. Conjugation of DNA to silanized colloidal semiconductor nanocrystalline quantum dots. Chem Mater 2002; 14:2113-2119.

36. Yeh S, Simone E, Zhang C et al. Specific single-biomolecule detection with quantum dots in a flow-controlled microchannel. 17th IEEE International Conference on Micro Electro Mechanical Systems (MEMS 2004).
37. Li HT, Zhou DJ, Browne S et al.Molecule by molecule direct and quantitative counting of antibody-protein complexes in solution. Anal Chem 2004; 76:4446-4451.
38. Castro A, Okinaka RT. Ultrasensitive, direct detection of a specific DNA sequence of Bacillus anthracis in solution. Analyst 1999; 125:9-11.
39. Marina O, Castro A. Applications of single-molecule detection to the analysis of pathogenic DNA. Curr Pharm Biotech 2004; 5:279-284.
40. Schwille P, MeyerAlmes FJ, Rigler R. Dual-color fluorescence cross-correlation spectroscopy for multicomponent diffusional analysis in solution. J Biophys 1997; 72:1878-1886.
41. Crut A, Geron-Landre B, Bonnet S et al. Detection of single DNA molecules by multicolor quantum-dot end-labeling. Nucleic Acids Res 2005; 33:e98.
42. Ho Y, Kung M, Yang S et al. Multiplexed hybridization detection with multicolor colocalization of quantum dot nanoprobes. Nano Lett 2005; 5:1693-1697.
43. Zander C, Enderlein J, Kabata H. Single molecule detection in solution: methods and applications. Wiley-VCH, 2002.
44. Anazawa T, Matsunaga H, Yeung ES. Electrophoretic quantitation of nucleic acids without amplification by single-molecule imaging. Anal Chem 2002; 74:5033-5038.
45. Foquet M, Korlach J, Zipfel W et al. DNA fragment sizing by single molecule detection in submicrometer-sized closed fluidic channels. Anal Chem 2002; 74:1415-1422.
46. Stavis SM, Edel JB Samiee KT et al. Single molecule studies of quantum dot conjugates in a submicrometer fluidic channel. Lab Chip 2005; 5:337-343.
47. Sauer M, Angerer B, Ankenbauer W et al. Single molecule DNA sequencing in submicrometer channels: state of the art and future prospects. J Biotechnol 2001; 86:181-2001.

CHAPTER 12

Nanochannels for Genomic DNA Analysis:
The Long and the Short of It

Robert Riehn,* Walter Reisner, Jonas O. Tegenfeldt, Yan Mei Wang,
Chih-kuan Tung, Shuang-Fang Lim, Edward Cox, James C. Sturm,
Keith Morton, Steven Y. Chou and Robert H. Austin

Abstract

This review will discuss the theory of confined polymers in nanochannels and present our experiments, which test the theory and explore use of nanochannels for genomic analysis. Genomic length DNA molecules contained in nanochannels, which are less than one persistence length in diameter, are highly elongated. Thus, nanochannels can be used to analyze genomic length DNA molecules with very high linear spatial resolution. Also, nanochannels can be used to study the position and dynamics of proteins such as transcription factors that bind to DNA with high specificity. In order to realize these goals not only must nanochannels be constructed whose radius is less than the persistence length of DNA, but it is also necessary to understand the dynamics of polymers within nanochannels and develop experimental tools to study the dynamics of polymers in such confined volumes, tools which we review here.

Introduction

The bending persistence length p of a polymer is basically a measure of the mean radius of curvature of the polymer in a solvent at temperature T and it is a function not only of T but also of the Young's modulus of the polymer E and the cross-section of the polymer as determined by the surface moment of inertia I_A.[34] Although in the case of a molecule it is a little risky to use a continuum approach to describe a molecular property, it is intuitively helpful to give here the continuum expression for p:

$$p \sim \frac{EI_A}{k_B T} \qquad (1)$$

where k_B is Boltzmann's constant. Double-stranded DNA (dsDNA) has a persistence length under normal 100 mM saline DNA analysis of about 50 nm,[66] which translates into a Young's modulus of about $10^8\,Nm^{-2}$,[26] about the same as a hard plastic such as nylon. Thus, although dsDNA is relatively stiff in terms of E, the 2 nm diameter of the dsDNA molecule makes for a relatively Small p compared to the enormous length which can approach centimeters in eukaryotic chromosomes.[74] Although dsDNA has an incredible length L to persistence length p ratio, with some notable exceptions such as FISH,[5] most techniques which have manipulated DNA molecules have traditionally concentrated on moderate length DNA molecules[62] of less than 20 kB (or about 6

*Corresponding Author: Robert Riehn—Department of Physics, North Carolina State University, Raleigh, North Carolina U.S.A. Email: rriehn@ncsu.edu

Integrated Biochips for DNA Analysis, edited by Robin Hui Liu and Abraham P. Lee.
©2007 Landes Bioscience and Springer Science+Business Media.

microns). Dealing with truly long intact genomic molecules poses a formidable challenge, since the genomic length molecule is easily sheared and fragmented.

Further, it is difficult to determine accurately the size of very long DNA molecules. Gel electrophoresis for short dsDNA molecules creates different mobilities for different DNA lengths by letting the molecules interact with obstacles (the gel components), giving rise to length-dependent mobilities.[61] The matrix is necessary because of the "free-draining" character of DNA, that is, the moving ions in solution create a hydrodynamic shear boundary at the Zeta potential surface, which is typically on the order of 10 Å or so.[38] Beyond that surface, all the long-range hydrodynamic interactions are effectively short-circuited and so electrophoretic mobilities are independent of the length of the polymer! Paradoxically the interaction with a sieving matrix leads to an elongation of the molecules even at relatively low field strengths and makes all the various lengths above a few tens of kbp move with the same electrophoretic mobility. Using pulsed electric fields with different field directions overcame that problem, but at the price of separation times of days. The apparent shortfalls of conventional gel electrophoresis that challenged attempts to sequence large genomes around 1990 were the large amount of DNA needed, the difficulty of integration of gel electrophoresis into an automated device, the reliance on the random nature of the sieving matrix, which prevented predictive calibration of gel mobilities and the long times needed for large DNA at 100 mM salt concentrations since elongated molecules in gels do not have a length-dependent mobility.

Fueled by those considerations, artificial gels consisting of regular arrays of obstacles were fabricated using semiconductor industry methods.[77] However, it quickly became apparent that even synthetic matrixes suffered from exactly the same problem that natural gels did: genomic length molecules of course became easily elongated even in synthetic matrices.[78] One shortfall of the first artificial separation matrices was their inability to create narrow bands of launched molecules. Later, the synthetic array structures were also adapted to large genomic DNA separations using pulsed field techniques that had previously been discovered to work in gels,[59,2] but by this point the battle had long been lost for fractionation of relatively short molecules: capillary gel electrophoresis using liquid polymer media had been proven to be a remarkably effective and rapid technique that could be automated easily and cheaply.[75] However, eventually a technology was evolved that used asymmetric pulsed fields and hexagonal arrays, which gave clear paths along the field directions and forced the molecules to back-track when the field direction was changed.[27] This technology could continuously separate different sizes without the need for a narrow launching line by creating a DNA "prism" which dispersed different DNA lengths into different angular deviations. While all of the above methods advanced the idea of measuring fewer and fewer molecules in smaller and smaller liquid volumes, other than the DNA Prism concept they suffer from the necessity to launch narrow lines of DNA into the structure and wait for the device to resolve the lengths of that line.

However, concurrent with the failure of the synthetic arrays to find an application confining structures resembling nanoslits were developed, which were used to determine molecule lengths by dynamically measuring the entropically driven recoil of DNA from a nanofluidic into a microfluidic region.[72] Conceptually somewhat similar devices utilizing entropic traps that yield different activation energies for DNA molecules of different lengths had been proposed earlier by Volkmuth[76] and the dynamics of DNA in nanoslits had been described earlier in detail,[3] but the first real demonstrations were by Han et al.[22,23] It was a short step from nanoslits to nanochannels, but there are major advantages to putting genomic length molecules in tubes rather than slits.[19]

The basic idea of DNA analysis in nanochannels is shown in Figure 1. A single double-stranded DNA molecule is introduced into a nanochannel with a cross-section that is much smaller than the radius of gyration of the unconfined DNA molecule. The channel's length is much greater than the DNA molecule's contour length. Typically the DNA molecule is brought in by electrophoretic or hydrodynamic forces.[8,64] The DNA molecule becomes stretched out by confinement forces in the process, the apparent end-to-end length r of the molecule in the channel is measured using fluorescence microscopy. When a genomic length DNA molecule is elongated in a nanochannel

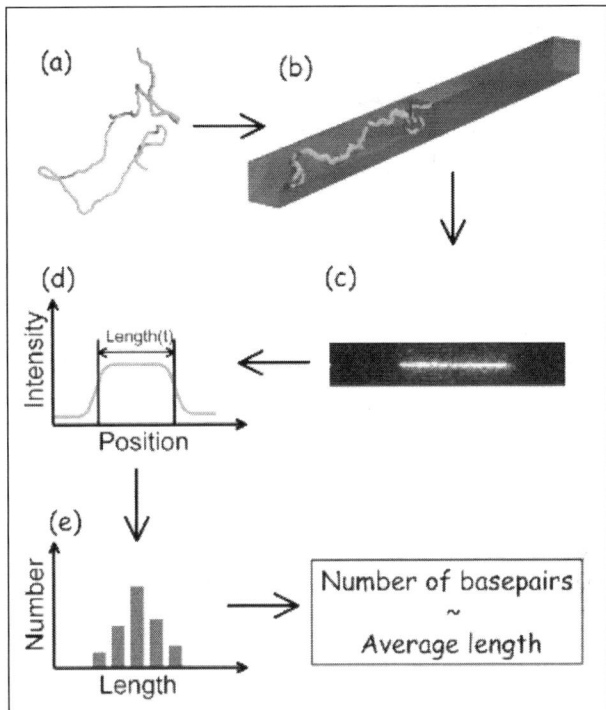

Figure 1. Schematic of basic nanochannel measurement. a) DNA molecules in solution form random coils. b) By applying an electric field single DNA molecules can be stretched out by bringing them into nanochannels. c) The a time-series of fluorescence images of the stretched molecule is recorded. d) For each time, the end-to-end length is determined. e) A histogram of instantaneous end-to-end lengths yields the average length. Using a calibration or a priori knowledge about the channels, the end to end length can be converted into an absolute basepair number.

the entire molecule is open for inspection. Nanochannel analysis of linear DNA does not need to launch narrow lines of DNA and it examines individual molecules.

We typically choose channels which have a cross-section of no more than 100 nm × 100 nm or (preferably) smaller, we strive for linear dimensions which are less than the persistence length of double stranded DNA (50 nm). Note that the stretching is not a dynamic effect of the introduction into the channel, but rather an equilibrium property of nanoconfinement. By making the reasonable assumptions that neither the basepair volume density nor the mechanical tension along the polymer backbone can diverge for very long DNA molecules, we conclude that r must be proportional to the contour length of the molecule in the case of self-avoidance.[69] It is important to point out that the DNA is free to fluctuate longitudinally around an equilibrium configuration, since it is only subject to lateral confinement. Hence, subsequent measurements if spaced far enough apart in time are statistically independent and the equilibrium configuration can be determined by repeated measurements to an arbitrary precision, at least in principle.

Of course, there are other ways to stretch DNA molecules than putting them into nanochannels. Popular methods for stretching are extension of a gel in which DNA has been immobilized, stretching DNA on a sticky surface using liquid flow, or stretching DNA by attaching handles on the ends and manipulating those handles using optical tweezers or similar. The achievements of those methods are impressive, with particular examples being the application to restriction mapping and the observation of protein and marker binding to single molecules. However, the initial promise of automatization of the stretch-and-fix method for commercial applications has not yet been fulfilled, partially due to the fact that the stretching has proven difficult to control. More specifically, inhomogeneous stretching, breaking of molecules and the need for sophisticated image and data analysis have been complications, although some of those problems have been overcome. On a more fundamental level, one has to note that the notion of single molecule measurement is not strictly fulfilled in the sense that multiple molecules are needed to construct a histogram. Because of that, multiple images of the same molecule in the stretched and fixed state are not statistically independent and hence averaging has to occur over multiple molecules.

Although the stretching and fixing of DNA prior to observation appeared to be a promising concept, it is a batch process and not amenable for integration into a μTAS environment. Hence our group developed an alternative approach by extending the polymer chains dynamically in a microfluidic device.[3,70] There has been a particularly high interest in the electrohydrodynamic flow concept, since it lends itself to automation and incorporation into an integrated measurement concept.[10] However, attribution of the original sources of ideas[70] appears to have been a problem for those authors.

Our initial interest in nanofluidic DNA analysis was sparked by the wish to improve on the dynamic stretching of DNA in microchannels. In particular, reducing the lateral dimensions of the microchannel to some hundreds of tens of nanometers was intended to slow or completely arrest the recoil that naturally occurs in microchannels. However, soon we realized that performing an equilibrium measurement of stationary or quasi-stationary stretched molecules is far more desirable, since averaging over multiple measurements is identical to simply taking a movie of the microscope image. We were even more attracted by the fact that multiple, statistically independent measurements of the same molecule become available, which opens the door to true single-molecule studies of heterogeneous populations.

The research path taken by our group can roughly be divided into three stages. Initially the focus was on designing a robust fabrication platform upon which the nanochannel stretching experiments could be based. While the initial demonstrations of stretching DNA in nanochannels were obtained early,[1,8,9] some effort had to be undertaken to obtain control of critical dimensions, reproducibility, integration with microfluidics and efficient microfluidic control. During the second stage, we assessed the basic equilibrium properties of DNA inside nanochannels and developed an assay that was able to demonstrate that the technique indeed was capable of sizing long molecules with a high precision. We further investigated the statistics of fluctuating molecules, determined the optimal dimensions for nanochannel measurements and confirmed the applicability of the established theory to our system.[51] The third stage in our work is the application of nanoconfinement to study functional regions along DNA molecules, both by attaching fluorescent markers and DNA modification.[81,53] Other groups have lately begun to explore the non-equilibrium dynamics of DNA in confining structures.[50,37,7] These dynamics are of considerable interest for the operation of a device, in particular the introduction of DNA into the nanochannels and can govern how rapidly a molecule can be consider fit for measurement. However, non-equilibrium processes do not appear to influence the measurement process itself.

Theory

Remarkably, the theoretical treatment of polymers in nanochannels predates the application to DNA analysis by about quarter of a century. Pierre de Gennes[14] and Odijk[44] made early predictions about the stretching ratio, which we have found to be in good agreement with our actual measurement.[51] The reason for the large temporal difference between the theoretical development and the emergence of the experimental technique are the stringent demands on device manufacturing technology that are laid out by the theory.

The point of departure for most introductory treatments of polymer physics is the ideal chain model: a polymer is envisioned as being composed of a series of freely rotating rigid links capable of self-intersection. In free solution such a chain is described by a simple random walk: it will form a coil that scales in size as the square root of the number of monomers (links) in the polymer. An ideal chain confined in a channel undergoes a 1-D random walk down the channel axis; the extension of the ideal polymer in the channel is just the projection of the unconfined chain's spatial extent along the channel axis. In contrast, DNA confined in a nanochannel is observed experimentally to stretch with an extension along the channel proportional to the contour length.[69] While the ideal model gives, in certain cases, a reasonable approximation to the behavior of DNA (for example, DNA in solution in the absence of confinement), the ideal model completely fails to describe DNA in nanoconfinement.

DNA Is a Semiflexible and Self-Avoiding Polymer

The stretching of DNA in nanochannels is ultimately a consequence of two physical properties of the "real" molecule missing from the ideal description: semiflexibility and self-avoidance. The DNA chain has an intrinsic stiffness. At Small enough length scales the molecule "finds it" difficult to bend. Moreover, the fact that monomers units have a physical diameter, i.e., they cannot occupy the same space, necessitates an alteration in the statistical description of the polymer that has a crucial effect when the molecule is confined. This self-avoidance effect is likely amplified in low molarity buffers by screened electrostatic repulsion between monomer units.

The semiflexibility and self-avoidance of DNA are quantified in the theoretical description by two key parameters: the persistence length P and the effective diameter of the diameter of the polymer w_{eff} (Fig. 2). The persistence length is set by a balance of thermal energy against the intrinsic rigidity of the DNA molecule: $P = \kappa/k_B T$ with κ the bending modulus of DNA (related to the Young's modulus). The quantity κ directly determines the free energy cost of bending a segment l of DNA through a radius of curvature R via,

$$\Delta F_{bend} = \frac{k}{2} \frac{l}{R^2}.$$ (2)

Thermal fluctuations tend to coil the DNA at large enough length scales, scales much larger than P, yet if we examine Small enough segments of the molecule—much below P—we observe that the DNA in locally rigid. In mathematical terms, the persistence length determines the scale over which the tangent-tangent correlations along the molecule contour decay,[15]

$$\langle t(0) \cdot t(l) \rangle = \exp\left(-\frac{l}{P}\right)$$ (3)

Screened electrostatic repulsion between neighboring charges on the DNA backbone creates a weak dependence of the persistence length on ionic strength.[60,44,21] In the roughly 50 mM buffers used the persistence length is around 50 nm.

The effective diameter w_{eff} gives a measure of the "thickness" of the DNA molecule at a local scale. At high salt concentrations, the effective diameter is simply the intrinsic diameter of 2 nm for B-form DNA.[4] At low salt, electrostatic repulsion between adjacent monomers becomes significant. These electrostatic interactions are often approximated by assigning an effective hard core diameter to the DNA chain. In other words, one approximates the charged molecule by an equivalent neutral chain of slightly larger effective diameter.[80,79] The effective diameter can be obtained experimentally by measuring the probability that larger DNA molecules form a knot during cyclization.[55] This knotting probability is sensitive to the strength of excluded volume interactions within the chain. These measurements suggest that the effective diameter has a pronounced dependence on NaCl concentration, ranging from the intrinsic diameter of 2 nm at 1 M salt to as high as 20 nm at 10 mM.[79]

DNA in Free Solution

For polymers in free solution semiflexibility and self-avoidance also affect the chain statistics, but only at a quantitative level. The key quantity describing the conformation of a polymer in free solution is the mean square end-to-end length $<R^2>$ or the radius of gyration R_g, related to the end-to-end length by a numerical factor.

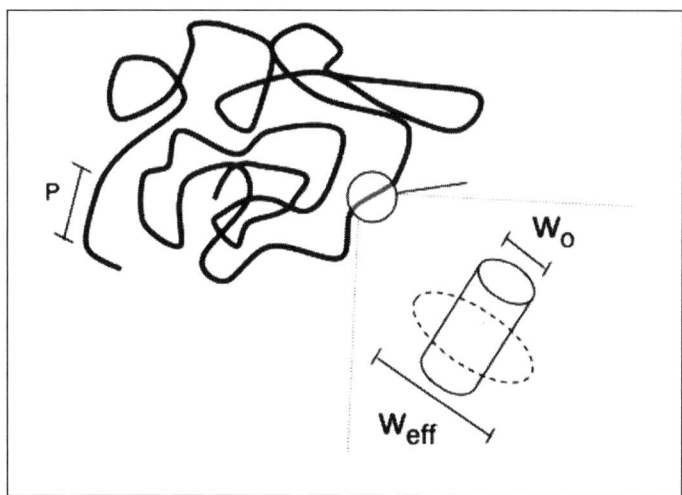

Figure 2. The ``Spaghetti'' model of DNA. P is the persistence length, the average scale over which the DNA remains rigid. On the magnified portion of the molecule are shown the definitions of w_o (the intrinsic diameter of 2 nm) and w_{eff} (the effective diameter taking into account the effect of electrostatic interactions). The effective diameter can range from 2 nm to as high as 20 nm for molarities ranging from 100 mM to 10 mM.[79] The contour length is just the total length of the spaghetti if it were stretched out and measured from one end to the other.

An ideal polymer is an energetic cipher: it consists purely of entropy. The entropy of an ideal chain is usually obtained from the distribution function of the end-to-end length.[14] The distribution function is given by the Gaussian

$$p(R,N)=\left(\frac{3}{2\pi\langle R^2\rangle}\right)^{3/2}\exp\left(-\frac{3R^2}{2\langle R^2\rangle}\right) \tag{4}$$

with a mean square average $<R^2>$ proportional to the number of links:

$$\langle R^2\rangle=\sum_{i,j=1}^{N}\langle t_i\cdot t_j\rangle=\sum_{i,j=1}^{N}l^2\delta_{ij}=l^2 N=lL. \tag{5}$$

The entropy of the chain is by definition given by the logarithm of the distribution function: $S=\ln(p(\mathbf{R},N))$. Consequently, $F=-k_B T S$ and denoting the polymer elongation r the free energy is given by,

$$F_{ideal}=\frac{3k_B T}{2\langle R^2\rangle}r^2. \tag{6}$$

This equation shows that an ideal chain responds to an externally applied force on both ends like a simple spring with spring constant

$$k_{ideal}=\frac{3k_B T}{\langle R^2\rangle}=\frac{3k_B T}{lL}. \tag{7}$$

This elasticity arises from the fact that stretching the coil tends to reduce the number of available conformations.

Entropy, which favors a Smaller coil, is in DNA balanced by excluded-volume interactions, which tend to swell the chain. Flory showed that it is surprisingly simple in mean field approximation to deduce the effect of excluded volume on the chain statistics.[14] The monomer concentration $c = N/r^3$, with N the total number of monomers. The free energy increase due to pairwise monomer-monomer interactions in this approximation is given by

$$F_{ex} \cong \frac{1}{2} k_B T \chi c^2 r^3 = \frac{1}{2} k_B T \chi \frac{N^2}{r^3} \tag{8}$$

The parameter χ has dimensions of volume and measures the strength of excluded volume interactions between the monomer pairs. The total free energy is given by adding to Equation. 8 the entropic term (Eq. 7),

$$\frac{F_{tot}}{k_B T} \cong \frac{3}{2lL} r^2 + \frac{1}{2} \chi \frac{N^2}{r^3} \tag{9}$$

Minimizing this result leads to a scaling relation for the end-to-end length of a Flory coil

$$R_F \cong \chi^{1/5} l^{2/5} N^{3/5}. \tag{10}$$

For a flexible polymer with isotropic monomer units, $\chi = l^3$. The prefactor in Equation 10 is then proportional to the linksize. For a persistent polymer, the link size is proportional to the persistence length and the excluded volume parameter has the classic form due to Onsager $\chi \approx w_{eff} P^2$ reflecting the volume occupied on average by a link of size P twirled into a disk. Equation 10 then becomes[58,41]

$$R_F \cong (Pw_{eff})^{1/5} L^{3/5} \tag{11}$$

The best existing theoretical estimate of the scaling exponent for L is around 0.588,[15] remarkably close to the Flory estimate.

Existing data on the coil size as a function of contour length suggest that the self-avoiding exponent holds down to several kbp.[63,73] Below ~1 kbp, the measured exponent actually increases due to the semiflexibility of the chain. Reference 73, a AFM study of DNA absorbed on glass, measures the end-to-end length over scales ranging from 10 μm to 2 nm. The data, when fit to a double power law, shows that the end-to-end length as a function of contour interpolates smoothly from a power law with Flory exponent of 0.589 (the precision was sufficient to obtain the exact Flory value!) to a form linear in L, the expected result in the sub persistence length limit. Lastly, nanopore translocation experiments indirectly access information about the polymer coil size. Storm, Dekker and Joanny have proposed that the translocation time of a DNA molecule in the pore should scale as the square of the coil size.[66] Measurements of the translocation time as a function of the coil size then yield a value of the scaling exponent of 0.64 ± 0.05 (the running buffer was 1 M KCl, 10mM Tris-HCl pH 8, 1mM EDTA), consistent again with the self-avoiding prediction. So far, contrary to various theoretical predictions and estimates,[18,41] experiment has not yet uncovered an ideal scaling regime.

Confined Self-Avoiding Polymers

In the limit $D >> P$ the polymer is free to coil in the channel as the energy to make a backbend is on order of $k_B T$. A merely semiflexible (but not self-avoiding polymer) would behave ideally and undergo a random walk down the channel axis. Self-exclusion, however, will raise the energetic cost of back-looping and give rise to an extension in the channel scaling linearly with contour.

De Gennes developed the classic model of a confined self-avoiding polymer.[13,14] He envisioned the confined chain as being divided into a series of blobs (see Fig. 3). He argued that self-exclusion has two effects: (1) the blobs repel like hard spheres, uniformly distributing the polymer contour along the channel and (2) each blob behaves as a "Flory coil", so that the blob radius R_b scales according to Equation 11. Confinement forces R_b to scale as D, i.e., $R_b \sim D$. The contour stored per blob, L_b, can then be found by back-solving for L_b as a function of D:

$$L_b \cong \frac{D^{5/3}}{(Pw_{eff})^{1/3}}. \tag{12}$$

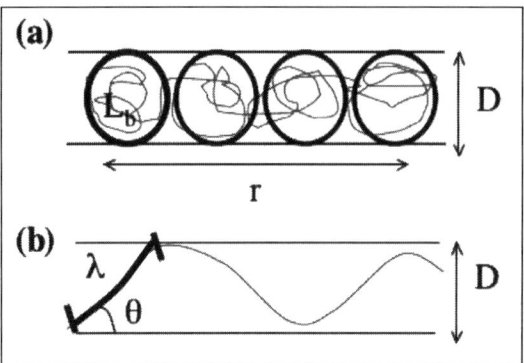

Figure 3. a) A confined polymer in the De Gennes regime: $D \gg P$. The molecule can be sub-divided equally into a series of blobs with contour length L_b, the stretch arises from the mutual repulsion of the blobs b) A confined polymer in the Odijk regime: $D \ll P$. The molecule undergoes a series of deflections with the wall. The deflections occur on average over the Odijk scale $\lambda \sim (PD^2)^{1/3}$. Reproduced from ref. 51.

The extension of the chain parallel to the tube axis is then simply the number of blobs L/L_b times the extension of each blob (D),

$$r \cong \frac{L}{L_b} D . \tag{13}$$

Using Equation 12 to eliminate L_b in terms of D, Equation 13 gives

$$r \cong L \frac{(w_{eff} P)^{1/3}}{D^{2/3}} . \tag{14}$$

This result can also be obtained by Monte Carlo[32,40] and Langevin simulation.[29,30]

It is also possible to obtain Equation 14 via a Flory free energy analogous to Equation 9.[6] The trick is to adjust the self-exclusion term to reflect confinement in a channel:

$$\frac{F_{tot}}{k_B T} \cong \frac{3}{2lL} r^2 + \frac{1}{2} \chi \frac{N^2}{rD^2} . \tag{15}$$

Minimizing this free energy will lead to Equation 14. This particular argument is useful because it shows how easily channels with non-unity aspect ratios can be accommodated within the scaling picture. Consider a channel with a height D_1 and width D_2. Then the self-exclusion term will $\sim \chi$ $N^2/(rD_1D_2)$, implying that the parameter D can be in general replaced by $D_{ave} = (D_1 D_2)^{1/2}$, the geometric average of the dimensions. It is also possible to obtain this result via blob arguments.[71]

Lastly, the de Gennes theory also yields the increase in free energy ΔF of a polymer due to confinement. Here the Flory model gives incorrect results, but the blob model gives the right answer.[14] One simply argues that each blob has around $k_B T$ of energy. The total free energy is then the sum over the free energy of all the blobs:

$$\Delta F \cong k_B T \frac{L}{L_b} \cong k_B T \frac{(P w_{eff})^{1/3}}{D^{5/3}} L . \tag{16}$$

For λ-DNA and 100 nm channels, this equation would estimate a confinement free energy in the range of 35-60 $k_B T$ for w_{eff} varying from 2 to 10 nm.

Confined Semiflexible Polymers

When the channel diameter drops below a persistence length the confined polymer can no longer coil. The free energy of making a coil in a channel is estimated using Equation 2. Setting the radius of curvature $\approx D/2$ and the arclength of the bend $\approx \pi D/2$:

$$F_{coil} = \pi k_B T \frac{P}{D}. \tag{17}$$

The probability of a coil being formed is then determined by the Boltzman factor $\exp(-F_{coil}/k_B T)$ so that

$$P_{coil} \cong \exp\left(-\pi \frac{P}{D}\right). \tag{18}$$

Clearly, as the channel diameter drops below the critical diameter $D_{crit} \approx \pi P \approx 150$ nm, P_{coil} rapidly becomes Small.

In the regime $D << D_{crit} \approx P$, the physics is dominated not by excluded volume but by the interplay of confinement and intrinsic DNA elasticity. As back-folding no longer occurs, contour length is stored exclusively in deflections made by the polymer with the walls (see Fig. 3). These deflections occur on average over a special scale λ_O called the "Odijk" length.[45,46] The Odijk length gives the average increment in contour between the successive "bumps" and plays the role of a blob in the strongly confined regime. The key to devising scaling relations for the extension is to determine how λ_O scales with P and D. The answer is rather non-intuitive. In his original argument Odijk envisioned a polymer, starting at the channel center and initially tangent to the channel axis, that gradually wends away from the channel axis. The key idea is to find the horizontal excursion of the polymer away from the channel axis as a function of the contour length s of the polymer segment. To do so, Odijk first cited the known moments for a semiflexible chain, in the approximation $s<<P$,

$$\left\langle R(s)^2 \right\rangle = s^2 \left(1 - \frac{s}{3P} + \cdots\right) \tag{19}$$

$$\left\langle (R(s) \cdot t(0))^2 \right\rangle = \left\langle z(s)^2 \right\rangle = s^2 \left(1 - \frac{s}{P} + \cdots\right) \tag{20}$$

The quantity $t(0)$ is the initial tangent vector the polymer, assumed to be parallel to the channel axis. The first of these relations can be obtained by expanding the Kratky-Porod formula for the end-to-end length of a semiflexible polymer to first order in s/P.[15] The second is rather obscure and is derived in reference 82. Odijk argued that the mean squared transverse deflection $<R_\perp(s)^2> = <x(s)^2> + <y(s)^2>$ is then given by $<R(s)^2> - <R(s)^2 \bullet t(0)^2>$. Taking the difference of Equation 19 and Equation 20 leads to

$$\left\langle R_\perp(s)^2 \right\rangle = \frac{2}{3} \frac{s^3}{P}. \tag{21}$$

The deflection length is defined as the mean contour stored between bumps with the wall. Consequently, $<R_\perp(\lambda_O)^2> = (2/3)(\lambda_O^3/P) \approx (1/4) D^2$. Backsolving for λ_O as a function of D leads to

$$\lambda_O \cong (PD^2)^{1/3}. \tag{22}$$

Odijk also obtained this scaling via more abstract similarity arguments[46] that do not depend on the little known Equation 20.

Once a scaling relation for λ_O is known, the calculation of the polymer extension along the channel axis becomes straightforward. The extension is the number of Odijk segments L/λ_O times the average projection of an Odijk segment on the channel axis. Assuming that the average deflection made by the polymer with the walls is Small, $\cos(\theta) \approx 1 - \frac{1}{2}\theta^2$, $\theta \approx D/\lambda$ and

$$r = Lcos(\theta) = L\left[1 - A\left(\frac{D}{P}\right)^{2/3}\right].$$
(23)

The value of the constant A will be estimated from experiment. This formula predicts a gradual approach to full stretching as the channel width decreases. From a device point of view, this suggests that in the sub-persistence length regime further decreases in the channel width will yield diminishing returns in terms of increasing the polymer extension further. Note that Equation 23 clearly is not valid in the limit that $D \gg P$.

The free energy increase due to confinement ΔF in the Odijk regime can be obtained via a similar argument. The total free energy increase of the polymer should scale as $k_B T$ times the total number of Odijk segments in the polymer, L/λ_O. This yields

$$\Delta F = Bk_B T \frac{L}{P^{1/3}D^{2/3}}.$$
(24)

with the parameter B another undetermined scaling constant.

Brownian Fluctuations of Confined Polymers

The equilibrium theory can be extended to describe Small longitudinal fluctuations of the confined polymer. We imagine, to first approximation, that these extension fluctuations δr are described by a damped oscillator equation for the extension alone,

$$\zeta \frac{d(\delta r)}{dt} = -k_{eff}\delta r + \xi(t)$$
(25)

$$\langle \xi(t) \rangle = 0$$
(26)

$$\langle \xi(t)\xi(t') \rangle = 2\zeta k_B T\delta(t-t')$$
(27)

The quantity k_{eff} is an effective spring constant for the confined polymer and ζ is a friction factor due to hydrodynamic drag. The second term in Equation 25, defined via Equation 26 and Equation 27, is a stochastic force experienced by the polymer due to collisions with the fluid molecules.[15] (The prefactor, $2\zeta k_B T$, in Equation 27, ensures that Equation 25 leads to a diffusion constant satisfying the Einstein relation $D = k_B T/\zeta$ in the case that the restoring term $-k_{eff}\delta r$ is removed). Such an approximation is equivalent to neglecting higher order internal modes of the confined polymer which the experiments are not sufficient to resolve. Equation 25, Equation 26 and Equation 27 lead to an exponentially damped autocorrelation function for δr,

$$\langle \delta r(t)\delta r(t') \rangle = \langle (\delta r)^2 \rangle \exp\left(-\frac{(t-t')}{\tau}\right).$$
(28)

The quantity τ is a relaxation time scale for the polymer fluctuations determined by the ratio of the friction factor to the spring constant: $\tau = \zeta/k_{eff}$. Informally speaking, τ sets the time scale required for the polymer to relax back to the equilibrium extension after it is driven away by a thermal fluctuation. The spring constant and friction factor are functions of the polymer parameters (L, P and w_{eff}) and depend also on the degree of confinement.

The spring constant can be obtained from the scaling relations for the extension and free energy of the polymer in equilibrium. Our argument is the following: first express the free energy is terms of r by eliminating D. The spring constant as a function of r can then be determined from the second derivative of the free energy with respect to r. Finally, eliminating r back again in terms of D gives the spring constant as a function of D. In the weakly confined $D \gg P$ regime, using Equation 14 and Equation 16, this argument leads to:

$$k_{eff} \simeq \frac{k_B T}{L(Pw_{eff})^{1/2}} \frac{(Pw_{eff})^{1/6}}{D^{1/3}} \qquad (29)$$

Analogously, in the strongly confined $D << P$ regime, Equation 23 and Equation 24 give

$$k_{eff} \cong \frac{k_B T}{PL}\left(\frac{P}{D}\right)^2 . \qquad (30)$$

Comparing Equation 29 and Equation 30 shows that a strongly confined polymer should be much stiffer than a weakly confined polymer. The same relation for the spring constant of a confined chain is used in studies of semiflexible polymer networks.[42,43] This work, while in a rheological rather than device context, is closely related: theoretical descriptions of the rheology of a polymer network are based upon the fluctuations of a polymer filament in an effective tube created by the constraining effect of the polymer network on an individual chain.

The friction factor can be obtained in the self-avoiding regime by an argument due to Brochard and de Gennes.[12] The hydrodynamic interactions of the polymer should be screened over scales greater than D. Consequently, the total friction factor of the polymer should be the sum of the individual friction factors of each blob. Estimating that a blob friction factor should be $\approx 6\pi\eta(s)$ D, where $\eta(s)$ is the solvent viscosity, we have:

$$\zeta = 6\pi\eta_s D\frac{L}{L_b} = 6\pi\eta_s r = 6\pi\eta_s L \frac{(Pw_{eff})^{1/3}}{D^{2/3}} . \qquad (31)$$

Intuitively, in the strong confinement regime, we would expect the friction to be dominated by the hydrodynamic interaction of a short segment of DNA with the channel wall. Consequently, we argue it should be estimated by the friction factor of a cylinder of diameter $w_o = 2$ nm moving inside a coaxial cylinder of diameter D. Classical hydrodynamics then implies,[33]

$$\zeta \cong \frac{2\pi\eta_s L}{log\left(\frac{D}{w_o}\right)} \qquad (32)$$

The same relation for the friction factor is used in reference 43.

The relaxation time is then given by the ratio of the friction factor to the spring constant. In the weakly confined regime,

$$\tau \cong \frac{6\pi\eta_s}{k_B T} \frac{(Pw_{eff})^{2/3}}{D^{1/3}} L^2 \qquad (33)$$

This suggests that the relaxation time increases very slowly with decreasing width in the de Gennes regime as the friction factor has a slightly stronger dependence on D than the spring constant. Equation 33 agrees with the formula proposed by de Gennes and Brochard for the relaxation time of the lowest collective mode of the confined polymer.[12] In the Odijk regime the situation is reversed:

$$\tau \cong \frac{2\pi\eta_s}{k_B TP} \frac{D^2}{log\left(\frac{D}{w_o}\right)} L^2 . \qquad (34)$$

Here the relaxation time actually decreases with decreasing width, due the spring constant's strong dependence on D. This formula agrees with the time scale derived by Morse for excess contour to diffuse the length of the polymer.[43,§] This behavior would be qualitatively consistent with experiments conducted by S. Quake on the relaxation dynamics of polymers under tension,

---

§We should note that there appears to be some controversy in the literature over the correct scaling exponent of D. While our result agrees with the exponent of 2 suggested by Morse, Maggs, using a different argument, obtains a factor of 4/3.[28]

which also show that the relaxation time decreases as extension increases.[25,48] Finally, note that Equation 33 and Equation 34 suggest that the behavior of the relaxation time of a confined polymer should have a curious nonmonotonic structure. The relaxation time should initially increase slowly as channel width is decreased and then rapidly decrease as the channel width drops below the critical width at which coiling is suppressed. The maximum relaxation time will occur in the cross-over regime.

Methods: Realizations of Nanochannels

While the idea of functional imaging of DNA in nanochannels in itself is conceptually not difficult, the actual fabrication of devices with dimensions below 50 nm has posed a significant obstacle. The demands on the device are size control of the channel widths, smoothness of the internal surface, fabrication of a sealed volume, low background fluorescence, electrically insulating surface and virtual elimination of DNA surface adhesion. The material selection appears to be of paramount importance for device performance. In general, we can classify possible materials into crystalline or glassy and thus "hard" and polymeric or "soft".

Soft materials initially appear very attractive because molding or imprinting provides rapid and low cost patterning.[49,56,20] However, problems of DNA adhesion have persistently prevented the use of those devices unless DNA was forced through channels using large electric fields.[56] We believe that the difficulty of controlling wall adsorption stems from the large surface to volume fraction, the entropic driving force for adsorption of polymers onto walls and the low negative, or even positive zeta potential of common moldable polymeric materials. We speculate here that carefully tailored surfactant systems may overcome the problem of DNA adhesion to polymer materials. In addition, most polymeric materials have a background fluorescence that makes them unsuitable for the observation of single chromophores.

Hard materials such as fused silica clearly provide superior background fluorescence, but initially pose a greater problem for patterning and formation of a closed fluidic system. However, the surfaces of silicon, silicon oxide, silicon nitride and silicon oxinitride bear a large number of hydrogen atoms when processed appropriately. Those surfaces deprotonate in neutral or basic buffers, leaving a negatively charged surface that actively suppresses double-stranded DNA adsorption.

Pattern formation for hard nanofluidics can utilize many techniques that have been developed for the semiconductor industry. Most prominent are nanoimprint lithography, electron beam lithography and focused ion beam milling. Each of those techniques has a particular field in which they excel. Nanoimprint lithography can achieve excellent resolution over large areas at a low cost per wafer. However, that is contrasted by the high initial cost for the master mold. Electron beam lithography also can achieve very good resolution, although the cost per wafer is higher than for imprint lithography. E-beam lithography can yield arbitrary patterns over large areas. FIB can afford fair resolution sufficient for 100 nm nanochannels, but is not able to pattern large areas. However, as a fast prototyping system it can be of value.

A second consideration for hard nanofluidic is the formation of a sealed fluidic system. The first concept is to use a sacrificial layer over which a conformal layer of silicon oxide or silicon nitride is deposited.[36,72] In a subsequent step the sacrificial layer is removed by a thermal or wet process. An example of a sacrificial-layer process is illustrated in Figure 4. The second pathway to a closed hard nanofluidic system uses a pattern transfer from resist into a polished wafer using reactive ion etching (RIE) and subsequent sealing of the structure using a second, transparent wafer.[69] An example of a process using planar, postpatterning sealing is shown in Figure 5. While both processes are capable of yielding sufficiently controlled nanofluidic systems, there are distinctive advantages and disadvantages. The first point of interest is how efficient the sealing is. The limitation of the sacrificial layer process is the quality and pinhole density of the conformal layer. In particular, silicon oxide layers grown by PECVD or thermal evaporation are known to have a problematic number of pinholes, limiting the list of useful materials to silicon nitride or sputtered silicon oxide. The limiting process parameter in planar sealing is the smoothness of the two relevant surfaces, since any roughness translates into a cavity between the wafers. The roughness of wafers commercially

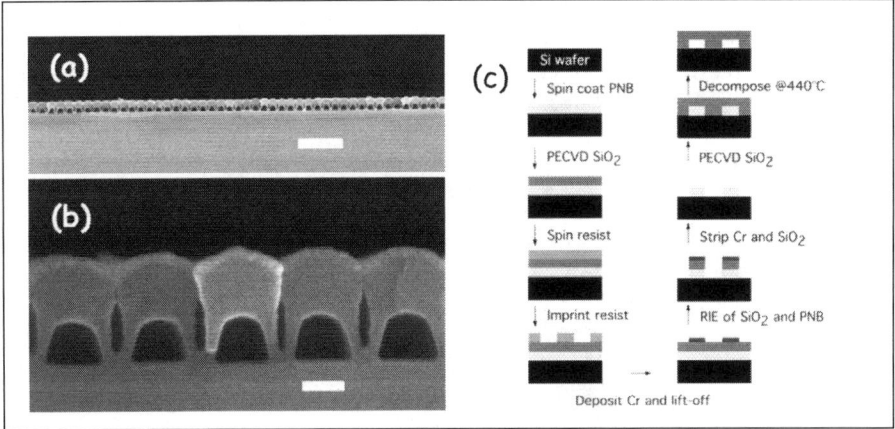

Figure 4. Sacrificial layer process for fabricating SiO_2 nanochannels after Li et al [36]. a) Using nanoimprinting lithography and the sacrificial resist poly(butylnorbornene) (PNB), nanoscale channels can be uniformly defined over large areas. The scale bar corresponds to 0.5 μm. b) Detail of the channels with a diameter of 100 nm resulting from nanoimprinting of PNB. The scale bar corresponds to 100 nm. c) Nanofabrication process diagram.

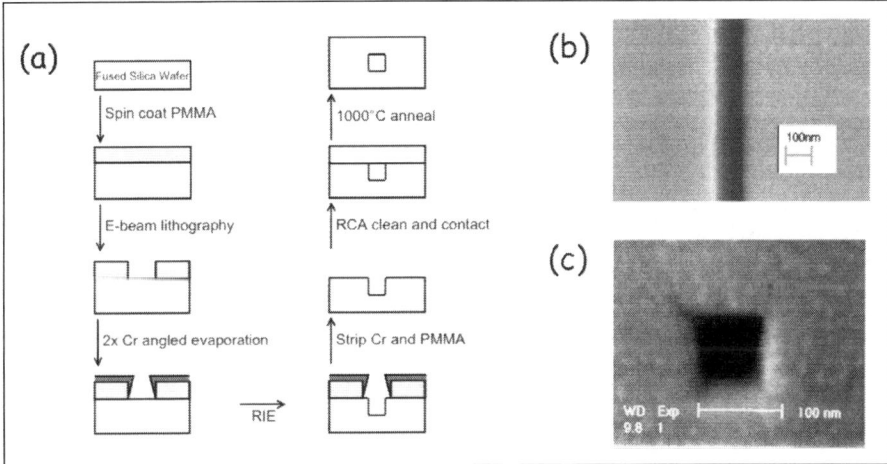

Figure 5. Fused silica device fabricated using planar sealing technique. a) Schematic of the process using electron beam lithography (EBL) and reactive ion etching (RIE) for pattern definition and planar high-temperature sealing. b) Topview SEM of channel after patterning. c) Cross-section of 80 nm x100 nm channel after capping and high-temperature annealing.

available is less than 1 nm and hence the sealing is sufficient for channels down to 30 nm or so. Furthermore, any presence of contaminations such as dust makes proper sealing impossible.

For both sacrificial-layer as well as planar sealing, material compatibility issues exist. In planar sealing both wafers have to have similar thermal expansion coefficients or thermal annealing steps will fail. Also, all materials used in the device have to be compatible with the highest temperature of the process, which can be as high as 1,000°C. For sacrificial-layer methods, the thermal removal or etching of the sacrificial layer has to be complete and sufficient etching resistance and thermal compatibility can be an issue. A recent interesting approach to hard nanofluidics is that of using self-organized inorganic nanotubes as a template for nanochannels. Initially the tubes would be hollowed out and then connected to microfluidics.[16]

Even when one successfully fabricates nanochannels, there may still be the problem of wetting the nanochannels. The problem typically occurs in regions where the cross-section of a fluidic device suddenly decreases, such as in a funnel with a high aspect ratio where gas bubbles can be trapped. This behavior is easily rationalized by noting that liquid wetting is a capillary-driven process and that it proceeds faster at corners than between parallel plates. The simple strategy for removing the gas bubble by flushing it out using pressure-driven flow is impeded by the high internal pressure of those bubbles. The Laplace equation for the pressure P inside a spherical bubble of radius R when the surface tension between the media has a value of γ dictates

$$P = \frac{2\gamma}{R}. \tag{35}$$

The problem is of course that as R goes to zero the internal pressure diverges and the hydraulic pressure needed to push a bubble into a nanochannel diverges.

We have found a way around this problem by simply making γ go to zero by wetting with a liquid that is above its critical point, in essence removing the difference between liquid and gaseous phases.[54] The critical point of water is at 374°C and a pressure of 3212 psi. The liquid form of water is obtained after the supercritical wetting by lowering the temperature. The process is thus conceptually related to critical point drying, which is often utilized in the fabrication of micro electromechanical systems (MEMS).[83]

The process was realized by using high-pressure cells made from two Conflat® flanges separated by a soft copper seal. Cells are loaded with the dry device with all parts submerged in deionized water. After closing the cell, the cell was placed in an oven and the temperature was ramped to 400°C and then slowly lowered down to room temperature. Figure 6 shows a device with a single-entrance nanofluidic system. Traditionally, such a system would be impossible to wet because the air has no way of escaping. The wetting of the nanofluidic channels was visualized by letting the laser dye Rhodamine 610 diffuse into the wet device overnight and then observing the fluorescence under 568 nm irradiation from a Kr+ laser. Note that because of the high process temperature, only "hard" nanofluidic devices are suitable for this method of wetting.

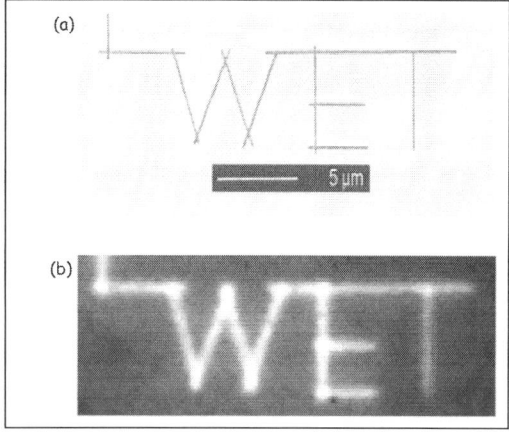

Figure 6. Single-inlet nanochannels. a) SEM image of nanochannels before sealing and wetting of the device. The region shown in the image is connected to the microfluidic channel via a 100 micron long nanochannel. b) Fluorescence micrograph after wetting and diffusion of Rhodamine 610 dye. Reproduced from ref. 54.

Experimental Results

This final section will show some experimental examples of how the technology and theory we have presented can be used. We will first test the theory, discuss how nanochannels can be used in conjunction with AC and DC fields to direct genomic length DNA in different directions and then give two biological examples of nanochannels.

Tests of Elongation Statics and Dynamics in Nanochannels

We will now turn to our experimental studies of DNA under nanoconfinement and the comparison to the theory outlined above. Work in this section is based on references 69 and 51.

The nanochannels used here were nanofabricated on fused silica substrates by means of RIE pattern transfer after nano-imprint or electron beam lithography. Closed fluidic volumes were formed by bonding the substrate to a second plate or coverslip of fused silica using a combination of the surface-cleaning protocol (RCA), room-temperature bonding and annealing at 1,000°C. The device structure also included microchannels, prepared by optical lithography and RIE, which where used for loading of the imaging solution (Fig. 7). Via holes for accessing the fluidic system were prepared using a air abrasion unit commonly used for dental work.

The sealed devices were wet with a loading buffer consisting of 0.045 M tris-base, 1 mM EDTA and 0.045 M boric acid (0.5xTBE). To suppress bleaching and photo-nicking of confined DNA, an oxygen scavenging system was added consisting of 4 mg/ml β-D+ glucose, 0.2 mg/ml glucose oxidase, 0.04 mg/ml catalase and 0.07 M β-mercaptoethanol. In some of the work, this system was replaced by adding 1 mM dithiothreitol (DTT) instead. DNA was moved through the devices and driven into the nanochannels using electrophoresis by means of platinum electrodes

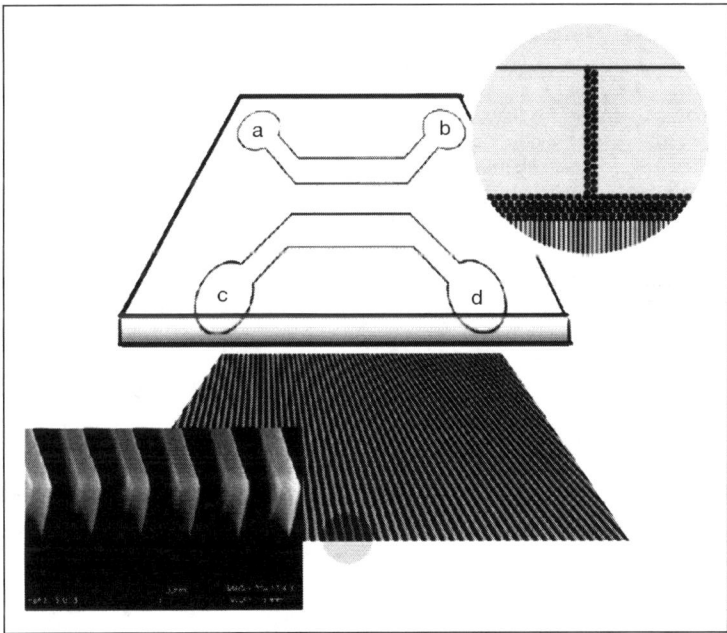

Figure 7. The assembly of a sealed 100-nm-wide nanochannel array with a microfabricated coverslip. The nanoimprinted chips were made in fused silica. The cover chips were patterned using standard lithographical techniques and reactive ion etching. DNA molecules from a gel plug were moved along the path from well a to well b and a driving voltage was used to transfer molecules into wells c and d through the nanochannels on the mating nanoimprinted quartz wafer. Posts of 1 μm in diameter that were separated by 2 μm were used to prestretch the genomic length molecules. Reproduced from ref. 69.

that were placed in the reservoirs interfacing the microchannels. No voltage was applied during the length measurement.

DNA was visualized by staining it with the bis-intercalating dye TOTO-1 (Molecular Probes) at a dye to basepair ratio of 1:10. The fluorescence under 488 nm laser illumination was observed using an optical microscope, oil-immersion microscope objectives with an numerical aperture of 1.4 (both Nikon) and intensified CCD camera (iPentamax, Ropert Scientific, Trenton, NJ). Previous experiments investigating the stretching of DNA stained with intercalating dyes suggest that the dye, to first approximation, just increases the contour length and persistence length up to a saturating value of 30%.[3,47] The saturating dye concentration is 1 dye molecule per 4 base pairs, so at our dye concentration (40% of full dying) we expect an increase of 13%, yielding a contour length of 18.6 μm and a persistence length of 57.5 ± 2 nm (using the value for the persistence length obtained in ref. 67).

The intensity $I(z)$ of the elongated molecule was assumed to be a convolution of a Boxcar function of width r with a Gaussian point-spread function of width σ_0. For a given molecule the intensity transverse to the channel axis was summed to obtain a 1-D intensity scan $I(z)$ along the channel axis. We obtained the length r by fitting $I(z)$ for each frame to

$$I(z) = \frac{I_0}{2}\left[Erf\left(\frac{z}{\sigma_0\sqrt{2}}\right) - Erf\left(\frac{z-r}{\sigma_0\sqrt{2}}\right)\right] \qquad (36)$$

where z is the length along the molecule, Erf is the error function and r and σ_0 are the fitting parameters (offset along the z-axis omitted here). σ_0 will in general not coincide with the Rayleigh criterion because (1) the point spread function is only approximately a Gaussian, (2) the DNA is not of uniform intensity, (3) the DNA fluctuates during exposure time per frame (on the order of 100 ms). For instance, for a x60 objective and 100 ms exposure we obtained a value of 0.4 μm by curve fitting. A typical intensity profile along nanoconfined DNA together with the corresponding curve-fit is shown in Figure 8.

For biologically relevant measurements, the most significant property of DNA stretching in nanochannels is the proportionality between the contour length and the observed length r along the channel. For this test we used nanochannels with a cross-section of 100 nm × 200 nm and a λ-DNA ladder consisting of concatamers of the 48.5-kbp-long monomer (cI857 ind1 Sam7) embedded in low-melting-point agarose (product no. N0340S, New England Biolabs). This sample is a well-established size standard for gel electrophoresis. The DNA was brought into the nano-channel region by electrophoresis and then observed for about 10 seconds. For each molecule the average length was determined over all frames and lengths were compiled into a histogram (Fig.

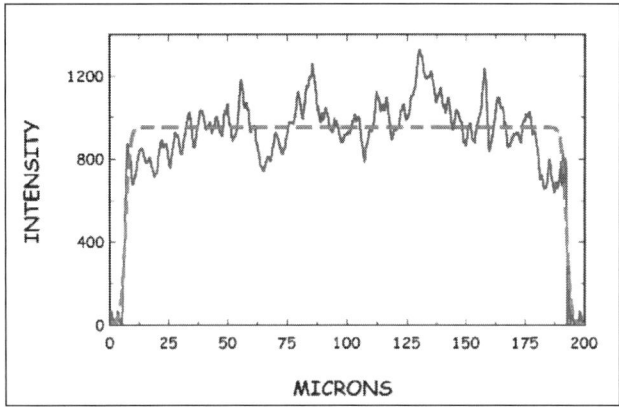

Figure 8. Intensity vs. length of a confined DNA molecule with r = 185 μm. The dashed line is the fit of the data. Reproduced from ref. 69.

9). As anticipated, the histogram consists of clear, discrete peaks, which we can be attributed to the different ligation numbers, i.e., monomers, dimers, etc. In the histogram we show the amount of DNA (number of molecules times average length) instead of the number of molecules in order to obtain a better comparison with the data obtained by pulsed-field gel electrophoresis (Fig. 10). We find the first four peaks observed at $r = 8 \pm 1$ μm (n = 1), 16 ± 1 μm (n = 2), 24 ± 1 μm (n = 3) and 32 ± 1 μm (n = 4).

A plot of average length per ligation number reveals a linear relationship, consistent with the theories introduced above (Fig. 11). From the slope of the curve we find an extension factor $\varepsilon = 0.36$, roughly consistent with expectation from the de Gennes theory.

While this important result can be reproduced by a number of theoretical descriptions, we expected a study of the extension as a function of the channel dimension to pose a far more stringent test. We conducted the following experiments in 35 nm- to 440 nm-channels either with

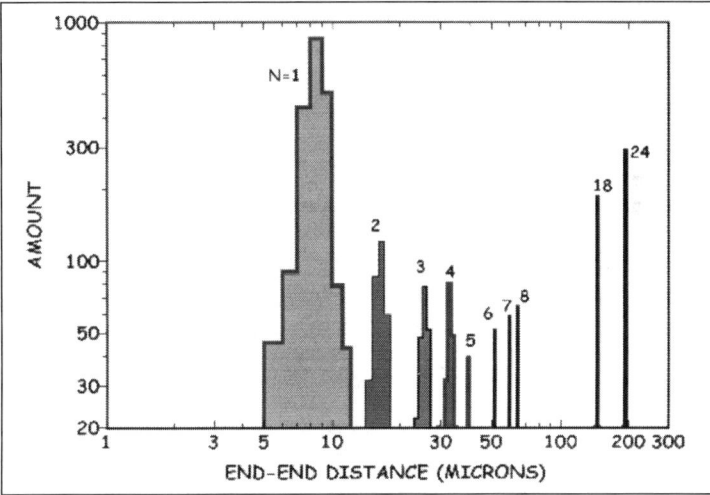

Figure 9. A histogram of end-to-end distances r of molecules observed in 2 min of running DNA molecules into 100-nm-width nanochannels vs. the amount of DNA is shown, as described in the text. The assignment of the single DNA molecules to n = 5-8 is based on the assumption that $r \sim L$. Reproduced from ref. 69.

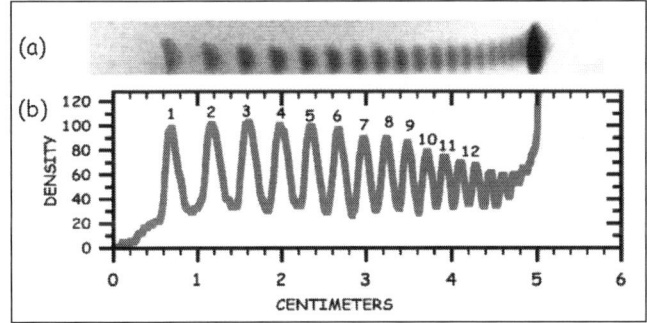

Figure 10. Analysis of pulsed-field gels. a) Gel of the λ-ladder used in this experiment. b) Scanned density of the gel lane, with N-mer labeling. An applied electric field of 5 V/cm was used, with the field direction switching 60° to the average direction, with a period that was linearly ramped from 5-120 s over the entire run (~18 h). Reproduced from ref. 69.

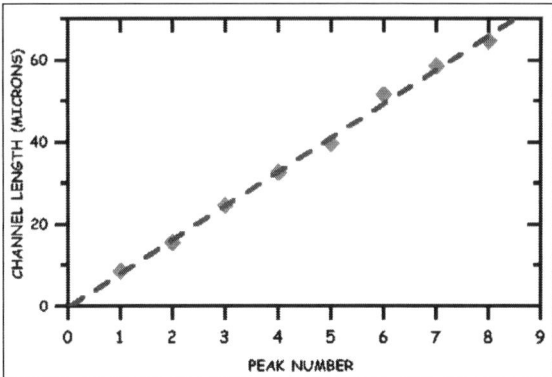

Figure 11. Observed end-to-end distance vs. the N-mer ligation value. The data are shown as diamonds and a linear fit is shown by the dashed line. Reproduced from ref. 69.

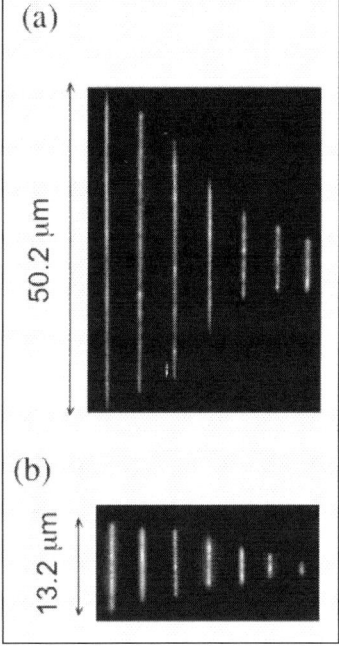

Figure 12. a) Averaged intensity of selected T2 DNA molecules in 30 × 40 nm, 60 × 80nm, 80 × 80 nm, 140 × 130 nm, 230 × 150 nm, 300 × 440 nm and 440 × 440 nm channels (left to right). b) Averaged intensity of selected λ-DNA molecules in the same channels. Reproduced from ref. 51.

λ-phage DNA (48.5 kbp, L = 16.5 μm, L_{dye} = 18.6 μm, New England Biosciences) or T2 DNA (164 kbp, L = 55.8 μm, L_{dye} = 63 μm, Sigma). Figure 12 shows images of λ-phage and T2 dsDNA molecules confined in the nanochannels. The stretching of the polymers is clearly a function of the channel width and plateaus as the width drops below the persistence length. Figure 13 shows a plot of the DNA extension versus the geometric average of the nanochannel dimensions. The data fits well to a power law of the form $\sim D^{-0.85}$ with the clear exception of the data point for the

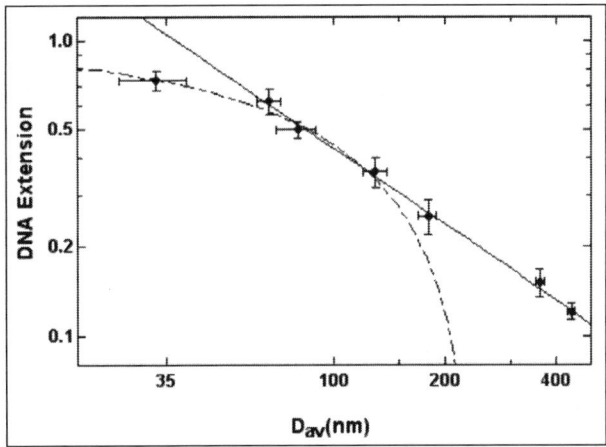

Figure 13. Log-log plot of λ-DNA extension as a function of D_{av}, the geometric average of the channel depth and height. The DNA extension is normalized to the (dye-adjusted) total contour length of 18.63 μm. The bold line is a best power-law fit to the data for the 440 nm, 300 nm, 230 nm, 140 nm, 80 nm and 60 nm wide channels (the best fit exponent is −0.85 ± 0.05). The dashed line is the Odijk prediction, which fits to the three smallest channels with a persistence length of 52 ± 5 nm, in agreement with the dye-adjusted persistence length of 57.5 ± 2 nm. The data points shown are averages over all the extensions measured for molecules in a given width (typically 10-40). The error-bars are the standard deviation taken from these measurements. Reproduced from ref. 51.

30 nm channel, suggesting that the smallest channel is in the Odijk regime. The transition scale can be precisely defined by requiring that Equation 23 and the power law fit merge continuously at a critical scale $D_{critical}$. This stipulation enforces $D_{critical} = \gamma P$, with the proportionality constant $\gamma = 1.93$ fixed entirely by the power law exponent and the value of A in Equation 23.

The behavior of the extension for widths greater than the cross-over scale differs somewhat from the classic de Gennes theory[14] which predicts the extension should scale as $D^{-2/3}$. We believe that this discrepancy arises because in the cross-over regime channel widths around 100 nm semiflex ibility and self-avoidance are equally important, so one would not expect the pure self-avoidance exponent to provide the full physical picture. However, the agreement of our measurements with de Gennes' and Odijk's treatments has to be considered largely satisfactory and we will use them for predictive estimates in planning and analyzing subsequent measurements.

We will now take a closer look at the fluctuations of the end-to-end length r of nanoconfined DNA. Figure 14 shows a plot of r as a function of the time, clearly revealing fluctuations in r. A histogram of the instantaneous value of r reveals an approximately Gaussian distribution, with a width of $\sigma_t = 0.6$ μm and an average length of 8.4 μm (σ_t refers to the thermal fluctuations). In Equation 29 of the theory section an effective spring constant k_{eff} was derived. The amplitude of the fluctuations can now be obtained from

$$\sigma_t \simeq \sqrt{\langle \delta r^2 \rangle} = \sqrt{\frac{k_B T}{k_{eff}}} = \sqrt{\frac{15}{4}} \sqrt{L} (P w D)^{1/6} \qquad (37)$$

The experiments using the λ-DNA ladder show a fair agreement with that prediction (Fig. 15). Note that this finding may be due to the limited sampling time during the experiment, or inhomogeneities in nanochannel widths.

It is obvious that in order to perform meaningful averaging and estimation of the error of the mean of the measurement, the time has to be found over which measurement of the same molecule can be considered statistically independent. We determined those times for the dataset

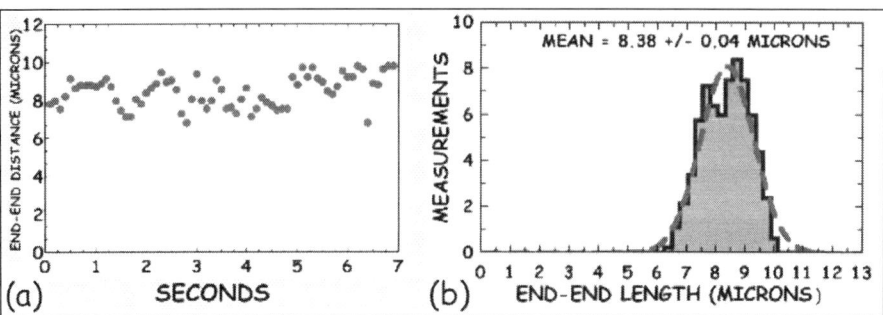

Figure 14. The end-to-end dynamics of a confined DNA molecule. a) End-to-end distance of a λ monomer confined in a 100-nm-wide channel as a function of time. b) Histogram of the observed end-to-end distances *r* of the monomer data. The SD of the mean length of 8.38 ± 0.15 μm as determined by the Gaussian curve fit (dashed line). Reproduced from ref. 69.

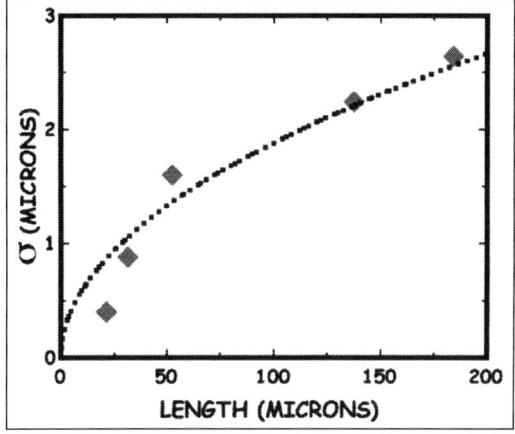

Figure 15. The observed SD in the length of a confined channel of cross-section 100 nm × 200 nm vs. the length of the molecule. Reproduced from ref. 69.

of Figure 13 by taking the extension fluctuations δr about the mean extension r and calculating the autocorrelation function $C_{\delta r} = <\delta r(t) \, \delta r(t + \delta t)>$. Exponential fits to $C_{\delta r}$ yield the relaxation time (see Fig. 16). The extension and relaxation time were then averaged over as many molecules as could be conveniently measured in a single experiment to obtain the best estimates of r and t for a given channel width and molecule size (typically, 10-30 molecules were used). The measured relaxation times as a function of the channel dimension are shown in Figure 17. These data provide added proof for the cross-over between de Gennes and Odijk regimes and show that the relaxation times are maximized between 80 and 130 nm, consistent with $D_{critical} = 1.9P = 110$ nm. To explain the existence of this maximum, note that the relaxation time scales as the ratio of a friction factor ξ to an effective spring constant k (for Small displacements about the equilibrium extension). The self-avoidance model predicts a friction factor that increases slightly faster than the spring constant, leading to a relaxation time that increases slowly with decreasing width.[51] The situation is reversed in the Odjijk regime: for $D << P$ the relaxation time rapidly decreases with decreasing width due to the strong scaling of the spring constant with width in this regime. Again, we find fair agreement between theory and experiment.

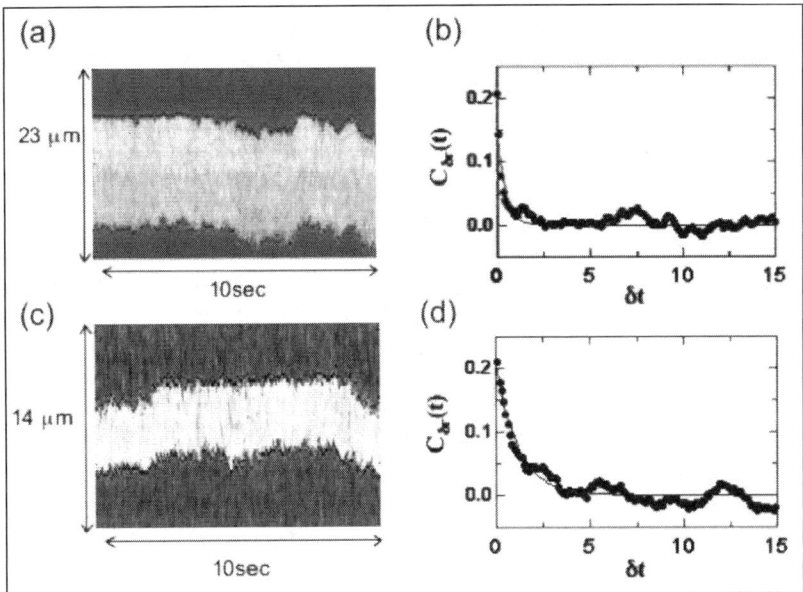

Figure 16. a) Time evolution of intensity stripe for a λ-phage DNA molecule in a 60 nm channel. b) Autocorrelated extension fluctuation extracted from molecule shown in (a) with exponential fit. c) Time evolution of intensity stripe for a λ-phage DNA molecule in a 180 nm channel. d) Autocorrelated extension fluctuation for molecule shown in (c) with exponential fit. The black dots at the stripe edges in figures (a) and (c) represent the edges of the molecules determined via an edge-finding algorithm. Reproduced from ref. 51.

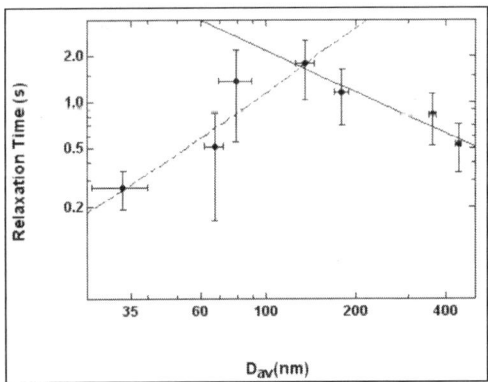

Figure 17. Log-log plot of λ-DNA relaxation time as a function of D_{av}, the geometric average of the channel depth and height. The data points shown are averages over all the relaxation times measured for molecules in a given width. The error-bars are taken from the standard deviation of these measurements. The large error bars result from two factors: (1) statistical variation resulting from the relatively Small amount of data used and (2) slow variations in the extension due to interaction of the DNA with channel defects. Shown superimposed is a best power law fit to the data taken for channels greater than 140 nm (bold curve) and a fit to the model $\tau \sim D^{\alpha}/\log(D/w)$ for channel widths less than 140 nm (dashed curve). The de Gennes theory underestimates the scaling exponent: the best fit exponent for the large channel widths is −0.9 ± 0.4. The best fit exponent for the Small widths is α = 1.6 ± 0.4. Reproduced from ref. 51.

With the time scale for statistically independent measurements established, we can now examine the resolution of the technique. The standard deviation of the mean extension $<r>$ of a given molecule should scale as $\sigma_r/\sqrt{M}$ after M independent measurements. For the channels the molecule shown in Figure 14, the time for independent measurements is about 1 s and hence the average length of the molecule is known with an uncertainty of 0.15 μm, or about 1 kbp.

We now can estimate the achievable resolution for a measurement over a given time. As in pulsed field electrophoresis, the resolving power R, given by the ratio of the obtained length and uncertainty of the measurement, is dependent both on the length of the measured molecule and the time over which the measurement was performed. For the de Gennes regime, we have previously derived all the necessary quantities for evaluating the resulting expression

$$R = \frac{r}{\sigma_r}\sqrt{\frac{t_{meas}}{2\tau}}. \tag{38}$$

That relationship obviously holds only for $\tau << t_{meas}$. Using this relationship we find an uncertainty of about 1 kbp for the molecule in Figure 14, in good agreement with the experiment.

We can also use this equation to obtain some intuitive understanding of the practical power of the nanochannel technique. In Figure 10 we show the results of a sizing experiment using pulsed field electrophoresis, which was run over the duration of 18 hours. We focus on the long molecules, in particular the concatamer with linking number 16. If we were to run a nanochannel experiment for the same time (18 hours), we would obtain an uncertainty on the order of 2 kbp (1 sigma), where the pulsed field gel barely can distinguish molecules of that length, which differ by 50 kbp. A nanochannel measurement of only one frame can equal that resolution (about 25 kbp, after equilibration for 5 minutes). If we were to size a λ-monomer for 18 hours, we would obtain a resolution of 34 bp.

In the last paragraph we mentioned the importance of thermalization before measurements are performed. The authors of references 50 and 37 have studied the entry and exit of DNA into nanochannels, as well as the relaxation of DNA after compression. Interestingly, Mannion et al find that the relaxation of mildly overstretched DNA in channels can be approximated by an exponential decay relationship. Hence we can speculate here that equilibration may not be a not be required for length measurements, but that a fit to an exponential decay can yield the desired equilibrium length just as well. Reccius et al find that for strongly compressed DNA the exponential assumption does not hold, although the model these authors provided does also not provide an excellent fit.

Nanochannel Railroad Switchyards

We have discussed up to this point simple linear arrays of nanochannels, in which DNA is confined to an essentially one-dimensional volume. However, the channel topologies can be made more complex with surprising and useful results.[52] As an example of this complexity, we discuss here a two-dimensional meta-material made of an asymmetric lattice of nanochannels and show that in two dimensions DNA can be transported in a fashion that is similar to propagation of light in the Pockels effect.[67]

The asymmetric lattice is formed by orthogonally intersecting two arrays of nanochannels that have equal pitch, but dissimilar channel cross-sections (Fig. 18). The width of the nanochannels was chosen so that DNA would be stretched to about 50% of its contour length, with channels running in one direction wider than in the other (100 nm and 130 nm wide, respectively, 100 nm deep, 2 micron period). For our experiments, we interfaced a stripe of the meta-material to two microfluidic channels, such that voltages could be applied at a 45° angle to both principal directions of the two-dimensional structure. The transport direction under a d.c. bias is not parallel to the applied external field, but parallel to one principal axis of the lattice. The transport direction can be switched by 90° by applying an a.c. modulation to the d.c. bias. This effect arises from an interplay of entropic and dielectrophoretic energy contribution. Our results imply that program-

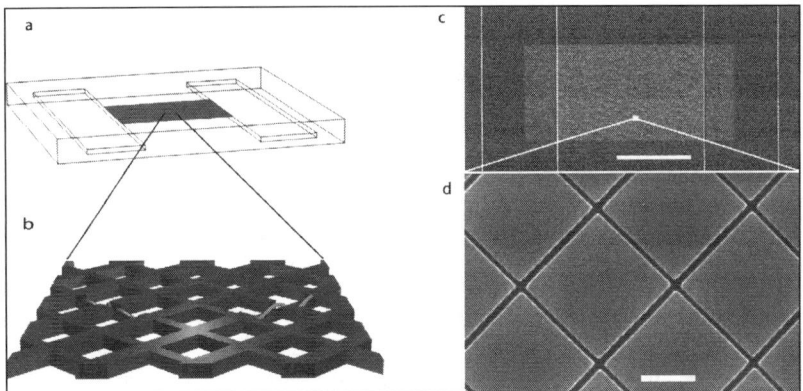

Figure 18. Schematic of the device and SEM micrographs of structures before sealing the fluidic system. a) Schematic overview of the device, which consists of one microchannel each on the left and the right, both 100 μm wide and 1 μm deep, linked by a 200 μm-long and 120 μm-wide nanofluidic region. b) Enlarged schematic of the nanofluidic region, consisting of a square lattice of nanochannels, with differing channels widths along the principal axes. The green DNA molecule is confined to a wide channel, as in the case of a purely d.c. field. The red DNA molecule is confined to a narrow channel and corresponds to a molecule that is subject to an appropriate a.c. voltage. c) Overview SEM of a device used in this publication. d) SEM of the local structure, showing a lattice with 2 μm period and nanochannel widths of 100 nm and 140 nm, respectively. During transport measurements, the electric field was applied horizontally in this figure. Reproduced from ref. 52.

mable DNA transport in nanofluidic devices is possible without the need for moving parts, external pumps, or an inflationary number of local electrodes.

The effects of anisotropic transport and switching of the transport direction are demonstrated in Figure 19. A λ-DNA molecule was observed moving along the direction of the wider channels

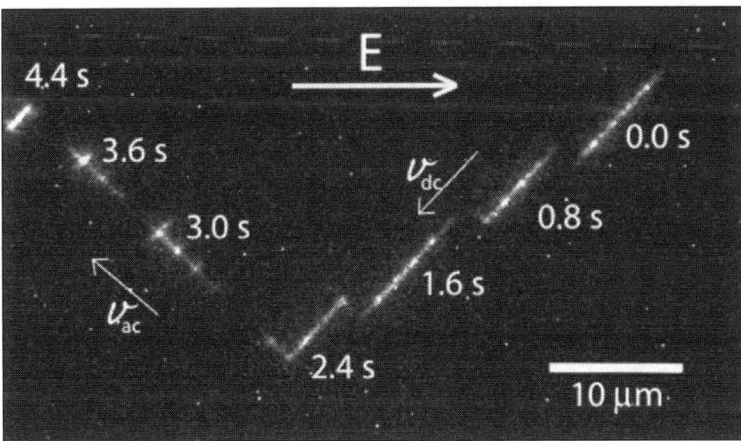

Figure 19. λ-DNA moving through the 2-d meta-material. The orientation of the lattice and the stripe direction are the same as in Figure 18. At times before 2.4 s, only a d.c. bias was applied and the DNA traveled mostly inside wide channel segments. At 2.4 s, an a.c. field (350 V/cm, 300 Hz) was added to the d.c. bias and the polymer subsequently aligned with the narrow channels. The molecule then moved along the narrow channels. Reproduced from ref. 52.

when only a d.c. bias was applied between the microchannels. The molecule was stretched out and aligned with the wide channels, at an angle to the direction of the externally applied field. However, when an a.c. voltage was added to the d.c. bias, the DNA molecule aligned along the direction of the narrow channels, orthogonal to the previous direction and proceeded to move along that axis. This effect is unique to the tailored energetic environment that the polymer experienced in our structure.

We can understand the effect of an asymmetric lattice in the absence of an applied electric field by considering the free energy of confinement U_S.[51] For use with a single nanochannel segment in a lattice, we rewrite de Gennes' result[14,57] in terms of the occupied length along the channel, r, which is set equal to the lattice constant:

$$U_S \cong k_B T \frac{r}{D}. \tag{39}$$

D is the width of the nanochannel segment, which can be replaced by the geometric average of width and height for rectangular channels. Remarkably, this relationship is independent of the structural details of the polymer, except for that the molecule has to be long enough to be treated as linearized inside the nanochannel segment. Note that U_S is of direct importance to the configuration of molecules in an asymmetric lattice, as it acts to drive DNA from narrow channels into wider channels.

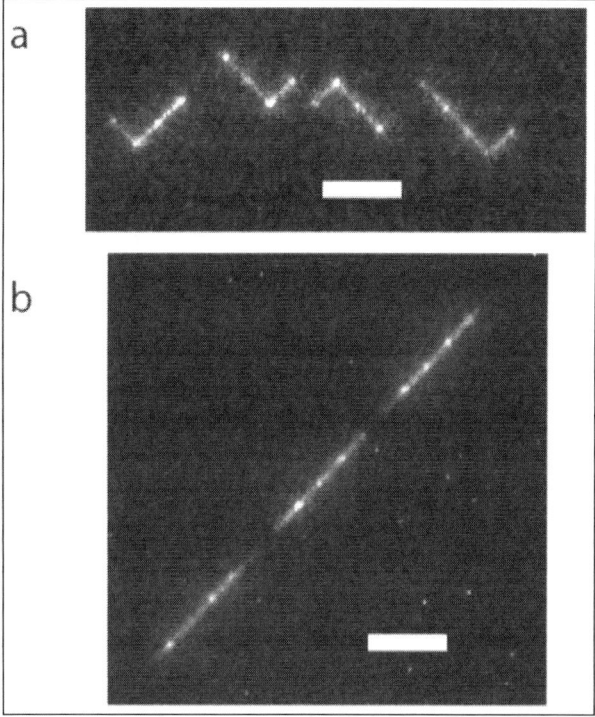

Figure 20. Movement of λ-DNA in lattices without an a.c. field. In order to show multiple configurations of the same molecule, a small d.c. bias was applied in both cases. a) Movement in a symmetric square lattice of 100 nm-wide nanochannels (scale bar 5 μm, frame spacing 8 s). The molecule has no preference for occupying a particular direction and the ratio of straight to kinked geometries at junctions is about one. b) Movement in an asymmetric lattice (scale bar 5 μm, frame spacing 1.5 s). The molecule is completely confined inside wide channels. Reproduced from ref. 52.

The effect of U_S is illustrated in Figure 20, which shows DNA molecules in a symmetric and an asymmetric lattice, respectively. In the symmetric lattice (Fig. 20A), the molecule formed kinked and straight conformations at junctions with about equal probability. On average, the molecule had no preferred orientation. In contrast to the symmetric lattice, an asymmetric structure (Fig. 20B) did result in almost exclusively straight configurations at channel junctions. More specifically, we observed that the molecule was contained within the wider channels, in agreement with the functional dependence of U_S on D. Molecules thus exhibited a net alignment.

Interestingly, the average migration direction of molecules under d.c. electrophoresis mirrored the alignment within both symmetric and asymmetric devices. On average, DNA moved parallel to the stripe direction (horizontal) in the symmetric structure and at a 45° angle to that in the asymmetric lattice. An asymmetric material thus remarkably leads to polymer transport that is not parallel to the average current density. Note that the entropic effect described here is substantially different from size-dependent sorting,[27] since DNA has no local intrinsic size and the phenomenon is independent of DNA length. We can estimate the effectiveness of alignment in the asymmetric lattice by using a Boltzmann relationship and Equation 39. For the structure presented in Figure 20, we conclude that the fraction of narrow channel segments out of all occupied segments should be about 8%. This fraction is roughly consistent with the experimental value of at most 10% that was observed using molecules with a drift velocity equal to or less than 2 μm/s.

We now consider a device with an applied a.c. voltage. The periodic polarization of the counter-ion cloud surrounding the DNA leads to an attractive dielectrophoretic potential U_E that is proportional to the square of the electric field.[11] An intuitive understanding of the electric field distribution inside the device can be obtained by the following Gedanken experiment. We consider an infinitely long stripe of the meta-material that has a finite width. Let us initially also assume that the electric field strength inside wide and narrow channels is equal. Since the resistance of the wide channels is lower than the one of the narrow channels, the current through wide channels is larger than through narrow channels and the combined average current flows not parallel to the stripe, but at an angle to it. The current component perpendicular to the stripe would however charge the insulating boundaries and that charge would give rise to an electric field component perpendicular to the stripe. In order to reach steady state the charging has to cease and that is exactly the case when the current is parallel to the stripe, with no perpendicular component. At that point an equal current has to flow through narrow and wide channels. Given equal buffer conductivity but different cross-sections of narrow and wide nanochannels, that in turn requires that the local field strengths inside the nanochannels are inversely proportional to their cross-sections. A detailed numerical calculation of the electric fields (Fig. 21A) assuming equal total currents through narrow and wide channels confirms this simple argument, but shows some added complexity at channel junction. We have also tested whether the assumption of an infinitely long stripe holds for our geometry using a mean-field model with an anisotropic conductivity and found a satisfactory agreement with the above argument in the center of the device and close to the insulating walls.

Because of the distribution of the electric field strength and functional dependence of the dielectrophoretic potential U_E on it, U_E will counteract the free energy of confinement U_S and drive DNA into narrower channel segments. On average, a high-strength a.c. electric field will orient DNA along the narrow channels. To rephrase the result, we are able to tune the strength of the confinement potential to the point of reversing the sign of the net force by applying a.c. electric fields.

We demonstrate the effect of an a.c. field in Figure 21B,C, using the same device as in Figure 20B. Upon application of the a.c. voltage, the molecule rapidly moved from wide channel segments into a narrow segment and contracted at the same time. When the a.c. voltage was turned off, a slow relaxation back into a wide channel followed. We then added a d.c. offset to the a.c. voltage and observed movement of DNA along the narrow channels, with occasional lane skips (Fig. 21D). The motion occurs not homogeneously, but rather in jumps. We can explain this by the pinning of DNA at junctions due to the local minima of the entropic and dielectrophoretic potentials there.

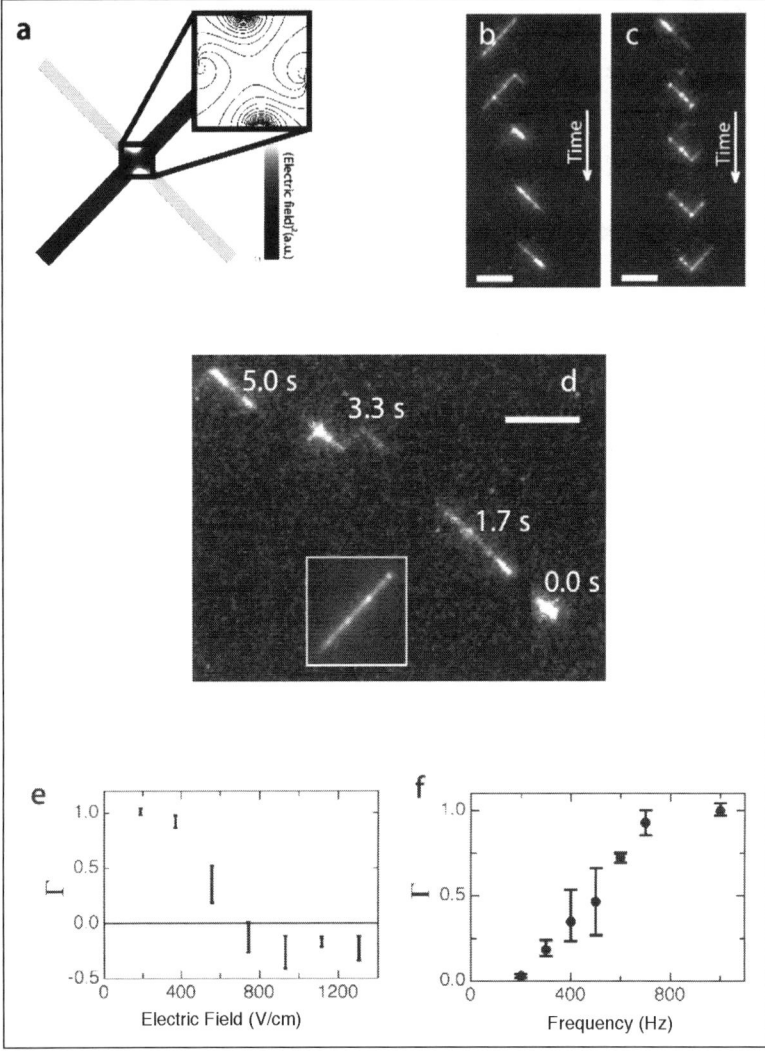

Figure 21. Alignment of DNA under an a.c. field. a) Electric fields squared in the unit cell of a square lattice with 100 nm and 140 nm wide nanochannels, in arbitrary units, calculated using FEMLab (Comsol, Burlington, MA). The inset shows lines of equal potential energy due to dielectrophoresis close the nanochannel junction. b) DNA moving from wide into narrow channels of an asymmetric lattice as an a.c. voltage is applied in the second frame (frame spacing 650 ms, scale bar 5 μm. c) DNA relaxing into wide nanochannel after removal of the a.c. voltage (frame spacing 2.6 s, scale bar 5 μm). d) Molecule moving through an asymmetric array in the presence of a.c. field with a d.c. offset. Exposure time 0.55 s, frame spacing 1.65 s, scale bar is 5 μm. The DNA is aligned with the narrow channels and moves by hopping between narrow segments. The inset shows the same molecule before applying any voltage and aligned with the wide channels. e) Alignment coefficient Γ for a λ-DNA molecule in a lattice with 100 nm and 130 nm-wide nanochannels as a function of the applied a.c. electric field at a frequency of 1 kHz. Positive values signify alignment along the direction of the wide channels and negative ones alignment along the direction of the narrow channels. Bars indicate minima and maxima over all measurements. f) Alignment coefficient Γ for a λ-DNA dimer versus frequency at an a.c. field strength of 200 V/cm. Reproduced from ref. 52.

A few points about DNA confinement in an a.c. electric field merit further attention. We quantified the effect of the a.c. voltage and frequency by calculating the alignment coefficient Γ, which is defined in the following (Fig. 21). Γ is based on the average spatial 2-dimensional fast Fourier transform (FFT) for a molecule over all frames of a movie. We took the difference between the components in the principal directions divided by the total intensity in the Fourier transform along the two axes. For each molecule we calculated this ratio with and without a.c. field and the ratio of those is the alignment coefficient Γ. Positive values signify alignment along the direction of the wide channels and negative ones alignment along the direction of the narrow channels.

The graph of Γ as a function of the applied a.c. voltage exhibits the signature of a phase transition with a transition point from alignment in wide channels to alignment in narrow channels at around 600 V/cm. When the frequency of the a.c. field was varied between 200 Hz and 1 kHz at a constant a.c. voltage, we found that the alignment in narrow channels decreased as the frequency was increased. Both observations are in agreement with the treatment presented by Chou et al[11] who describe the dielectric response as arising from the polarization and relaxation of the counter-ion cloud around a polyelectrolyte in a salt solution. However, the precise mechanism may be more intricate since for nanoconfined DNA the interaction of counter-ions, channel walls and DNA may become inextricably linked. Note that in Figure 21, Γ did not fully reverse sign for the following reasons. A) DNA gets compressed and especially at very high voltages DNA is concentrated at nodes. B) Because of the electric field distribution at junctions, kinked configurations are preferred over straight ones at very high fields.

While all the observations reported in this work were obtained using double stranded DNA, the results are expected to hold in a more general sense. All semiflexible, self-excluding polymers are expected to align in an asymmetric matrix, given that the polymer strands are long enough to fill a full period of the lattice. Polymer transport can occur by fluid flow, or any other mechanism such as electrophoresis. The electrical switching of the anisotropy is dependent on the a.c. response of the polymer, but we expect the basic behavior to be maintained for all polyelectrolytes. We also expect that the basic principles demonstrated here will be used in integrated devices for biological and biopolymer analysis.

Restriction Mapping

To this point we have discussed the behavior of DNA in nanochannels and how this knowledge can be used to size DNA by observing the end-to-end length. However, we have neglected that the stretching is uniform along all molecules and each point of a molecule in a micrograph can be attributed to a specific genetic location. Location-specific functional imaging thus becomes possible if a means of visualization for the functionality can be found. We have achieved that visualization by observing DNA modification through restriction enzymes.

Restriction endonucleases, also simply called restriction enzymes, cleave DNA at specific sequences that are characteristic of the particular enzyme. Hence, a cut by a restriction enzyme can serve as a positive confirmation that a certain sequence exists at the cut position. Restriction mapping, the measurement of landmarks along genome-sized molecules using restriction enzyme, has been a very successful application of the stretch-and-fix approach to single-molecule DNA measurements. In particular, that technique has made obtaining ordered restriction maps of megabasepair-molecules possible for the first time.

At first glance, restriction mapping is a technique at which nanochannel devices should excel: the stretching for fragments is homogeneous and independent of each other, each fragment can be measured multiple times and the measurement is easily automated. It hence overcomes many of the problems that have impeded the commercialization of the stretch-and-fix approach. However, the apparent challenge is that the cutting must not occur before the molecule is fully stretched within the nanochannel. We solve that problem by manipulating the concentration of the necessary enzyme cofactor Mg^{2+} throughout the device.

In order to understand how Mg^{2+} is used to control the restriction process, we first have to look at the steps involved in the function of restriction enzymes. In the first step, the enzyme attaches unspecifically to the DNA strand. In the second phase, the molecule finds its specific target site. These two steps can proceed without divalent ions for selected restriction enzymes. The final steps of enzymatic cleavage of the DNA molecule and release of the cleavage products are dependent on the presence of Mg^{2+}. Our strategy is to first decorate DNA fully with the desired restriction enzyme, to introduce it into a nanochannel that contains a Mg^{2+} gradient and to move it electrophoretically to a high-Mg^{2+} region within the nanochannel where the restriction reaction proceeds.

The device, shown in Figure 22 consists of a microfluidic "loading" channel containing the DNA to be analyzed, the restriction enzyme and EDTA; and a microfluidic "exit" channel containing Mg^{2+} and the restriction enzyme. The two microfluidic channels are linked by 10 nanofluidic channels of about 100 nm in diameter. We expect that, in the absence of an applied voltage along the nanochannels, a Mg^{2+} gradient will be established as the result of Mg^{2+} and EDTA diffusion and chelation of Mg^{2+} by EDTA. With an applied voltage, we expect the higher mobility of the Mg^{2+} ions to result in a constant Mg^{2+} concentration equal to that on the "exit" side. Note that the particular shape of the Mg^{2+} concentration is not crucial to the functioning of the device, only the facts that there is a concentration sufficient for restriction in the active region, i.e., the nanochannels and a negligible concentration in the "loading" microchannel.

We performed restriction mapping of λ-DNA (48.5 kbp) using the enzymes *Sma* I and *Sac* I in nanochannels where the DNA was stretched to about 40% and 30% of its full length, respectively (Fig. 23). Molecules were observed for about one minute and both typical frames as well as time traces are shown. The latter is a stack of lines, each of which shows the intensity along the molecule for a single 10-ms frame. *Sma* I cuts λ-DNA into fragments of 19.4 kbp, 12.2 kbp, 8.3 kbp and 8.6

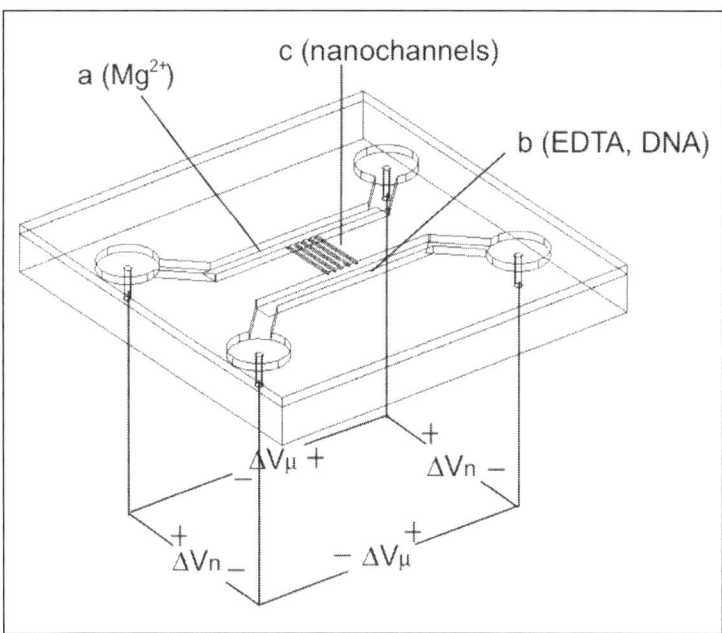

Figure 22. Schematic of the device employed in the experiments. We linked two microfluidic channels (a, b, 1 μm × 100 μm cross-section) with 10 nanochannels of about 100 nm × 100 nm (c). The solution in the "loading" microchannel (b) contained DNA and EDTA, while the "exit" microchannel (a) contained Mg^{2+}. Both channels contained restriction enzyme. Both DNA and Mg^{2+} were moved through the device by electrophoresis using four electrodes. The voltage applied across the length of the nanochannels is marked ΔV_n (~ 2 V) and the voltage across the microchannels is ΔV_μ (~ 2 V). During DNA imaging, no voltages were applied. Reproduced from ref. 53.

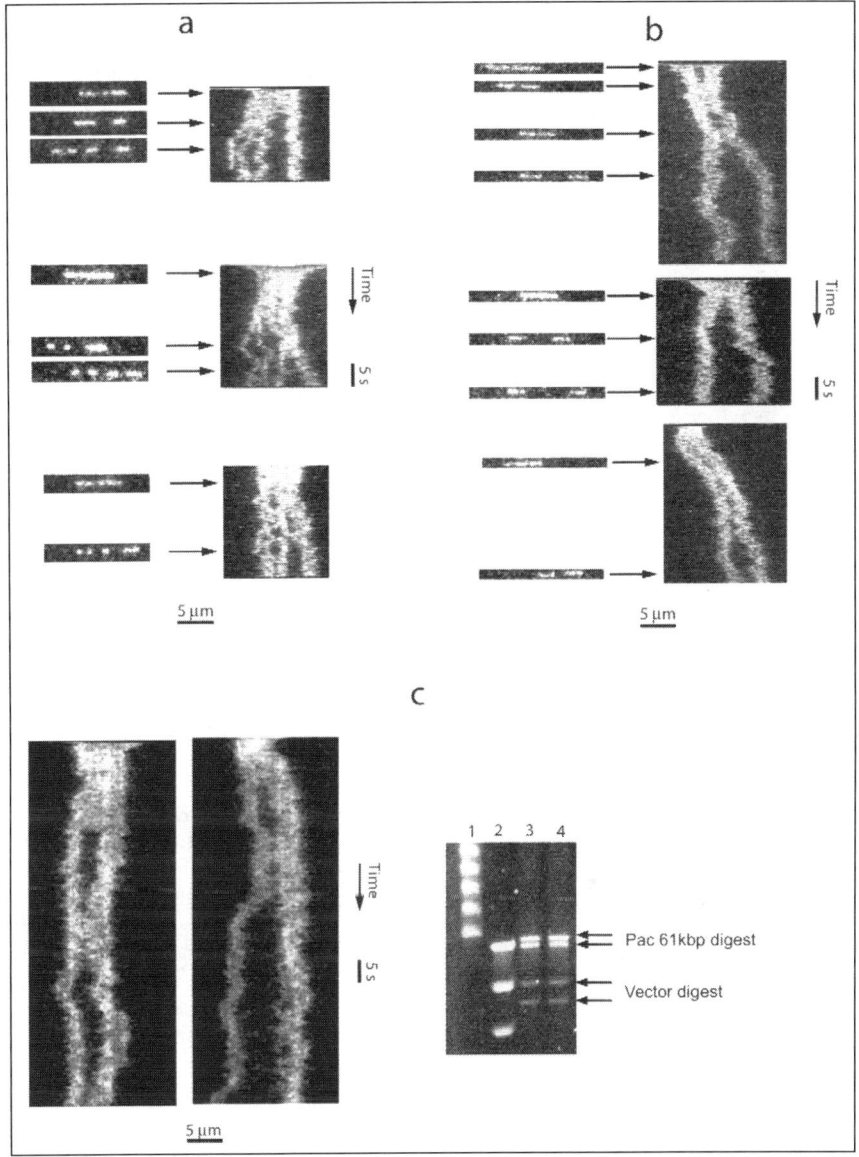

Figure 23. a) Restriction of three λ-DNA (48.5 kbp) molecules using *Sma* I in channels of about 120 nm × 120 nm cross section. On the left, individual 10-ms frames are shown, while the right panels contain timetraces, in which each line corresponds to intensity along the nano-channel in a single frame. b) Restriction of three λ-DNA molecules using *Sac* I in channels of 140 nm × 180 nm crosssection. On the left are individual 10-ms frames and on the right are the timetraces. c) Cutting 61 kbp DNA with *Pac* I. The top panels are time traces of cutting in roughly 120 nm wide nano-channels, where we expect a stretch of 40%. The right panel shows a pulsed-field gel electrophoresis separation of the digestion product of an unpurified 61 kbp DNA and cloning vector. 1) λ-DNA ladder, (2) long range PFGE ladder, (3) and (4) digestion product after *Pac* I. Reproduced from ref. 53.

Table 1. *Cutting positions (A) and fragment lengths (B) of Sma I on λ-DNA in kilobasepairs (kbp) derived from molecules with two and three cuts.* A)

Sequence	19.4	31.6	39.9	
Histogram	19.3 ± 1.2	32.1 ± 1.0	40.6 ± 2.0	
Weighted Average	19.9 ± 1.3	32.8 ± 1.3	40.7 ± 1.7	
B)				
Sequence	19.4	12.2	8.3	8.6
Histogram	19.3 (−0.1)	12.8 (+0.6)	8.5 (+0.2)	7.9 (−0.7)
Weighted Average	19.9 (+0.5)	12.9 (+0.7)	8.0 (−0.3)	7.8 (−0.8)

The values in the second line are obtained from a three-Gaussian fit to the histogram in Figure 24. The values in the third line were calculated from the average values for the individual molecules, by forming the weighted mean of the individual-molecule averages with the observation times for the individual molecules as weights. The errors in the upper panel are the statistical variances and 29 molecules were used for the statistical analysis. In the lower panel we have indicated the difference between the experimental and the sequencing values in brackets. Reproduced from ref. 53.

kbp, in that order. All 4 fragments were clearly observed. *Sac* I cuts λ-DNA into three fragments 22.6 kbp, 0.9 kbp and 24.8 kbp long. Only the two longer fragments could be clearly observed. Since cuts were easily observed in Figure 23 we can reexamine the question why cuts do become visible. We did not observe obvious recoil of the ends at the cutting positions. That is because the thermalized polymer has a constant stretching independent of its length and no intrinsic tension is acting along the backbone. Thus we have to conclude that the gaps in enzymatically cleaved DNA molecules become visible because the diffusion of restriction fragments inside the channels.

We have used the curve-fitting method outlined above to size the restriction fragments from the restriction mapping of λ−DNA using *Sma* I. Figure 24 is a histogram of calculated cut positions of *Sma* I on λ−DNA, constructed from all molecules with 2 or 3 observed cuts. There are three known restriction sites and the positions derived from the histogram correspond to the positions in the sequence (Table 1). We believe that the incomplete digestion is due to the staining of the DNA with an intercalating dye, which is know to interfere with sequence-recognizing enzymes.[39] Furthermore, the recognition sequence for *Sma* I contains a known preferred binding site of the intercalating dye TOTO-1.[24] We find that Figure 24 shows no signs of photo-induced or unspecific enzymatic cutting. Note also that cuts by *Sma* I on λ−DNA appear within, at most, a few seconds (Fig. 23).

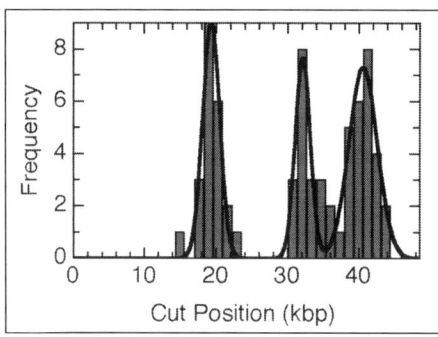

Figure 24. The absolute cut positions from 29 molecules with 2 and 3 cuts of *Sma* I on λ−DNA. The line is a fit to the histogram using the sum of three Gaussian distributions. Reproduced from ref. 52.

Interestingly, the idea of triggering an enzymatic reaction inside a nanochannel using a divalent ion is not restricted to restriction endonucleases, but is expected to apply to a much wider class of enzymes. We expect that many DNA and RNA polymerases, exonucleases, helicases and other ATP/NTP-dependent enzymes can be triggered in a similar fashion. We believe the nanochannel technology could be used to study the site-selective interaction between DNA and a range of metal-ion induced enzymes.

Imaging Transcription Factors in Nanochannels

While the imaging of DNA modification by enzymes is certainly a powerful method, it is not general in the sense that one needs to find enzymes that fulfill a rather long list of requirements, i.e., type-II endonuclease, nonsticking, nonstaring, high activity and reaction buffers amenable to electrophoresis and fluorescence imaging. That naturally puts strong limits on the number of sequences that can be tested for. A far more general approach to functional imaging of DNA inside nanochannels is to observe fluorescently labeled proteins, or artificial fluorescent probes. The necessary custom oligomers and probes have become commercially available and hence do not add significantly to the experimental overhead.

We developed a platform utilizing nano-fabrication and internal reflection fluorescence (TIRF) microscopy to localize the bound protein along DNA. The device, which is designed to manipulate and enable direct imaging of LacI-DNA, is shown schematically in Figure 25A. The idea is to drive DNA molecules into the micro-channel first and then into the nano-channels using electrophoresis. The one-micron deep micro-channels were fabricated using photolithography and reactive-ion-etching; the nano-channels were fabricated using focused ion beam milling (FIB) into fused silica (amorphous quartz). Figure 25B shows the Scanning Electron Microscopy (SEM) image of the 120 nm x 150 nm nano-channel. All images of DNA in nano-channels shown in this report were observed in nano-channels of the same dimension—120 nm × 150nm. A prism was placed on top of the nano-channel region for TIRF microscopy.

Since proteins tend to stick to quartz surfaces, it is necessary to treat the channel surfaces with anti-sticking reagents before injecting LacI-DNA solution. The channels were wetted by capillary force with a surface treating solution of 0.5xTBE, 1 mg/ml Bovine serum albumin (BSA) protein and 0.1% (weight/volume) POP-6 (Performance-optimized linear polyacrylamide, Applied Biosystems, Foster City, CA). BSA and POP-6 together were used to prevent the sticking of proteins to quartz surfaces, while POP-6 functions to prevent electroosmosis. A few hours of soaking are sufficient for the surface treatment. The surface treatment solution was then removed from the

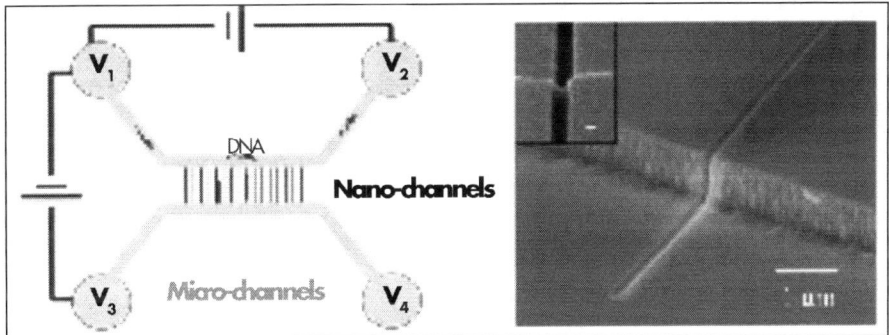

Figure 25. Schematics of the micro- and nano-fluidic device. Blue regions are micro-channels and the bridging black lines are nano-channels where the DNA molecules (red) are elongated. DNA molecules are guided consecutively into micro- and nano-channels using electrophoresis. b) Scanning electron microscope images of a 120nm × 150nm channel made using FIB. The inset is the top slanted view (at 52°) of the nano-channel showing its dimensions. The scale bar for the inset is 100 nm wide. Reproduced from ref. 21.

reservoir and replaced with a solution LacI-DNA in 0.5 × TBE. To prevent photo-bleaching of DNA dye, Glucose (β-D-glucose, 4 mg/ml, Sigma G7528), glucose oxidase (0.1 mg/ml, Sigma G-7016), catalase (20 μg/ml, Sigma C-40) and 100 mM β -mercaptoethanol were added in the solution.

To drive DNA into the micro-channels, a voltage drop was applied across the micro-channel (between V_1 and V_2). After a sufficient amount of DNA appeared in the micro-channel, a different voltage drop was applied across the nano-channels (between V_1 and V_4) and across the micro-channel (between V_1 and V_2) simultaneously. The purpose of the voltage drop across the nano-channels is to drive DNA into the nano-channels and the voltage drop across the micro-channel is to ensure a continue DNA flow. Once a DNA molecule had entered a nano-channel, the voltage was turned off and the molecule came to rest. We then imaged the DNA in nano-channel with one excitation wavelength (568 nm for BOBO-3 dye) and the bound LacI-GFP protein with another wavelength (488 nm) using TIRF microscopy.

Figure 26 is an image of LacI-GFP bound to lacO$_{256}$-DNA elongated in a nano- channel. The molecule enters from the right micro-channel region into the nano-channel driven by an electric field of 5V/50 μm. The number of protein bound is characterized using a photon counting method. There are ~ 20 LacI-GFP bound onto the lacO$_{256}$. LacI-GFP proteins distribute across the 10k bp long tandem lacO$_{256}$. After protein binding, the length of tandem lacO segment is of the same as that without the bound protein, indicating that the binding of 20 LacI over 10K bp doesn't show obvious effect on DNA properties such as persistence length. DeGennes' scaling law for polymers in a tube can thus be applied to protein-bound-DNA complexes.

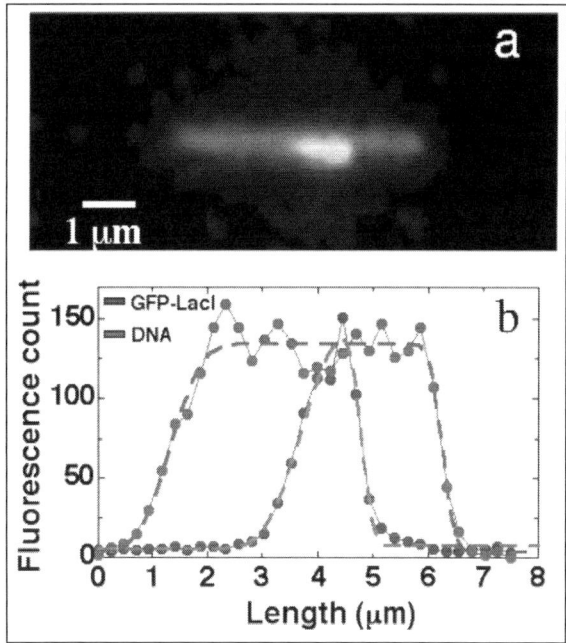

Figure 26. a) Time-averaged image of LacI-GFP bound to lacO$_{256}$-DNA elongated in a 130 nm x 150nm channel. There are 20 LacI-GFP molecules bound to this lacO$_{256}$-DNA molecule. The right half of the protein-bound region is brighter; it contains "2.5" more LacI-GFP than the left half. This molecule travels from right to left into the nano-channel driven by an electric field of 5V/50 μm. b) Fluorescence intensity profiles of the DNA and LacI-GFP. Reproduced from ref. 81.

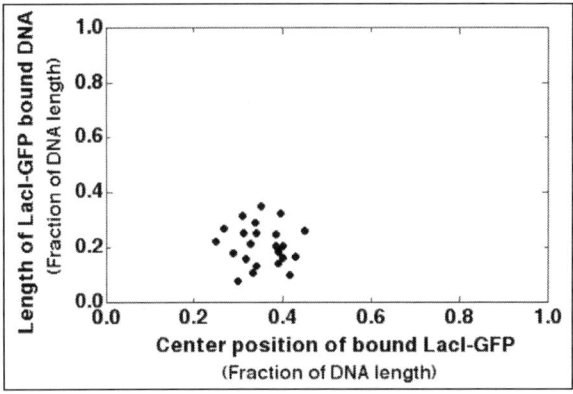

Figure 27. Statistics of center location and occupation length of bound LacI-GFP to lacO$_{256}$-DNA. The cell length is considered as unit one, the center position and the length of the protein bound DNA segment are expressed as fraction of the cell length. The center position is the fractional distance from the center of the LacI bound DNA segment to the closest end of the DNA molecule comparing to the length of the whole DNA molecule. Reproduced from ref. 81.

Figure 27 is a statistics plot of LacI-GFP distribution along lacO$_{256}$-DNA elongated in nano-channels. The lacO$_{256}$ insert is between 24.02K bp and 33.236K bp and is 9.2K bp long. Its center locates at lacO$_{256}$ insert is at 15K bp from the closer end (0.319 of the fractional cell length position). The length of the 10K bp lacO$_{256}$ is 25% of the whole length; the bound length is below 0.35 on the scaled length, corresponding well with bound protein number distribution. Few bound proteins occupy a shorter region of DNA and of order 10 and more proteins distribute across lacO$_{256}$.

Conclusions

We hope we have provided the reader a fairly complete "state of the art" report of how nanochannels can be used in molecular biology in 2006. In many respects the technology is still in its infancy. In terms of becoming truly nano in size, since the channels are still about half the persistence length and so the fluctuation amplitude and relaxation time can be significantly improved. Also, Smaller channels are needed in order to elongate ssDNA. The optical techniques that we discuss will fail in resolution as the channel sizes shrink, so electronic technologies will have to be developed if we are to achieve single basepair resolution. However, there is a firm theoretical background in place, so we believe that in the coming years nanochannels will become a very important tool in genomic analysis, certainly at the single cell level.

Over the last few years another nanofluidic tool for DNA analysis, namely nanopores, have been developed.[35,17,65] The core idea behind that technique is to electrophoretically thread a single DNA molecule through a short pore that has a cross-section on the order of the diameter of the polymer. The presence of DNA in the pore is detected by monitoring the current through the pore. Fundamentally, nanopore measurements are thus dynamic measurements since the DNA molecule must pass through the pore and will entropically coiled on either side of the pore, with the attendant polymer dynamics and relation time scales that we have discussed which must be dealt with.[31] However, the detection mechanism is certainly elegant since no labeling of the DNA is needed and single-stranded molecules can be measured. An integration off nanochannels and nanopores may become possible, with the nanochannel wringing out the entropy and controlling the relaxation dynamics and some form of nanopore controlling electronic readout of the DNA length and ultimately, someday, basepair sequence.

Acknowledgements

This work was supported by grants from DARPA (MDA972-00-1-0031), NIH (HG01506), NSF Nanobiology Technology Center (BSCECS9876771, the State of New Jersey (NJCST 99-100-082-2042-007) and US Genomics. It was also performed in part at the Cornell Nano-Scale Science and Technology Facility (CNF) which is supported by the National Science Foundation under Grant ECS-9731293, its users, Cornell University and Industrial Affiliates.

References

1. Austin RH, Tegenfeldt JO, Cao H et al. Scanning the controls: Genomics and nanotechnology. IEEE Transactions on Nanotechnology 2002; 1(1):12-18.
2. Bakajin O, Duke TAJ, Tegenfeldt J et al. Separation of 100-kilobase DNA molecules in 10 seconds. Anal Chem 2001; 73(24):6053-6056.
3. Bakajin OB, Duke TAJ, Chou CF et al. Electrohydrodynamic stretching of DNA in confined environments. Phys Rev Lett 1998; 80(12):2737-2740.
4. A. D. Bates and A. Maxwell. DNA Topology. Oxford University Press 2005.
5. Bolzan AD, Bianchi MS. Telomeres, interstitial telomeric repeat sequences and chromosomal aberrations. Mutat Res 2006; 612(3):189-214.
6. Brochard-Wyart F, Tanaka T, Borghi N et al. Semi-flexible polymers confined in soft tubes. Langmuir 2005; 21:4144.
7. Campell LC, Wilkinson MJ, Manz A et al. Electrophoretic manipulation of single DNA molecules in nanofabricated capillaries. Lab Chip 2004; 3:225-229.
8. Cao H, Tegenfeldt JO, Austin RH et al. Gradient nanostructures for interfacing microfluidics and nanofluidics. Appl Phys Lett 2002; 81(16):3058-3060.
9. Cao H, Yu ZN, Wang J et al. Fabrication of 10 nm enclosed nanofluidic channels. Appl Phys Lett 2002; 81(1):174-176.
10. Chan EY, Goncalves NM, Haeusler RA et al. DNA mapping using microfluidic stretching and single-molecule detection of fluorescent site-specific tags. Genome Res 2004; 14(6):1137-1146,
11. Chou CF, Tegenfeldt JO, Bakajin OB et al. Electrodeless dielectrophoresis of single- and double-stranded DNA. Biophys J 2002; 83(4):2170- 2179.
12. Daoud M, de Gennes PG. Dynamics of Confined Polymer Chains. J Chem Phys 1977; 67:52.
13. Daoud M, de Gennes PG. Statistics of Macromolecular Chains Trapped in Small Pores. J Physique 1977; 38:85.
14. P.G. de Gennes. Scaling Concepts in Polymer Physics. Cornell University Press 1979.
15. M. Doi and S. F. Edwards. The Theory of Polymer Dynamics. Oxford University Press 1986.
16. Fan R, Karnik R, Yue M et al. DNA translocation in inorganic nanotubes. Nano Lett 2005; 5(9):1633-1637.
17. Fologea D, Gershow M, Ledden B et al. Detecting single stranded DNA with a solid state nanopore. Nano Lett 2005; 5(10):1905-1909.
18. Ghosh K, Carri G, Muthukumar M. Configurational Properties of a Single Semiflexible Polyelectrolyte. J Chem Phys 1991; 115:4367-4375.
19. Godde R, Akkad DA, Arning L et al. Electrophoresis of DNA in human genetic diagnostics-state-of-the-art, alternatives and future prospects. Electrophoresis 2006; 27(5-6):939-946.
20. Guo LJ, Cheng, X, Chou CF. Fabrication of size-controllable nanofluidic channels by nanoimprinting and its application for DNA stretching. Nano Lett 2004; 4(1):69-73.
21. Ha B, Thirumalai D. Persistence Length of Flexible Polyelectrolyte Chains. J Chem Phys 1999; 110:7533-7541.
22. Han J, Craighead HG. Separation of long DNA molecules in a microfabricated entropic trap array. Science 2000; 288(5468):1026-1029.
23. Han JY, Craighead HG. Characterization and optimization of an entropic trap for DNA separation. Anal Chem 2002; 74(2):394-401.
24. Hansen LF, Jensen LK, Jacobsen JP. Bis-intercalation of a homodimeric thiazole orange dye in DNA in symmetrical pyrimidinepyrimidine-purine-purine oligonucleotides. Nucleic Acids Res 1996; 24(5):859-867.
25. Hatfield J, Quake S. Dynamic Properties of an Extended Polymer in Solution. Phys Rev Lett 1998; 82:3548.
26. Hogan M, LeGrange J, Austin B. Dependence of DNA helix flexibility on base composition. Nature 1983; 304:752-754.
27. Huang LR, Tegenfeldt JO, Kraeft JJ et al. A DNA prism for high-speed continuous fractionation of large DNA molecules. Nature Biotechnol 2002; 20(10):1048-1051.
28. Isambert H., Maggs H. Dynamics and Rheology of Actin Solutions. Macromolecules 1996; 29:1036.

29. Jendrejack R, Dimalanta E, Schwartz D et al. DNA Dynamics in a Microchannel. Phys Rev Lett 2003; 91:038102.
30. Jendrejack R, Schwartz D, Graham M et al. Effect of Confinement on DNA Dynamics in Microfluidic Devices. J Chem Phys 2003; 119:1165.
31. Keyser UF, Koeleman BN, Van Dorp S et al. Direct force measurements on DNA in a solid-state nanopore. Nature Phys 2006; 2(7):473-477.
32. Kremer K, Binder K. Dynamics of Polymer Chains Confined in Tubes: Scaling Theory and Monte Carlo Simulations. J Chem Physics 1984; 81:6381.
33. Landau L and Lifshitz E. Fluid Mechanics. Butterworth Heinmann 1998.
34. Landau L and Lifshitz E. M. Statistical Physics. Addison-Wesley, Reading, PA 1958.
35. Li JL, Gershow M, Stein D et al. JA DNA molecules and configurations in a solid-state nanopore microscope. Nature Mater 2003; 2(9):611-615.
36. Li WL, Tegenfeldt JO, Chen L et al. Sacrificial polymers for nanofluidic channels in biological applications. Nanotechnology 2003; 14:578-583.
37. Mannion JT, Reccius CH, Cross JD et al. Conformational analysis of single DNA molecules undergoing entropically induced motion in nanochannels. Biophys J 2006; 90(12):4538-4545.
38. Martin-Molina A, Quesada-Perez M, Galisteo-Gonzalez F et al. Electrophoretic mobility and primitive models: Surface charge density effect. J Phys Chem B 2002; 106(27):6881-6886.
39. Meng X, Cai WW, Schwartz DC. Inhibition of restriction endonucleases activity by DNA binding fluorochromes. J Biomol Struct Dyn 1996; 13(6):945-951.
40. Milchev A, Paul W, Binder K. Polymer Chains Confined into Tubes with Attractive Walls: A Monte Carlo Simulation. Macromol. Theory Simul 1994; 3:305.
41. Moon J, Nakanishi H. Onset of the Excluded Volume Effect for the Statistics of Stiff Chains. Phys Rev A 1991; 44:6427-6442.
42. Morse D. Viscoelasticity of Concentrated Isotropic Solutions of Semiflexible Polymers. 1. Model and Stress Tensor. Macromolecules 1998; 31:7030.
43. Morse D. Viscoelasticity of Concentrated Isotropic Solutions of Semi- flexible Polymers. 2. Linear Reponse. Macromolecules 1998; 31:7044.
44. Odijk T. Polyelectrolytes near the Rod Limit. J Polym Phys Ed 1977; 15:477-483.
45. Odijk T. On the Statistics and Dynamics of Confined or Entangled Stiff Polymers. Macromolecules 1983; 16:1340.
46. Odijk T. Similarity Applied to the Statistics of Confined Stiff Polymers. Macromolecules 1984; 17:502.
47. Perkins TT, Quake SR, Smith DE et al. Relaxation of a single DNA molecule observed by optical microscopy. Science 1994; 264(5160):822-826.
48. Quake S, Babcock H, Chu S. The Dynamics of Partially Extended Single Molecules of DNA. Nature 1997; 388:151.
49. Quake SR, Scherer A. From micro- to nanofabrication with soft materials. Science 2000; 290(5496):1536-1540.
50. Reccius CH, Mannion JT, Cross JD et al. Compression and free expansion of single DNA molecules in nanochannels. Phys Rev Lett 2005; 95(26):268101.
51. Reisner WW, Morton K, Riehn R et al. Statics and Dynamics of Single DNA Molecules Confined in Nanochannels. Phys Rev Lett 2005; 94:196101.
52. Riehn R, Austin RH, Sturm JC. A nanofluidic railroad switch for DNA. Nano Lett 2006; 6(9):1973-1976.
53. Riehn R, Lu M, Wang Y et al. Restriction Mapping in Nanofluidic Devices. Proc Natl Acad Sci USA 2005; 102:10012-10016.
54. Riehn R, Austin RH, Wetting Micro- and Nanofluidic Devices Using Supercritical Water. Anal Chem 2006; 78(16):5933-5944.
55. Rybenkov V, Cozzarelli N, Vologodskii A. Probability of DNA Knotting and the Effective Diameter of the DNA Double Helix. Proc Natl Acad Sci USA 1993; 90:5307-5311.
56. Saleh OA, Sohn LL. An artificial nanopore for molecular sensing. Nano Lett 2003; 3(1):37-38.
57. Schaefer DW, Curro JG. Statistics of a single polymer-chain. Ferroelectrics 1980; 30(1-4):49-56.
58. Schaefer DW, Joanny JF, Pincus P. Dynamics of Semiflexible Polymer Chains in Solution. Macromolecules 1980; 13:1280-1289.
59. Schwartz DC, Cantor CR. Separation of yeast chromosome-sized DNAs by pulsed field gradient gel-electrophoresis. Cell 1984; 37(1):67-75.
60. Skolnick J, Fixman M. Electrostatic Persistence Length of a Wormlike Polyelectrolyte. Macromolecules 1977; 10:944-948.
61. Slater GW, Desrulsseaux C, Hubert SJ et al. Theory of DNA electrophoresis: A look at some current challenges. Electrophoresis 2000; 21(18):3873-3887.

62. Slater GW, S. Guillouzic S, Gauthier MG et al. Theory of DNA electrophoresis. Electrophoresis 2002; 23(22-23):3791-3816.
63. Smith D, Perkins TT, Chu S. Dynamical Scaling of DNA Diffusion Coefficients. Macromolecules 1996; 29:1372-1373.
64. Stein D, van der Heyden FHJ, Koopmans W et al. Pressure-driven transport of confined DNA polymers in fluidic channels. Proc Natl Acad Sci USA 2006; In Press.
66. Storm AJ, Storm C, Chen J et al. Fast DNA Translocation through a Solid-State Nanopore. Nano Lett 2005; 5:1193-1197.
67. Strick T, Allemand JF, Croquette V et al. Twisting and stretching single DNA molecules. Prog Biophys Mol Biol 2000; 74(1-2):115-140.
68. Takada T. Acoustic and optical methods for measuring electric charge distributions in dielectrics. IEEE Transactions on Dielectrics and Electrical Insulation 1999; 6(5):519-547.
69. Tegenfeldt JO, Prinz C, Cao H et al. The Dynamics of Genomic-Length DNA Molecules in 100nm Channels. Proc Natl Acad Sci USA 2004; 101:10979-10983.
70. Tegenfeldt JO, Bakajin O, Chou CF et al. Near-field scanner for moving molecules. Phys Rev Lett 2001; 86(7):1378-1381.
71. Turban, L. Conformation of Confined Macromolecular Chains: Cross- Over Between Slit and Capillary. J Physique 1984; 45:347.
72. Turner SWP, Cabodi M, Craighead HG. Confinement-induced entropic recoil of single DNA molecules in a nanofluidic structure. Phys Rev Lett 2002; 88(12):128103.
73. Valle F, Favre M, Rios P et al. Scaling Exponents and Probability Distributions of DNA End-to-End Distance. Phys Rev Lett 2005; 95:158105.
74. Venter JC et al. The sequence of the human genome. Science 2001; 291(5507):1304-+ .[75] Viovy JL. Electrophoresis of DNA and other polyelectrolytes: Physical mechanisms. Rev Modern Phys 2000; 72(3):813-872.
76. W.D. Volkmuth. Ph.D. Thesis. Princeton University, Princeton, NJ 1994.
77. Volkmuth WD, Austin RH. DNA Electrophoresis in Microlithographic Arrays. Nature 1992; 358:600-602.
78. Volkmuth WD, Duke T, Wu MC et al. DNA electrodiffusion in a 2-d array of posts. Phys Rev Lett 1994; 72:2117-2120.
79. Vologodskii AV, Cozzarelli NR. Modeling of Long-Range Electrostatic Interactions in DNA. Biopolymers 1994; 35:289-296.
80. Vologodskii AV, Cozzarelli NR. Conformational and Thermodynamic Properties of Supercoiled DNA. Annu Rev Biophys Biomol Struct 1994; 23:609-643.
81. Wang Y, Tegenfeldt JO, Reisner WW et al. Single-Molecule Studies of Repressor-DNA Interactions Show Long-Range Interactions. Proc Natl Acad Sci USA 2005; 102:9796-9801.
82. Yamakawa H, Fujii M. Wormlike Chains near the Rod Limit: Path Integral in the WKB Approximation. J Phys Chem 1973; 59:6641.
83. Zhu ZQ, Zhang J, Zhu JZ. An overview of Si-based biosensors. Sens Lett 2005; 3(2):71-88.

CHAPTER 13

Beyond Microtechnology— Nanotechnology in Molecular Diagnosis

Paolo Fortina,* Joseph Wang, Saul Surrey, Jason Y. Park and Larry J. Kricka

Introduction

Advances in technology have led to significant improvements in our ability to detect and diagnose disease. In particular, the advent of nucleic acid technology has allowed unparalleled insight into the genetic basis of disease and a highly specific means to detect and diagnose infectious disease. A further technological advance, dating back to the 1980s, has been the use of microfabrication techniques widely utilized in the microelectronics industry to fabricate microminiaturized analyzers.[1,2] These devices have simplified and improved various steps in typical genetic test procedures, such as cell isolation and selection, nucleic acid extraction, amplification (PCR, RTR-PCR, LCR) and quantitation of nucleic acids (e.g., quantitative microchip capillary electrophoresis of PCR amplicons).[3-5] More importantly, integration of the analytical steps in nucleic acid assays can be achieved in a microchip format (lab-on-a-chip).[6] For example, an integrated microchip was developed that performs cell isolation, cell lysis, nucleic acid purification and recovery in nanoliter volumes.[7,8] Other microchips combine various steps such as DNA amplification and capillary electrophoresis[9] or electrochemical detection.[10]

Nanotechnology is an emerging technology, which will further simplify nucleic acid technology, as well as accelerate and extend its acceptance and utilization. It is defined as "technology development at the atomic, molecular or macromolecular range of approximately 1-100 nanometers to create and use structures, devices and systems that have novel properties" (www.nano.gov). The current range of nanostructures includes nanoarrays, nanocantilevers, nanodisks, nanofibers, nanoparticles, nanoprisms, nanorods, nanoribbons, nanoshells, nanotransistors, nanotubes, nanowires and many of these structures have found application in nucleic acid technology for direct and indirect detection of nucleic acid sequences as well as for DNA sequence analysis. This chapter explores the use of selected nanostructures in tests for specific nucleic acid sequences and in nucleic acid sequence analysis.

History of Nanotechnology

The word nanotechnology was coined by Norio Taniguchi[11] in 1974, but its origins can be traced back to an after-dinner lecture in 1959 by the Nobel laureate Richard Feynman.[12] It was entitled "There's Plenty of Room at the Bottom" and it explored the problem of manipulating and controlling things on a small scale and the possibility of atomic-scale manufacturing (http://www.zyvex.com/nanotech/feynman.html). In the 1980s, Eric Drexler provided an important impetus

*Corresponding Author: Paolo Fortina—Kimmel Cancer Center, Thomas Jefferson University, 1009 Bluemle Life Sciences Building, 233 South 10th Street, Philadelphia, PA 19107, U.S.A. Email: paolo.fortina@jefferson.edu

Integrated Biochips for DNA Analysis, edited by Robin Hui Liu and Abraham P. Lee. ©2007 Landes Bioscience and Springer Science+Business Media.

to the field through his books "Engines of Creation" and "Nanosystems"[13,14] that outlined approaches and benefits of nanoscale technology (e.g., molecular manufacturing). Further momentum was provided by sizeable government funding of nanotechnology research in the United States (e.g., National Nanotechnology Initiative), Japan and Europe (http://www.nano.gov/html/res/IntlFundingRoco.htm). An early demonstration of the possibility of atomic scale manipulation was provided in 1990 by scientists at IBM who spelled-out "IBM" using 35 atoms of xenon.[15] In the following year, the first carbon nanotube was synthesized and this was the first of many new nanoscale structures and nanomaterials that have found numerous beneficial applications (e.g., sunscreen, coatings, composites, etc.) (http://www.nanotechproject.org).

Safety and Nanotechnology

Apprehension and unease often accompany the emergence of a new technology, as was the case for recombinant DNA technology in the 1970s.[16] Currently, there is growing concern over safety of nanotechnology and in particular, hazards associated with nanoscale materials such as nanoparticles and nanotubes. Indeed, current research into aerosolized nanoparticles indicates that both the size as well as the surface chemistry of particles determine their toxicity.[17,18] The properties of increased absorbability and reactivity that make nanoparticles desirable for in vivo diagnostics and therapeutics are the same properties that have already demonstrated potential toxicity.[19,20] The true bounds of safety and toxicity of nanoscale particles have not yet been determined and remain an active area of research. The safety issues that nanomaterials present can be generally categorized into: purposeful administration; inadvertent exposure through the workplace; environmental accumulation.

Purposeful administration encompasses the use of nanoscale therapeutics or diagnostics in patients. Although this has the potential for toxicity because of direct dosing, it is also the most controllable from a toxicity standpoint. Indeed, from a clinical perspective, numerous pharmacologic compounds are studied each year that have a high potential for toxicity. In a similar manner to the development of pharmacologic agents with multiple steps of both preclinical animal models and human studies, the purposeful exposure to nanoscale materials can be studied and assessed for toxicity. Once the toxicity has been assessed, then the risk of administering a nanoscale compound can be weighed against its possible benefits.

The safety of nanoscale materials in the workplace is somewhat more difficult to determine than in the purposeful administration of nanomaterials. In the latter, there is a clear risk versus benefit that can be determined for each individual. In the workplace, exposure to nanoscale materials may be the result of an industrial process that creates nanoscale materials as a by-product or the result of an industrial process that is designed to create nanoscale materials. In either case, the nanoscale materials in the work environment may be at high concentrations. The United States National Institute for Occupational Health and Safety (NIOSH) is currently seeking to address the concerns of nanoscale materials in the work environment. NIOSH has posted a pre-dissemination summary of current considerations in nanotechnology as they apply to workers (http://www.cdc.gov/niosh/topics/nanotech/nano_exchange.html); this document outlines the potential health and safety concerns that may apply to individuals working with engineered nanomaterials. In addition, NIOSH has made publicly available a searchable database of nanomaterials that includes their corresponding health and safety considerations (Nanoparticle Information Library; http://www2a.cdc.gov/niosh-nil/index.asp).

The end result of the manufacture and use of all nanotechnology products is the accumulation of nanoscale waste in the environment. This impact of engineering and mass producing nanomaterials upon the environment is not clearly understood, but is being actively studied. The United States Environmental Protection Agency (EPA) has prepared an external review draft Nanotechnology White Paper summarizing their current findings and proposed areas of activity regarding nanotechnology (http://www.epa.gov/osa/nanotech.htm). It is important to note that the EPA is funding research not only for determining the impact of nanomaterials upon the environment, but also for the utilization of nanomaterials and nanotechnology for the clean-up and improvement of environmental conditions.

Although there are few definitive answers regarding the safety of nanotechnology and nano-materials, the impact and utilization of nanotechnology in society is undeniable and already has begun. Nanomaterials already are integrated into hundreds of consumer products that are used at all levels of society. A partnership between the Woodrow Wilson International Center for Scholars and the Pew Charitable Trusts was established to study Emerging Nanotechnologies (http://www. nanotechproject.org/). A component of this partnership is an inventory that catalogs currently available consumer products that are known or declared by their manufacturer to contain a compo-nent of nanotechnology. This catalog lists hundreds of currently available consumer products that include cosmetics, clothing, nutraceuticals, mobile phones, cleaning agents and cooking products. Although there needs to be continued studies into the toxicological affect of nanomaterials, the research and development of novel diagnostic technologies should be framed in the context of the current high level of penetration of nanomaterials into everyday life.

Nanofabrication Methods

Nanofabrication methods can be divided into two categories:[21] (1) Top-down methods in which the nanostructure is created from a larger piece of material by removal of unwanted material; and, (2) Bottom-up methods in which the nanostructure is assembled from a set of component atomic or molecular parts. Examples among the former of these different types of methods in-clude atomic force microscopy (AFM), electron beam lithography and electron beam patterning, extreme ultraviolet lithography, laser-assisted direct imprinting, nano-imprint and nanosphere lithography and soft litography. Among the bottom-up methods, which include self-assembly and scanning tunneling microscopy (STM), one that has considerable analytical potential and will be discussed in detail is dip-pen nanolithography (DPN). DPN is a new technology that utilizes precise positioning of molecules by an atomic force microscope (AFM) tip onto a substrate at feature sizes less than 100 nm.[22,23] This scanning probe nanopatterning technique utilizes a tip coated with molecules to deliver molecules via capillary transport from the tip to a surface via a solvent meniscus. The pen-like deposition of molecules provides resolution of 10 nm; this is significantly better than standard photolithography (90 nm), micro-contact printing (100 nm), nanoimprint lithography (20 nm) and is only rivaled by the much more expensive and resource intensive electron-beam lithography (15 nm).

The scanning probe tip patterns a variety of "inks" (e.g., ranging from small organic molecules to organic and biological polymers and from colloidal particles to metals ions and sols) onto the surface of a variety of substrates including metals, semiconductors, insulators and of functional monolayers adsorbed on a variety of surfaces.

There are several attributes of DPN that make it unique among nanofabrication techniques. First, it is a one-step process, which does not require a mask. Instead, the pattern is created by movement of the pen or pens writing on the surface of the substrate. Second, multiple molecular inks can be precisely patterned on the same substrate without contamination and without loss of alignment between consecutively written nanoscale patterns. This can be done using a single pen and exchanging pens, or using parallel, multi-pen arrays.

DPN has been used to prepare both protein and nucleic acid arrays. Protein nanoarrays have been created by attaching enzymes and antibodies to gold surfaces.[24,25] Indeed, a nanoscale array of monoclonal antibodies for the HIV-1 p24 antigen was fabricated at a 100 nm feature size on gold surfaces for the sensitive and specific detection of the human immunodeficiency virus type 1 (HIV-1) in blood samples.[24] HIV-1 p24 antigen in plasma from HIV-1 infected patients was bound in a sandwich format to both the arrayed antibody as well as an antibody functional-ized with a gold nanoparticle. AFM topographical measurements demonstrated a height of the monoclonal antibody to be 6.4 nm. The addition of p24 antigen resulted in a specific increase of approximately 2.3 nm (total height from substrate surface increased from 6.4 nm to 8.7 nm). This system exceeded the sensitivity of conventional enzyme-linked immunosorbent assay (ELISA) and was capable of detecting the presence of p24 antigen in patients with less than 50 copies of HIV-1 mRNA per mL of serum. Thus, it is comparable to current HIV-1 nucleic acid detection systems that have a detection limit of 50 copies/mL of HIV-1 mRNA.[26] This increased sensitivity for HIV has significance in the detection and prevention of viral transmission in the perinatal period, needle-stick injuries and blood transfusion therapy.

Nucleic acid arrays have been created on both metallic and insulating substrates.[27,28] Specifically, DNA modified with hexanethiol groups were patterned on gold and DNA modified with 5′-terminal acrylamide groups were patterned on silicon oxide. In addition to using gold-thiol chemistry for nanopatterning, DNA was patterned onto oxidized silicon wafers.[27] The biological activity of the patterned oligonucleotides was verified by applications of both complementary and noncomplementary-labeled DNA. The DNA used in this study was either conjugated to a fluorophore to be assessed by epi-fluorescence microscopy or alternatively, the labeled DNA was conjugated to gold nanoparticles to be assessed by AFM in tapping-mode. The gold nanoparticle labels allowed detection of DNA features that were approximately 50 nm in diameter. This gold nanoparticle, size-based detection is an interesting method of detecting DNA based on topography rather than standard luminescent, colorimetric or radioactive labeling.

The perceived advantages of DPN fabricated arrays are their high density and possible greater simplicity in sample and reagent requirements. However, the practical application of these analytical systems cannot be determined from the relatively few studies published thus far. Currently, there is a limited range of clinical applications for this emerging technology and thus far, no devices produced by DPN have progressed to routine clinical application.

Selected Applications in Molecular Diagnosis

Several nanoscale structures show promise in molecular diagnostics and these include nanoparticles, nanopores, nanochannels, nanowires and nanotubes. Most work has focused on uses of nanoparticles.

Nanoparticles

A range of metallic and inorganic nanoparticles has been used as labels in nucleic acid assays[29-33] in conjunction with optical[32] and microgravimetric sensing.[33] The enormous signal enhancement associated with the use of nanoparticle labels provides the basis for ultrasensitive electrochemical detection with PCR-like sensitivity.[31] Most of these schemes have relied on a highly sensitive electrochemical stripping measurement of the metal tag. The remarkable sensitivity of such stripping measurements is attributed to the 'built-in' accumulation step, during which the target metals are electrodeposited onto the working electrode.[34,35]

DNA can be detected in the picomolar and sub-nanomolar levels using nanoparticle-based electrochemical bioaffinity assays based on capturing gold[36-38] or silver nanoparticles[39] to the bound target. This then is followed by acid dissolution and anodic-stripping electrochemical measurement of the solubilized metal tracer (e.g., hybridization detection of the 406-base human cytomegalovirus DNA sequence).[38] Further sensitivity enhancement can be achieved by catalytic enlargement of the gold tag in connection to nanoparticle-promoted precipitation of gold or silver.[37-40] Combining such enlargement of the metal-particle tracers, with the effective 'built-in' amplification of electrochemical stripping analysis paved the way to sub-picomolar detection limits.[41]

A simplified gold-nanoparticle based protocol achieved a detection limit of 0.78 fmol for PCR amplicons.[42] The method relies on pulse-voltammetric monitoring of the gold-oxide wave at ~1.20 V employing a disposable pencil graphite electrode to detect PCR amplicons bound to the pencil electrode and hybridized to oligonucleotide-nanoparticle conjugates.

It is possible to enhance sensitivity beyond the one nanoparticle reporter per one binding event by capturing multiple nanoparticles per binding event. For example, an electrochemical triple-amplification hybridization assay was developed which combines carrier-sphere amplifying units (loaded with numerous gold nanoparticles tags) with the 'built-in' preconcentration of the electrochemical stripping detection and catalytic enlargement of the multiple gold-particle

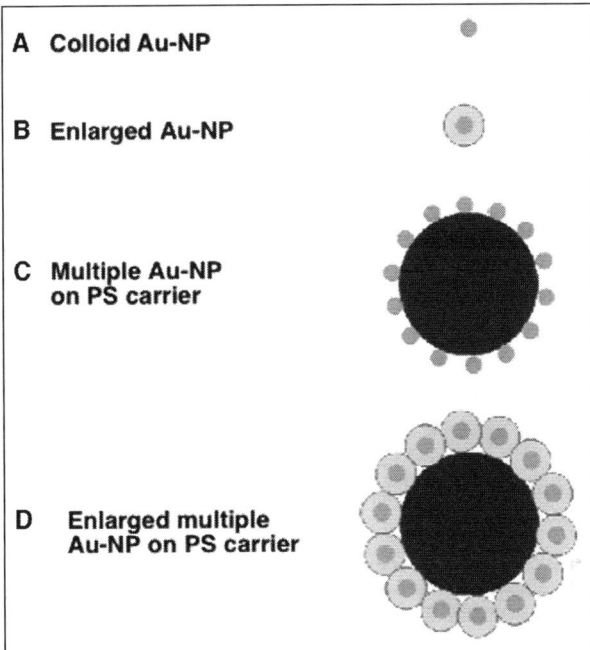

Figure 1. Different generations of amplification platforms for bioelectronic detection based on gold nanoparticle tracers: A) A single nanoparticle tag; B) catalytic enlargement of the nanoparticle tag; C) polymer carrier bead loaded with numerous gold nanoparticle tags; D) catalytic enlargement of multiple tags on the carrier bead.

tags.[43] The gold-tagged spheres were prepared by binding biotinylated metal nanoparticles to streptavidin-coated polystyrene spheres (~80 gold particles/polystyrene bead). Such multiple amplifications offered a dramatic enhancement in sensitivity. Enlargement of numerous gold nanoparticles tags (on a supporting sphere carrier) represents the 4th generation of an amplification strategy (Fig. 1) that started with the use of single gold nanoparticle tags.[36-38] In addition, it is possible to use gold nanoparticles as carriers of redox markers for amplified biodetection.[44] Gold nanoparticles covered with 6-ferrocenylhexaenthiol were used with a sandwich DNA hybridization assay. Due to elasticity of DNA strands, the ferrocene/Au-nanoparticle conjugates were positioned in close proximity to the underlying electrode to facilitate a facile electron-transfer reaction. A detection limit of 10 amol was observed with linearity up to 150 nm. The system was applied to detecting PCR products derived from hepatitis B virus.

Nanoparticle-induced changes in the conductivity across a microelectrode gap afford a highly sensitive (0.5 picomolar level) and selective detection of DNA hybridization.[45] Nanoparticle-tagged DNA targets are captured by probes immobilized in the gap between the two closely-spaced microelectrodes. Subsequent silver precipitation (Fig. 2) results in a conductive metal layer bridging the gap and a measurable conductivity signal.

The unique (tunable-electronic) properties of semiconductor (quantum dots) nanocrystals have generated considerable interest for optical DNA detection.[46] For example, detection of DNA hybridization using cadmium-sulfide nanoparticle tags and electrochemical stripping measurements of the cadmium has been described.[34] Nanoparticle-promoted cadmium precipitation was used to enlarge the nanoparticle tag and amplify the stripping DNA hybridization signal. In addition to measurements of the dissolved cadmium ions, it is possible to perform solid-state measurements following "magnetic" collection of the magnetic-bead/DNA-hybrid/CdS-tracer assembly onto a

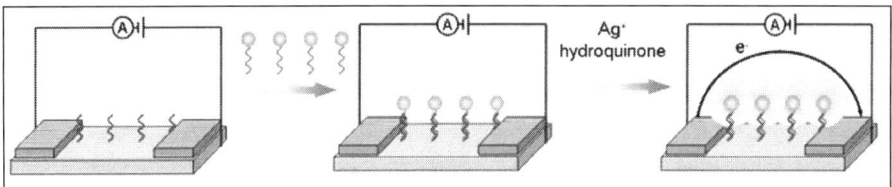

Figure 2. Conductivity detection of nanoparticle-based microelectrodes arrays. The capture of the nanoparticle-tagged DNA targets by probes confined to the gap and a subsequent silver enlargement, electrically short the gap and lead to a measurable conductivity signal.

screen-printed electrode transducer. Such protocols couple amplification features of nanoparticle/polynucleotide assemblies and highly sensitive potentiometric stripping detection of cadmium with magnetic isolation of the duplex. The low detection limit (100 fmol) was coupled to good precision (RSD = 6%). A substantially enhanced signal was obtained by encapsulating multiple CdS nanoparticles onto the host bead or by loading onto carbon-nanotube carriers.[47]

An interesting aspect of these inorganic nanocrystals is the possibility to combine different semiconductor tags linked to different DNA probes for simultaneous high-throughput analysis of different DNA targets.[48,49] Four encoding nanoparticles (cadmium sulfide, zinc sulfide, copper sulfide and lead sulfide) were used to differentiate signals of four different DNA or protein targets using hybridization or a sandwich immunoassay, respectively, with stripping voltammetry of the corresponding metals. Each recognition event yielded a distinct voltammetric peak. The size and position of the peak reflected the level and identity, respectively, of the corresponding antigen or DNA target (Fig. 3).

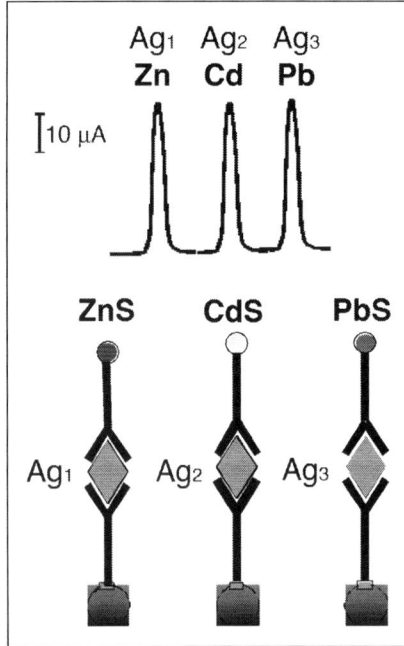

Figure 3. Use of different quantum-dot tracers for electrical detection of multiple protein targets. Top: stripping voltammogram for a solution containing dissolved ZnS, CdS and PbS nanoparticle tracers, corresponding to the three protein targets (Ag_1-Ag_3).[51]

Figure 4. Use of 'Electroactive' particles for ultrasensitive DNA detection, based on polystyrene beads impregnated with a redox marker. SEM image of the resulting DNA-linked particle assembly.

Metal nanoparticles also are useful for electrical coding of single nucleotide polymorphisms (SNP).[50] Such procedures rely on the hybridization of monobase-modified gold nanoparticles to the mismatched bases. The binding event leads to changes in the gold oxide peak and offers great promise for coding all mutational changes. Analogous SNP coding protocols based on different inorganic nanocrystals are currently being examined. This protocol involves the addition of ZnS, CdS, PbS and CuS crystals linked to adenosine, cytidine, guanosine and thymidine mononucleotides, respectively. Each mutation captures via base pairing different nanocrystal-mononucleotide conjugates yielding a distinct electronic fingerprint.

It is possible to further enhance sensitivity by employing multiple tracers per binding event, using for example, polymeric microbeads carrying multiple redox tracers externally (on their surface) or internally (via encapsulation). For example, a sensitive electrochemical detection of DNA hybridization was reported based on polystyrene beads impregnated with a ferrocene marker.[44] The resulting "electroactive beads" are capable of carrying a large number of the ferrocyanocarboxaldehyde-tagged molecules; and, hence, offer a remarkable amplification of single hybridization events. This allowed chronopotentiometric detection of target DNA in the 5.1×10^{-21} mol range (~31,000 molecules) in conjunction with a 20 min hybridization and "release" of the marker in an organic medium. The signal amplification advantage is combined with excellent discrimination in the presence of a large excess (10^7-fold) of non-target nucleic acids. The DNA-linked particle assembly that resulted from the hybridization event is shown in Figure 4. This image indicates that the 10 μm electroactive beads are cross-linked to smaller (~ 0.8 μm) magnetic spheres through the DNA hybrid. Current efforts are aimed at encapsulating different ferrocene markers in the polystyrene host beads in preparation for multi-target DNA detection. Other marker encapsulation routes hold promise for electrochemical bioassays. Particularly attractive are the recently developed nano-encapsulated microcrystalline particles, prepared by the layer-by-layer (LBL) assembly technique, that offer large marker/biomolecule ratios and superamplified bioassays.[51-53]

Nanopores and Nanochannels

A premier example of the promise of nanotechnology in nucleic acid analysis is provided by the nanopore and its potential for use in DNA sequence analysis.[54-57] The original work on nanopore sequencing utilized a 1.5 nm diameter alpha-hemolysin pore.[58] Attention now has shifted to solid-state nanopores (e.g., 3 to 10 nm in diameter) fabricated on a silicon nitride membrane[59] or a p-type silicon layer. A single strand of DNA can be drawn through the pore by an electrical field and as the DNA traverses the pore, it modulates the ionic current flowing through the pore. The amplitude and duration of the ion current blockade were found to depend on the base composition of the DNA. Currently, only ion current signatures characteristic of a nucleotide sequence have been obtained and a vital step will be refinement of the technique to distinguish individual nucleotides. Other nanopore devices integrate n-type sensor regions implanted in a p-type silicon layer sensors around a pore through the p-type silicon layer.[57]

Attractive features of this technology are simplicity of the instrumentation and the sequencing rate, which is estimated to be as high as 10,000 bases/sec (cf., 30,000 bases/day using conventional sequencers and 1235 bases/sec using the 454 Life Sciences technology).[60] A nanochannel also has been proposed as a sequencing tool with the advantage that it could extend the nucleic acid molecule and simplify DNA sequence analysis.[61]

Nanowires

Nanowires have received extensive attention owing to their great promise for sensing and coding applications. One unique and important application of nanowires is their use as tagging systems for a variety of bioanalytical applications. Striped metal nanowires are useful for electronic coding[48] and by incorporating different predetermined levels (or lengths) of multiple metal markers, such wires can lead to a large number of recognizable voltammetric signatures and hence to a reliable identification of a large number of biomolecules.

For example, multi-metal wires can be prepared by template-directed electrochemical synthesis, by plating indium, zinc, bismuth and copper onto a porous membrane template. Capping the wire with a gold end facilitates its functionalization with a thiolated oligonucleotide probe. Each nanowire thus yields a characteristic multi-peak voltammogram, whose peak potentials and current intensities reflect the identity of the corresponding DNA target. Thousands of usable codes could be generated in connection to 5-6 different potentials and 4-5 different current intensities.

The template-directed electrochemical route also can be used for preparing micrometer-long metal nanowire tags for ultra-sensitive detection.[48] The linear relationship between charge passed during the preparation and the size of the resulting wire allows tailoring of the sensitivity of the electrical DNA assay. For example, nanowire labels prepared by plating of indium into pores of a host membrane offered a lower detection limit (250 zmol) compared to bioassays based on spherical nanoparticle labels. Indium offers an attractive electrochemical stripping behavior and is not normally present in biological samples or reagents. Solid-state chronopotentiometric measurements of the indium microwire have been realized through a 'magnetic' collection of the DNA-linked particle assembly onto a screen-printed electrode transducer.

Because of the high surface-to-volume ratio and novel electron transport properties of semiconductor nanowires, their electronic conductance is strongly influenced by minor surface perturbations (such as those associated with the binding of macromolecules). Lieber's team recently described a label-free, multiplexed detection of cancer markers (prostate specific antigen (PSA), PSA-alpha1-antichymotrypsin, carcinoembryonic antigen and mucin-1) down to 0.9 pg/mL in undiluted serum using silicon-nanowire field-effect devices in which distinct nanowires and surface receptors are incorporated into arrays.[62] Nanowire arrays with different nucleic acid receptors enabled real-time monitoring of multiple binding events and this was successfully applied to the measurement of telomerase activity.

Nanotubes

The unique chemical and physical properties of carbon nanotubes (CNT) have paved the way to new and improved sensing devices, in particular, electrochemical biosensors. An array of vertically aligned MWCNT, embedded in SiO_2, has been shown useful for ultrasensitive detection of DNA hybridization.[59] Sub-attomoles of DNA targets have been measured by combining the CNT nanoelectrode array with $Ru(bpy)_3^{+2}$-mediated guanine oxidation. Similar to one-dimensional nanowires, nanotubes functionalized with different receptors also offer the prospect of rapid (real-time) and sensitive label-free bioelectronic transduction of binding events.

Conclusions

The applications of nanotechnology in molecular diagnostics are still at an early stage of development. However, encouraging results already have been obtained in assays that utilize a diverse range of nanostructures. In particular, nanoparticle labels have proved to be the basis of a variety of detection strategies and are particularly suited to multiplexed assays. Further development work will clarify which of the nanotechnology-based assays have the potential to enter the routine molecular diagnostic laboratory.

Acknowledgements

This work was supported in part by grants from the NIH (NCI R33-CA83220-PF, SS, NCI RO1-CA 78848-04-LJK and R01A 1056047-01 and R01 EP 0002189-JW), the PA Department of Health (SAP 4100026302-PF) as well as from National Science Foundation (CHE 0506529-JW).

References

1. Harrison DJ, Manz A, Fan ZH et al. Capillary electrophoresis and sample injection systems integrated on a planar glass chip. Anal Chem 1992; 64:1926-1932.
2. Cheng J, Kricka LJ eds. Biochip Technology. Philadelphia, Pa: Harwood Academic, 2001:1-372.
3. Dolnik V, Liu S. Applications of capillary electrophoresis on microchip. J Sep Sci 2005; 28:1994-2009.
4. Kricka LJ, Wilding P. Microchip PCR. Anal Bioanal Chem 2003; 377:820-825.
5. Lin YW, Huang MF, Chang HT. Nanomaterials and chip-based nanostructures for capillary electrophoretic separations of DNA. Electrophoresis 2005; 26:320-330.
6. Anderson RC, Su X, Bogdan GJ et al. A miniature integrated device for automated multistep genetic assays. Nucleic Acids Res 2000; 28:e60.
7. Hong JW, Quake SR. Integrated nanoliter systems. Nat Biotechnol 2003; 21:1179-1183.
8. Hong JW, Studer V, Hang G et al. A nanoliter-scale nucleic acid processor with parallel architecture. Nat Biotechnol 2004; 22:435-439.
9. Lagally ET, Emrich CA, Mathies RA. Fully integrated PCR-capillary electrophoresis microsystem for DNA analysis. Lab Chip 2001; 1:102-107.
10. Lee TM, Carles MC, Hsing IM. Microfabricated PCR-electrochemical device for simultaneous DNA amplification and detection. Lab Chip 2003; 3:100-105.
11. Taniguchi N. On the basic concept of nano-technology. Proc Intl Conf Prod Eng 1974; Tokyo, Part II, Japan Soc Precision Engineering.
12. Keiper A. Nanotechnology History: A Non-technical Primer. The New Atlantis 2003; Summer:17-34 (www.TheNewAtlantis.com).
13. Drexler KE. Engines of Creation: The Coming Era of Nanotechnology, New York: Anchor Books, 1986.
14. Drexler KE. Nanosystems: Molecular Machinery, Manufacturing and Computation. New York: Wiley, 1991.
15. Eigler DM, Schweizer EK. Positioning single atoms with a scanning tunneling microscope. Nature 1990; 344:524-526.
16. Berg P, Baltimore D, Brenner S et al. Summary statement of the Asilomar Conference on recombinant DNA molecules. Proc Natl Acad Sci USA 1975; 72:1981-1984.
17. Oberdorster G, Oberdorster E, Oberdorster J. Nanotoxicology: an emerging discipline evolving from studies of ultrafine particles. Environ Health Perspect 2005; 113:823-839.
18. Nel A, Xia T, Madler L et al. Toxic potential of materials at the nanolevel. Science 2006; 311:622-627.
19. Oberdorster E. Manufactured nanomaterials (fullerenes, C60) induce oxidative stress in the brain of juvenile largemouth bass. Environ Health Perspect 2004; 112:1058-1062.
20. Brown DM, Wilson MR, MacNee W et al. Size-dependent pro-inflammatory effects of ultrafine polystyrene particles: A role for surface area and oxidative stress in the enhanced activity of ultrafines. Toxicol Appl Pharmacol 2001; 175:191-199.
21. Gates BD, Xu Q, Stewart M et al. New approaches to nanofabrication: molding, printing and other techniques. Chem Rev 2005; 105:1171-1196.
22. Ginger DS, Zhang H, Mirkin CA. The evolution of dip-pen nanolithography. Angew Chem Int Ed 2004; 43:30-45.
23. Tang Q, Shi SQ, Zhou L. Nanofabrication with atomic force microscopy. J Nanosci Nanotech 2004; 4:948-963.

24. Lee KB, Kim EY, Mirkin CA et al. The use of nanoarrays for highly sensitive and selective detection of human immunodeficiency virus type I in plasma. Nano Lett 2004; 4:1869-1872.
25. Lee KB, Park SJ, Mirkin CA et al. Protein nanoarrays generated by pip-pen nanolithography. Science 2002; 295:1702-1705.
26. Johanson J, Abravaya K, Caminiti W et al. A new ultrasensitive assay for quantitation of HIV-1 RNA in plasma. J Virol Methods 2001; 95:81-92.
27. Demers LM, Ginger DS, Park SJ et al. Direct patterning of modified oligonucleotides on metals and insulators by dip-pen nanolithography. Science 2002; 296:1836-1838.
28. Zhang H, Li Z, Mirkin CA. Dip-pen nanolithography-based methodology for preparing arrays of nanostructures functionalized with oligonucleotides. Adv Mater 2002; 14:1472-1474.
29. Katz E, Willner I. Integrated nanoparticle-biomolecule hybrid systems: synthesis, properties and applications. Angew Chem Int Ed 2004; 43:6042-6108.
30. Rosi NL, Mirkin CA. Nanostructures in biodiagnostics. Chem Rev 2005; 105:1547-1562.
31. Wang J. Nanomaterial-based amplified transduction of biomolecular interactions. Small 2005; 1:1036-1043.
32. Storhoff JJ, Elghanian R, Mucic R et al. One-pot colorimetric differentiation of polynucleotides with single base imperfections using gold nanoparticle probes. Am Chem Soc 1998; 120:1959-1964.
33. Willner I, Patolsky F, Weizmann Y et al. Amplified detection of single-base mismatches in DNA using microgravimetric quartz-crystal-microbalance transduction. Talanta 2002; 56:847-856.
34. Wang J, Liu G, Polsky R. Electrochemical stripping detection of DNA hybridization based on cadmium sulfide nanoparticle tags. Electrochem Comm 2002; 4:722-726.
35. Wang J, Xu D, Kawde AN et al. Metal nanoparticle-based electrochemical stripping potentiometric detection of DNA hybridization. Anal Chem 2001; 73:5576-5581.
36. Dequaire M, Degrand C, Limoges B. An electrochemical metalloimmunoassay based on a colloidal gold label. Anal Chem 2000; 72:5521-5528.
37. Wang J, Polsky R, Danke X. Silver-enhanced colloidal gold electrochemical stripping detection of DNA hybridization. Langmuir 2001; 17:5739-5741.
38. Authier L, Grossiord C, Brossier P. Gold nanoparticle-based quantitative electrochemical detection of amplified human cytomegalovirus DNA using disposable microband electrodes. Anal Chem 2001; 73:4450-4456.
39. Cai H, Wang Y, He P et al. Electrochemical detection of DNA hybridization based on silver-enhanced gold nanoparticle label. Anal Chim Acta 2002; 469:165-172.
40. Lee TM, Li L, Hsing IM. Enhanced electrochemical detection of DNA hybridization based on electrode-surface modification. Langmuir 2003; 19:4338-4343.
41. Wang J, Polsky R, Merkoci A et al. Electroactive beads for ultrasensitive DNA detection. Langmuir 2003; 19:989-991.
42. Ozsoz M, Erdem A, Kerman K et al. Electrochemical genosensor based on colloidal gold nanoparticles for the detection of Factor V Leiden mutation using disposable pencil graphite electrodes. Anal Chem 2003; 75:2181-2187.
43. Kawde A, Wang J. Amplified electrical transduction of DNA hybridization based on polymeric beads loaded with multiple gold nanoparticle tags. Electroanalysis 2004; 16:101-107.
44. Wang J, Li J, Baca AJ et al. Amplified voltammetric detection of DNA hybridization via oxidation of ferrocene caps on gold nanoparticle/streptavidin conjugates. Anal Chem 2003; 75:3941-3945.
45. Park SJ, Taton TA, Mirkin CA. Array-based electrical detection of DNA with nanoparticle probes. Science 2002; 295:1503-1506.
46. Han M, Gao X, Su JZ et al. Quantum-dot-tagged microbeads for multiplexed optical coding of biomolecules. Nat Biotechnol 2001; 19:631-635.
47. Wang J, Liu G, Jan R et al. Electrochemical detection of DNA hybridization based on carbon-nanotubes loaded with CdS tags. Electrochem Comm 2003; 5:1000-1004.
48. Wang J, Liu G, Merkoci A. Electrochemical coding technology for simultaneous detection of multiple DNA targets. J Am Chem Soc 2003; 125:3214-3215.
49. Liu G, Lee TM, Wang J. Nanocrystal-based bioelectronic coding of single nucleotide polymorphisms. J Am Chem Soc 2005; 127:38-39.
50. Kerman K, Saito M, Morita Y et al. Electrochemical coding of single-nucleotide polymorphisms by monobase-modified gold nanoparticles. Anal Chem 2004; 76:1877-1884.
51. Liu G, Wang J, Kim J et al. Electrochemical coding for multiplexed immunoassays of proteins. Anal Chem 2004; 76:7126-7130.
52. Trau D, Yang W, Seydack M et al. Nanoencapsulated microcrystalline particles for superamplified biochemical assays. Anal Chem 2002; 74:5480-5486.
53. Mak WC, Cheung KY, Trau D et al. Electrochemical bioassay utilizing encapsulated electrochemical active microcrystal biolabels. Anal Chem 2005; 77:2835-2841.

54. Deamer DW, Branton D. Characterization of nucleic acids by nanopore analysis. Acc Chem Res 2002; 35:817-825.
55. Deamer DW, Akeson M. Nanopores and nucleic acids: prospects for ultrarapid sequencing. Trends Biotechnol 2000; 18:147-151.
56. Li J, Gershow M, Stein D et al. DNA molecules and configurations in a solid state nanopore microscope. Nat Mater 2003; 2:611-615.
57. Sauer JR, Zeghbroeck J van. Ultra-fast nucleic acid sequencing device and a method for making the same. US Patent 2002; 6:413-792.
58. Kasianowicz JJ, Brandin E, Branton D et al. Characterization of individual polynucleotide molecules using a membrane channel. Proc Natl Acad Sci USA 1996; 93:13770-13777.
59. Li J, Ng T, Cassell A et al. Carbon nanotube nanoelectrode array for ultrasensitive DNA detection. Nano Lett 2003; 3:597-602.
60. Margulies M, Eghold M, Altman WE et al. Genome sequencing in microfabricated high-density picolitre reactors. Nature 2005; 437:326-330.
61. Austin R. Nanopores: the art of sucking spaghetti. Nat Mater 2003; 2:567-568.
62. Zheng GF, Patolsky F, Cui Y et al. Multiplexed electrical detection of cancer markers with nanowire sensor arrays. Nature Biotechnol 2005; 23:1294-1301.

INDEX